Alexander P. Rotshtein and Hanna B. Rakytyanska

Fuzzy Evidence in Identification, Forecasting and Diagnosis

Studies in Fuzziness and Soft Computing, Volume 275

Editor-in-Chief

Prof. Janusz Kacprzyk
Systems Research Institute
Polish Academy of Sciences
ul. Newelska 6
01-447 Warsaw
Poland
E-mail: kacprzyk@ibspan.waw.pl

Further volumes of this series can be found on our homepage: springer.com

Vol. 261. Chris Cornelis, Glad Deschrijver, Mike Nachtegael, Steven Schockaert, and Yun Shi (Eds.)
35 Years of Fuzzy Set Theory, 2010
ISBN 978-3-642-16628-0

Vol. 262. Zsófia Lendek, Thierry Marie Guerra, Robert Babuška, and Bart De Schutter
Stability Analysis and Nonlinear Observer Design Using Takagi-Sugeno Fuzzy Models, 2010
ISBN 978-3-642-16775-1

Vol. 263. Jiuping Xu and Xiaoyang Zhou
Fuzzy-Like Multiple Objective Decision Making, 2010
ISBN 978-3-642-16894-9

Vol. 264. Hak-Keung Lam and Frank Hung-Fat Leung
Stability Analysis of Fuzzy-Model-Based Control Systems, 2011
ISBN 978-3-642-17843-6

Vol. 265. Ronald R. Yager, Janusz Kacprzyk, and Prof. Gleb Beliakov (eds.)
Recent Developments in the Ordered Weighted Averaging Operators: Theory and Practice, 2011
ISBN 978-3-642-17909-9

Vol. 266. Edwin Lughofer
Evolving Fuzzy Systems – Methodologies, Advanced Concepts and Applications, 2011
ISBN 978-3-642-18086-6

Vol. 267. Enrique Herrera-Viedma, José Luis García-Lapresta, Janusz Kacprzyk, Mario Fedrizzi, Hannu Nurmi, and Sławomir Zadrożny
Consensual Processes, 2011
ISBN 978-3-642-20532-3

Vol. 268. Olga Poleshchuk and Evgeniy Komarov
Expert Fuzzy Information Processing, 2011
ISBN 978-3-642-20124-0

Vol. 269. Kasthurirangan Gopalakrishnan, Siddhartha Kumar Khaitan, and Soteris Kalogirou (Eds.)
Soft Computing in Green and Renewable Energy Systems, 2011
ISBN 978-3-642-22175-0

Vol. 270. Christer Carlsson and Robert Fullér
Possibility for Decision, 2011
ISBN 978-3-642-22641-0

Vol. 271. Enric Trillas, P. Bonissone, Luis Magdalena and Janusz Kacprzyk
Combining Experimentation and Theory, 2011
ISBN 978-3-642-24665-4

Vol. 272. Oscar Castillo
Type-2 Fuzzy Logic in Intelligent Control Applications, 2012
ISBN 978-3-642-24662-3

Vol. 273. Rudolf Seising and Veronica Sanz (Eds.)
Soft Computing in Humanities and Social Sciences, 2012
ISBN 978-3-642-24671-5

Vol. 274. Irina Georgescu
Possibility Theory and the Risk, 2012
ISBN 978-3-642-24739-2

Vol. 275. Alexander P. Rotshtein and Hanna B. Rakytyanska
Fuzzy Evidence in Identification, Forecasting and Diagnosis, 2012
ISBN 978-3-642-25785-8

Alexander P. Rotshtein and
Hanna B. Rakytyanska

Fuzzy Evidence in Identification, Forecasting and Diagnosis

Springer

Authors
Prof. Alexander P. Rotshtein
Department of Industrial Engineering and Management
Jerusalem College of Technology - Lev Institute
Jerusalem
Israel

Dr. Hanna B. Rakytyanska
Department of Software Design
Vinnytsia National Technical University
Vinnytsia
Ukraine

ISSN 1434-9922
e-ISSN 1860-0808
ISBN 978-3-642-44421-0
ISBN 978-3-642-25786-5 (eBook)
DOI 10.1007/978-3-642-25786-5
Springer Heidelberg New York Dordrecht London

Softcover reprint of the hardcover 1st edition 2012

Printed on acid-free paper

Springer is part of Springer Science+Business Media (www.springer.com)

In memory of Dr. Igor Rotshtein (1971–2006)

Preface

The identification of an object refers to the construction of its mathematical model which determines the interrelationship between input and output variables by experimental data. The presence of the "input – output" model provides the possibility of solving two classes of problems; direct problems and inverse problems.

A direct problem is understood as a look into the future; i.e., a prediction of the effects (outputs) as a result of the observed causes (inputs).

An inverse problem is understood as a look into the past; i.e., a renewal of the causes (inputs) through the observed effects (outputs).

The resolution of these problems plays an important role in automatic and situational control, technical and medical diagnostics, pattern recognition, prediction, many-factor analysis, multi-criteria estimation and other *decision making tasks.*

As a rule, the task of identification is performed in two stages. At the first stage, called *structural identification*, some coarse model of an object is formed which approximates the input-output interconnection and contains adjustable parameters. At the second stage, called *parametric identification*, parameter values are chosen such that they minimize the distances between the model and experimental outputs of the object.

The stage of parametric identification is sufficiently formalized, because it amounts to the use of various optimization methods. The only difficulty is in finding the global minimum of nonlinear functions with divergence of theory and experimentation, and in computing complexity growth with the growth of the adjustable parameters number.

The stage of the structural identification is perceived to be more an act of art rather than that of science. The choice of the adjustable model considerably depends upon "the starting capital of the researcher", his/her qualifications, expertise, object essence understanding, bias in favor of one of many mathematical apparatuses and upon other subjective factors. In modern theory of identification, some quantitative relations in the form of various types of equations (algebraic, differential, difference, integral and others) are used. This apparatus is more intrinsically applied to those objects which are described by the laws of physics, mechanics, thermodynamics and electromagnetism. Classic theory leads to catastrophic complex models while identifying dependencies in the so-called *intellectual tasks* which are traditionally performed by people. Man walks, swims, does the most complex physical exercises, drives a car, recognizes familiar objects, conceives regularities in experimental data, solves other complicated mathematical tasks of control and decision making without resorting to strict quantitative relations.

Two unique qualities play fundamental role in solving the task of identification and decision making by man. They are known as: the *capability to learn* defined as the capability to minimize divergence of the actual activity result from some desired standard, and *linguistic capability*, which is the capability to express learned knowledge in a natural language.

Therefore, while simulating intellectual activity it is quite natural to employ such a mathematical apparatus which, in contrast to classical methods, is adjusted to employ learning and linguistic capabilities.

Fuzzy set theory, proposed by L. Zadeh in 1965, is a handy tool for natural language statements formalization. The first works on practical application of fuzzy logic in the direct problems of simulation and control performed in 1980 – 1990 are due to E. Mamdani, T. Takagi and M. Sugeno.

Quality of the fuzzy model directly depends on the parameters of fuzzy rules, fuzzy relations and fuzzy terms membership functions, which are chosen by employing expert methods at the stage of structural identification. Therefore, the stage of the fuzzy model tuning by using experimental data is required. The stage of tuning is connected with the statement and resolution of the problem of nonlinear optimization.

Investigations relative to fuzzy models tuning have been intensely developing since the end of the last century. Some combination of the genetic algorithm and the neural network appears to be the efficient means of solving tuning problems. The genetic algorithm provides a quick hitting into the area of the global minimum, while the neural network is then used for successive adjustment of the parameters of the fuzzy model in real time mode. One of the first monographs dedicated to the resolution of the direct problems of simulation on the basis of the complex use of fuzzy sets, genetic algorithms and neural networks is the work "Intellectual Technologies of Identification" by A. Rotshtein, 1999 (http:/matlab.exponenta.ru/fuzzy_logic/book5/index.php).

Solving inverse problems of fuzzy inference is connected with the problem of solving fuzzy logical equations. We propose a method for numerical resolution of the system of fuzzy logical equations by reducing this problem to the search for the minimum with the help of the genetic algorithm. A combination of the genetic algorithm with the neuro-fuzzy network is helpful in solving the inverse problem simultaneously with design and tuning of the fuzzy model on the basis of readily available expert and experimental information.

This monograph is written on the basis of the author's originally suggested investigations, devoted to the resolution of the direct and inverse problems of fuzzy inference with the use of genetic and neural algorithms.

The book consists of nine chapters:

Chapter One is a short introduction into intellectual technologies and contains the main knowledge of the theory of fuzzy sets, genetic algorithms and neural nets necessary to understanding the following chapters.

Chapter Two contains nonlinear objects approximation models on the basis of linguistic expressions joined in fuzzy knowledge bases. Some object models with continuous and discrete outputs are considered here. The approach to the linguistic approximation is based on the method of fuzzy logic equations suggested in the work "Medical Diagnosis based on Fuzzy Logic" by A. Rotshtein, 1997 (http:/matlab.exponenta.ru/fuzzy_logic/book7/index.php).

Chapter Three describes some methods of linguistic models tuning using genetic algorithms and neural nets. The tasks of optimal tuning are formulated in terms of mathematical programming for the objects with the continuous and discrete output as well as for the generalized object "many inputs - many outputs". Membership functions forms of fuzzy terms and weights of rules are considered as adjustable parameters.

In *Chapter Four*, the methods of IF-THEN rules extraction from experimental data using genetic algorithms and neural networks are described. The problem of linguistic knowledge extraction is formulated as the optimization problem, where the synthesis of the knowledge base amounts to finding the matrix of membership functions parameters interpreted as fuzzy terms.

In *Chapter Five*, we propose some procedures of numerical solution of the fuzzy relational equations using genetic algorithms. The procedures envisage the optimal solution growing from a set of primary variants using genetic cross-over, mutation and selection operations. To serve as an illustration of the procedures and genetic algorithm's effectiveness we present an example of the diagnosis problem.

In *Chapter Six,* we propose an approach for building fuzzy systems of diagnosis, which enables solving fuzzy relational equations together with design and tuning of fuzzy relations on the basis of expert and experimental information. The essence of tuning consists of the selection of such membership functions of the fuzzy causes and effects and such "causes-effects" fuzzy relations, which provide minimal difference between theoretical and experimental results of diagnosis. Genetic algorithms and neural networks are used for solving the optimization problems.

In *Chapter Seven*, an approach for an inverse problem solution based on the description of the interconnection between unobserved and observed parameters of an object with the help of fuzzy IF-THEN rules is proposed. The essence of the proposed approach consists of formulating and solving the optimization problems, which, on the one hand, find the roots of fuzzy logical equations, corresponding to IF-THEN rules, and, on the other hand, tune the fuzzy model using the readily available experimental data. The hybrid genetic – neuro approach is proposed for solving the formulated optimization problems.

In *Chapter Eight*, we consider a problem of multiple inputs – multiple outputs object identification expressed mathematically in terms of fuzzy relational equations. The identification problem consists of the extraction of an unknown relational matrix which can be translated as a set of fuzzy IF-THEN rules. In fuzzy relational calculus this type of the problem relates to the inverse problem and requires resolution for the composite fuzzy relational equations. The resulting solution is linguistically interpreted as a set of possible rules bases discovering the structure of the given experimental data.

The efficiency of the models and algorithms suggested in Chapters 3 – 8 is illustrated by computer experiments with standard objects as well as real examples of forecasting and diagnosis.

Chapter Nine describes the results of application of intellectual technologies of identification in system control problems, sport forecasting, automobile design, projects creditworthiness evaluation and system reliability analysis.

The presented bibliography does not in any way reflect the overwhelming majority of works in the given domain. Only the works which were used by the authors in carrying out the investigations are referred to.

While writing this book, we used the results of investigations carried out together with Dr. Denis Katelnikov and Dr. Yuriy Mityushkin, which remained the best reminiscences during our team work.

The first author would like to express his thanks to the colleagues from Jerusalem College of Technology (JCT – Machon Lev) Prof. Yaakov Friedman, Prof. Usiel Sandler, Prof. Morton Posner, Prof. Alan Stulman, Prof. Joseph M. Steiner and Mrs. Ariella Berkowitz for many useful discussions and support. Finally, we thank the JCT student Shaya Rubinshtein for the linguistic editing.

Contents

Chapter 1
Fundamentals of Intellectual Technologies

Intellectual technologies which are used to do the tasks of identification and decision making in this book represent a combination of three independent theories:

- *of fuzzy sets* - as a means of natural language expressions and logic evidence formalization;
- *of neural nets* - artificial analogs of the human brain simulating the capability to learn;
- *of genetic algorithms* - as a means of optimal decision synthesis from a multiplicity of initial variants on which the operations of crossing, mutation and selection are performed.

The concept of the linguistic variable underlies natural language expressions formalization [1, 2]. According to Zadeh [1], such a variable whose values are words or sentences of the natural language, that is the qualitative terms, is called the linguistic variable. Using the notion of membership function, each of the terms estimating a linguistic variable can be formulated in the form of a fuzzy set defined on a corresponding universal set [2]. Fuzzy logic apparatus does not contain learning mechanisms. That is why the results of fuzzy logic evidence strongly depend on the membership functions type used to formalize fuzzy terms: "small", "large", "cool", "hot" and alike.

The main feature of neural networks is their learning ability. This is realized by special algorithms among which the back-propagation algorithm is the most popular [3, 4]. There is no need for prior information about the structure of the sought functional dependence to train the neural network. Only the training data in the form of experimental "input – output" pairs are needed, and the price for it is the fact that a trained neural network – a graph with weighted edges – doesn't yield to semantic interpretation.

Optimization is the most important stage in solving identification problems [5 – 7]. A task of nonlinear optimization can be solved by various methods among which the gradient descent [8] is the most universal. However, when there is a great number of input variables the gradient descent method requires finding the minimum from various initial points that substantially increases computer time expenses. Genetic algorithms represent the powerful apparatus of optimal decision synthesis [9, 10]. These algorithms are analogues of random search [8], which is carried out simultaneously from various initial points, cutting the time of search for optimal solutions.

A.P. Rotshtein et al.: Fuzzy Evidence in Identif., Forecast. and Diagn., STUDFUZZ 275, pp. 1–37.
springerlink.com

1.1 Fuzzy Sets

This section is written on the basis of the works [1, 2, 11, 12]. The additional information relative to fuzzy sets and decision making under uncertainty can be found in the works [13 – 21].

The concept of a *set*, and set theory, are powerful tools in mathematics. Unfortunately, a *sin qua non* condition underlying set theory, i.e. that an element can either belong to a set or not, is often not applicable in real life where many vague terms as "large profit", "high pressure", "moderate temperature", "reliable tools", "safe conditions", etc. are extensively used. Unfortunately, such imprecise descriptions cannot be adequately handled by conventional mathematical tools.

If we wish to maintain the very meaning of imprecise (vague) terms, a crisp differentiation between elements (e.g., pressure values) that are either high or not high may be artificial, and some values may be perceived high to some extent, not fully high and not fully not high.

An attempt to develop a formal apparatus to involve a partial membership in a set was undertaken in the mid-1960's by Zadeh [1]. He introduced the concept of a *fuzzy set* as a collection of objects which might "belong" to it to a degree, from 1 for full belongingness to 0 for full nonbelongingness, through all intermediate values. This was done by employing the concept of a *membership function*, assigning to each element of a *universe of discourse* a number from the unit interval to indicate the *intensity (grade) of belongingness*. The concept of a membership function was evidently an extension of that of a *characteristic function* of a conventional set assigning to the universe of discourse either 0 (nonbelongingness) or 1 (belongingness). Then, basic properties and operations on fuzzy sets were defined by Zadeh (and later by his numerous followers) being essentially extensions (in the above spirit) of their conventional counterparts.

Since its inception, fuzzy sets theory has experienced an unprecedented growth of interest in virtually all fields of science and technology.

1.1.1 Fundamentals of Fuzzy Set Theory

Suppose that $X = \{x\}$ is a *universe of discourse*, i.e. the set of all possible (feasible, relevant, ...) elements to be considered with respect to a fuzzy (vague) concept (property). Then a *fuzzy subset* (or a *fuzzy set*, for short) A in X is defined as a set of ordered pairs $\{(x, \mu^A(x))\}$, where $x \in X$ and $\mu^A : X \to [0,1]$ is the *membership function* of A; $\mu^A(x) \in [0,1]$ is the *grade of membership* of x in A, from 0 for full nonbelongingness to 1 for full belongingness, through all intermediate values. In some contexts it may be expedient to view the grade of membership of a particular element as its degree of compatibility with the (vague) concept represented by the fuzzy set. Notice that the degrees of membership are clearly subjective.

Many authors denote $\mu^A(x)$ by $A(x)$. Moreover, a fuzzy set is often equated with its membership function so that both A and $\mu^A(x)$ are often used interchangeably.

Notice that if $[0, 1]$ is replaced by $\{0, 1\}$, this definition coincides with the characteristic function based description of an ordinary (nonfuzzy) set. Moreover, the original Zadeh's unit interval is chosen for simplicity, and a similar role may be played by an ordered set, e.g., a lattice.

It is convenient to denote a fuzzy set defined in a finite universe of discourse, say A in $X = \{x_1, x_2, ..., x_n\}$ as

$$A = \mu^A(x_1)/x_1 + \mu^A(x_2)/x_2 + ... + \mu^A(x_n)/x_n = \sum_{i=1}^{n} \mu^A(x_i)/x_i ,$$

where " $\mu^A(x_i)/x_i$ " (called a singleton) is a pair "grade of membership – element" and "+" is meant in the set-theoretic sense.

Example 1.1. If $X = \{1, 2, ..., 10\}$, then a fuzzy set "large number" may be given as

$$A = \text{" large number"} = 0.2/6 + 0.5/7 + 0.8/8 + 1/9 + 1/10$$

to be meant as: 9 and 10 are surely (to degree 1) "large numbers", 8 is a "large number" to degree 0.8, etc. and 1,2,...,5 are surely not "large numbers". Notice that the above degrees of membership are subjective (a "large number" is a subjective concept!) and context-dependent, and - by convention - the singletons with $\mu^A(\bullet) = 0$ are omitted.

In practice it is usually convenient to use a piecewise linear representation of the membership function of a fuzzy set as shown in Fig. 1.1 since only two values, $\bar{a}$ and $\underline{a}$, are needed.

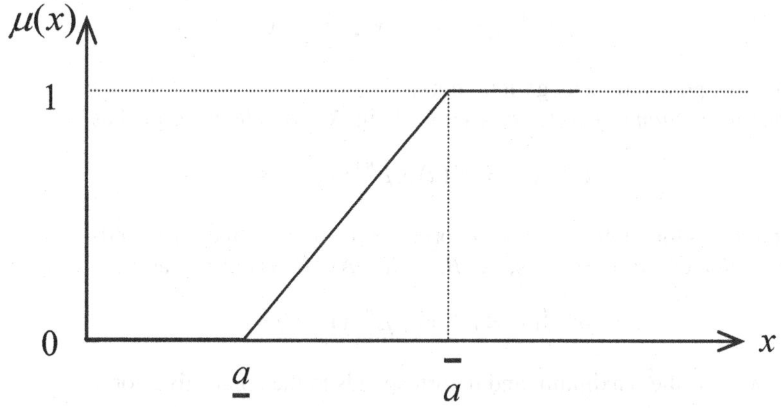

Fig. 1.1. Membership function of a fuzzy set

1.1.2 Basic Properties of Fuzzy Sets

A fuzzy set A in X is *empty*, $A = \varnothing$, if and only if $\mu^A(x) = 0$, $\forall x \in X$.

Two fuzzy sets A, B in X are *equal*, $A = B$, if and only if $\mu^A(x) = \mu^B(x)$, $\forall x \in X$.

A fuzzy set A in X is *included* in (is a *subset* of) a fuzzy set B in X, $A \subseteq B$, if and only if $\mu^A(x) \le \mu^B(x)$, $\forall x \in X$.

Example 1.2. Suppose $X = \{1, 2, 3\}$ and $A = 0.3/1 + 0.5/2 + 1/3$ and $B = 0.4/1 + 0.6/2 + 1/3$; then $A \subseteq B$.

An important concept is the *cardinality* of a fuzzy set. If $X = \{x_1, x_2, ..., x_n\}$, and

$$A = \mu^A(x_1)/x_1 + \mu^A(x_2)/x_2 + ... + \mu^A(x_n)/x_n = \sum_{i=1}^{n} \mu^A(x_i)/x_i ,$$

then the (nonfuzzy) cardinality of A is defined as

$$\text{card } A = |A| = \sum_{i=1}^{n} \mu^A(x_i) .$$

Example 1.3. If $X = \{1, 2, 3, 4\}$ and $A = 0.1/1 + 0.4/2 + 0.7/3 + 1/4$, then card A =2.2.

1.1.3 Basic Operations on Fuzzy Sets

The basic operations here are naturally the complement, intersection and union, as in the conventional set theory.

The *complement* of a fuzzy set A in X, $\neg A$, is defined as

$$\mu^{\neg A}(x) = 1 - \mu^A(x), \ \forall x \in X .$$

and it corresponds to the negation «not».

The *intersection* of two fuzzy sets A, B in X, $A \cap B$, is defined as

$$\mu^{A \cap B}(x) = \mu^A(x) \wedge \mu^B(x), \ \forall x \in X ,$$

where « $\wedge$ » is the minimum, and it corresponds to the connective «and».

The *union* of two fuzzy sets, A, B in X, $A \cup B$, is defined as

$$\mu^{A \cup B}(x) = \mu^A(x) \vee \mu^B(x), \ \forall x \in X ,$$

where « $\vee$ » is the maximum, and it corresponds to the connective «or».

Example 1.4. If $X = \{1, 2, ..., 10\}$,

A = "small number" = 1/1 + 1/2 + 0.8/3 + 0.5/4 + 0.3/5 + 0.1/6,
B = "large number" = 0.1/5 + 0.2/6 + 0.5/7 + 0.8/8 + 1/9 + 1/10,
then $\neg A$ = "not small number" = 0.2/3 + 0.5/4 + 0.7/5 + 0.9/6 + 1/7 + 1/8 + 1/9 + 1/10
$A \cap B$ = "small number" and "large number" = 0.1/5 + 0.1/6
$A \cup B$ = "small number" or "large number" = 1/1 + 1/2 + 0.8/3 + 0.5/4 + 0.3/5 + 0.2/6 + 0.5/7 + 0.8/8 + 1/9 + 1/10.

The above definitions are classic, and have been commonly employed though they are evidently by no means the only ones. For instance, the use of a t-norm for the intersection and an s-norm for the union has often been advocated. They are defined as follows:

a t-norm is defined as $t : [0, 1] \times [0, 1] \to [0, 1]$ such that:

a) a t 1 = a
b) a t b = b t a
c) a t b ≥ c t d, if a ≥ c, b ≥ d
d) a t b t c = a t (b t c).

Some more relevant examples of t-norms are:

$a \wedge b = \min(a, b)$ - this is the most popular t-norm,

$a \cdot b$,

$1 - [1 \wedge ((1-a)^p + (1-b)^p)^{1/p}]$, $p \geq 1$.

an s-norm (t-conorm) is defined as $s : [0, 1] \times [0, 1] \to [0, 1]$ such that:

a) a s 0 = a
b) a s b = b s a
c) a s b ≥ c s d, if a ≥ c, b ≥ d
d) a s b s c = a s (b s c).

Some examples of more popular s-norms are:

$a \vee b = \max(a, b)$ - this is the most popular s-norm,

$a + b - a \cdot b$

$1 \wedge \left(a^p + b^p\right)^{1/p}$, $p \geq 1$.

1.1.4 Further Properties and Related Concepts

An α*-cut* (α*-level set*) of a fuzzy set A in X is defined as the ordinary set $A_\alpha \subseteq X$ such that

$$A_\alpha = \{ x \in X : \mu^A(x) \geq \alpha \} , \quad \forall \alpha \in [0, 1] .$$

Example 1.5. If A = 1/1 + 0.8/2 + 0.5/3 +0.1/4, then $A_{0.1} = \{ 1,2,3,4 \}$, $A_{0.5} = \{ 1, 2, 3 \}$, $A_{0.8} = \{ 1,2 \}$ $A_1 = \{ 1 \}$.

The concept of an α-cut of a fuzzy set is crucial for the so-called *decomposition theorem* which states that any fuzzy set A in X may be represented as some (equivalent) operation on conventional sets (subsets of X).

Of fundamental importance here is the so-called *extension principle* [1] which gives a formal apparatus to carry over operations (e.g., arithmetic or algebraic) from sets to fuzzy sets. Namely, if $f{:}X \to Y$ is a function (operation) and A is a fuzzy set in X, then A induces via f a fuzzy set B in Y given by

$$\mu^B(y) = \begin{cases} \sup\limits_{y=f(x)} \mu^A(x) & , f^{-1}(y) \neq \varnothing \\ 0 & , f^{-1}(y) = \varnothing \end{cases} . \tag{1.1}$$

Example 1.6. Let $X = \{ 1, 2, 3, 4 \}$, $Y = \{ 1, 2, 3, 4, 5, 6 \}$ and $y = x + 2$. If now A = 0.1/1 + 0.2/2 + 0.7/3 + 1/4, then B = 0.1/3 + 0.2/4 + 0.7/5 + 1/6.

1.1.5 Fuzzy Relations

Fuzzy relations - exemplified by «much larger than», «more or less equal», etc. - are clearly omnipresent in human discourse. Formally, if $X = \{x\}$ and $Y = \{y\}$ are two universes of discourse, then a *fuzzy relation* R is defined as a fuzzy set in the Cartesian product $X \times Y$, characterized by its membership function $\mu^R{:}X \times Y \to [0, 1]$; $\mu^R(x, y) \in [0, 1]$ reflects the *strength of relation* between $x \in X$ and $y \in Y$.

Example 1.7. Suppose that X = {horse, donkey} and Y = {mule, cow}. The fuzzy relation «similar» may then be defined as

R = «similar» = 0.8/(horse, mule) + 0.4/(horse, cow) +
+ 0.9/(donkey, mule) + 0.5/(donkey, cow)

to be read that, e.g., a horse and a mule are similar to degree 0.8, a horse and a cow to degree 0.4, etc.

Notice that for finite, small enough X and Y, a fuzzy relation may be evidently shown in the matrix form.

A crucial concept related to fuzzy relations is their *composition*. If we have two fuzzy relations R in $X \times Y$ and S in $Y \times Z$, then their (*max-min*) *composition* is a fuzzy relation $R \circ S$ in $X \times Z$ defined by

$$\mu^{R \circ S}(x, z) = \sup_{y \in Y} [\mu^R(x, y) \wedge \mu^S(y, z)] .$$

Fuzzy relations play a crucial role in virtually all applications, notably in decision making and control.

1.1.6 Fuzzy Numbers

The extension principle defined by (1.1) is a very powerful tool for extending non-fuzzy relationships to their fuzzy counterparts. It can also be used, e.g., to devise *fuzzy arithmetic*.

A *fuzzy number* is defined as a fuzzy set in the real line, A in R, which is normal (i.e. $\sup_{x \in R} \mu^A(x) = 1$) and bounded convex (i.e. whose all α-cuts are convex and bounded). A fuzzy number may be exemplified by «about five», «a little more than 7», «more or less between 5 and 8», etc.

Notice that function f in (1.1) may be, say, the sum, product, difference and quotient, and we can extend via (1.1) the four main arithmetic operations: addition, multiplication, subtraction and division to fuzzy sets, hence obtaining fuzzy arithmetic.

Namely, for the basic four operations we obtain:

* addition

$$\mu^{A+B}(z) = \max_{z=x+y}[\mu^A(x) \wedge \mu^B(y)] \ , \ \forall x, y, z \in R ;$$

* subtraction

$$\mu^{A-B}(z) = \max_{z=x-y}[\mu^A(x) \wedge \mu^B(y)] \ , \ \forall x, y, z \in R ;$$

* multiplication

$$\mu^{A*B}(z) = \max_{z=x*y}[\mu^A(x) \wedge \mu^B(y)] \ , \ \forall x, y, z \in R ;$$

* division

$$\mu^{A/B}(z) = \max_{z=x/y,\, y\neq 0}[\mu^A(x) \wedge \mu^B(y)] \ , \ \forall x, y, z \in R .$$

Unfortunately, the use of the extension principle to define the arithmetic operations on fuzzy numbers is in general numerically inefficient, hence it is usually assumed that a fuzzy number is given in the so-called $L-R$ *representation* whose essence is that the membership function of a fuzzy number is

$$\mu^A(x) = \begin{cases} L\left(\dfrac{m-x}{\alpha}\right) , & \alpha > 0, \ \forall x \le m \\ R\left(\dfrac{m-x}{\beta}\right) , & \beta > 0, \ \forall x \ge m \end{cases} ,$$

where function L is such that:

a) $L(-x) = L(x)$,
b) $L(0) = 1$,
c) L is increasing in $[0, +\infty]$,

and similarly function R .

Here m is the *mean value* of the fuzzy number A , α is its *left spread*, and β is its *right spread*; notice that when α , β =0, then the fuzzy number A boils down to a real number m .

A fuzzy number A can now be written as $A = (m_A, \alpha_A, \beta_A)$, and the arithmetic operations may be defined in terms of the m 's, α 's and β 's. For instance, in the case of *addition*:

$$A + B = (m_A, \alpha_A, \beta_A) + (m_B, \alpha_B, \beta_B) = (m_A + m_B, \alpha_A + \alpha_B, \beta_A + \beta_B),$$

and similarly for the other arithmetic operations.

In practice, however, the $L-R$ representation is further simplified in that the functions L and R are assumed to be linear which leads to *triangular fuzzy numbers* exemplified by the one shown in Fig. 1.2a, and whose membership function is generally given by

$$\mu^A(x) = \begin{cases} (x - a^-)/(a - a^-) \ , \ a^- \le x \le a \\ (a^+ - x)/(a^+ - a) \ , \ a \le x \le a^+ \end{cases} .$$

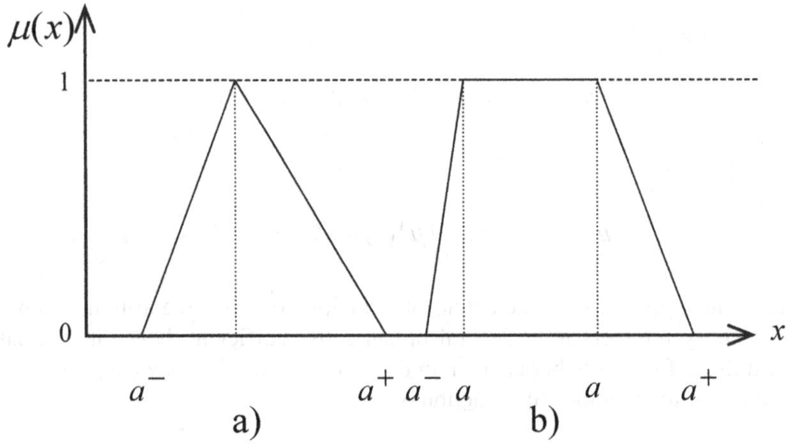

Fig. 1.2. Membership functions of triangular and trapezoid fuzzy numbers

Notice that a triangular fuzzy number may be adequate for the formalization of such terms as, say, *around 5* or *much more than 10* (in this case, evidently, a^+ must be a very large number). For the representation of such fuzzy numbers as, e.g., *more or less between 5 and 7* the *trapezoid fuzzy numbers* may be used which are exemplified in Fig. 1.2b and whose membership function is generally written as

$$\mu^A(x) = \begin{cases} (x-a^-)/(\underline{a}-a^-) \ , & a^- \le x \le \underline{a} \\ 1 & , \ \underline{a} \le x \le \overline{a} \ . \\ (a^+ - x)/(a^+ - \overline{a}) \ , & \overline{a} \le x \le a^+ \end{cases}$$

1.1.7 Fuzziness and Probability

Novices at fuzzy set theory very often try to compare it with theory of probability. However, both theories are hardly comparable because they treat uncertainty differently. Some statistical uncertainty is considered in theory of probability, e.g., *a probability of hitting the target is equal to* 0.9. Fuzzy set theory allows us to operate with linguistic uncertainty, e.g., *good shot.* These types of uncertainty can be formalized with the help of:

- distribution functions – for theory of probability,
- membership functions – for fuzzy set theory.

The founder of fuzzy set theory L. Zadeh gives the following example to illustrate the crucial difference between the two distributions [17].

Example 1.8. Let us consider the assertion «The author eats X eggs at breakfast»,

$$X = \{1, 2, 3, ... \} .$$

Some possibility and probability distributions correspond to the value of X, which can be considered as an uncertain parameter.

The possibility distribution $\pi_X(u)$ can be interpreted as a degree (a subjective measure) of easiness, with which the author eats u eggs. To define the probability distribution $P_X(u)$, it is necessary to observe the author over a period of 100 days.

Both distributions are presented below.

Possibility and probability distribution

u	1	2	3	4	5	6	7	8
$\pi_X(u)$	1	1	1	1	0.8	0.6	0.4	0.2
$P_X(u)$	0.1	0.8	0.1	0	0	0	0	0

It is seen, that the high degree of possibility $\pi_X(u)$ does not mean in any case the same high degree of probability $P_X(u)$. There is no doubt: *if an event is impossible, then it is also improbable.*

The crucial difference between theory of probability and theory of possibility is apparent wherein the axiom of complement is treated differently in these two theories:

$$P(A)+P(\overline{A})=1 \quad \text{- for theory of probability,}$$

$$\pi(A)+\pi(\overline{A})\neq 1 \quad \text{- for theory of possibility.}$$

1.2 Genetic Algorithms

As mentioned in the preface, optimization is the most important stage in solving identification problems [5 – 7]. The main difficulties in the application of the classical methods of nonlinear functions optimization [8] are related to the problems of finding a local extremum (Fig. 1.3) and overcoming of the "dimension curse" (Fig. 1.4).

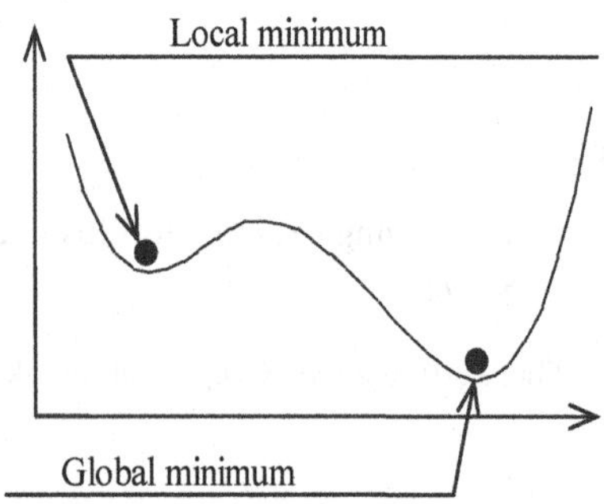

Fig. 1.3. Problem of local extremum

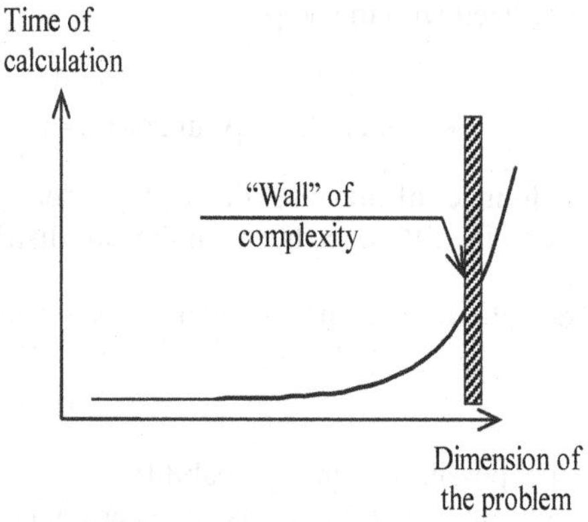

Fig. 1.4. Problem of "dimension curse"

The attempts to overcome these problems resulted in the creation of a special theory of genetic algorithms, which grow the optimal solution by crossing-over the initial variants with consequent selection using some criterion (Fig. 1.5). The general information about genetic algorithms presented in this chapter is based on the works [9, 10, 22, 23].

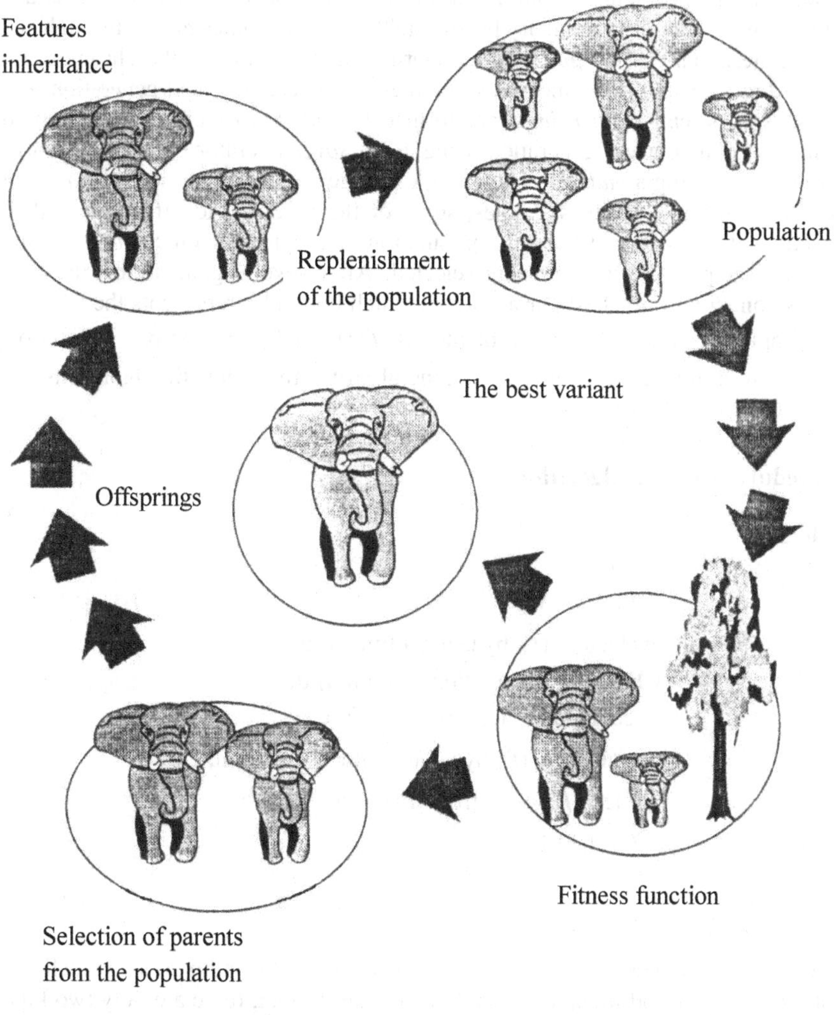

Fig. 1.5. Idea of genetic algorithm
(In: Goldberg D. Genetic Algorithms in Search, Optimization and Machine Learning, Addison Wesley, 1989)

1.2.1 General Structure of Genetic Algorithms

Genetic algorithms are stochastic search techniques based on the mechanism of natural selection and natural genetics. Genetic algorithms, differing from conventional search techniques, start with an initial set of random solutions called a *population*. Each individual in the population is called a *chromosome*, representing a solution to the problem at hand. A chromosome is a string of symbols; it is usually, but not necessarily, a binary bit string. The chromosomes *evolve* through successive iterations, called *generations*. During each generation, the chromosomes are *evaluated*, using some measures of fitness. To create the next generation, new chromosomes, called *offsprings*, are formed by either (a) merging two chromosomes from the current generation using a *crossover* operator or (b) modifying a chromosome using a *mutation* operator. A new generation is formed by (a) selecting, according to the fitness values, some of the parents and offsprings and (b) rejecting others so as to keep the population size constant. Fitter chromosomes have higher probabilities of being selected. After several generations, the algorithms converge to the best chromosome, which hopefully represents the optimum or suboptimal solution to the problem. Let $P(t)$ and $C(t)$ be parents and offsprings in current generation t; the general structure of genetic algorithms (see Fig. 1.6) is described as follows:

Procedure: Genetic Algorithm

```
begin
        t := 0;
        initialize P(t);
        evaluate P(t) by using a fitness function;
        while (not termination condition) do
          recombine P(t) to yield C(t);
          evaluate C(t) by using a fitness function;
          select P(t+1) from P(t) and C(t);
          t := t+1;
        end
    end.
```

Usually, initialization is assumed to be random. Recombination typically involves crossover and mutation to yield offspring. In fact, there are only two kinds of operations in genetic algorithms:

1. Genetic operations: crossover and mutation.
2. Evolution operation: selection.

The genetic operations mimic the process of heredity of genes to create new offspring at each generation. The evolution operation mimics the process of *Darwinian evolution* to create populations from generation to generation.

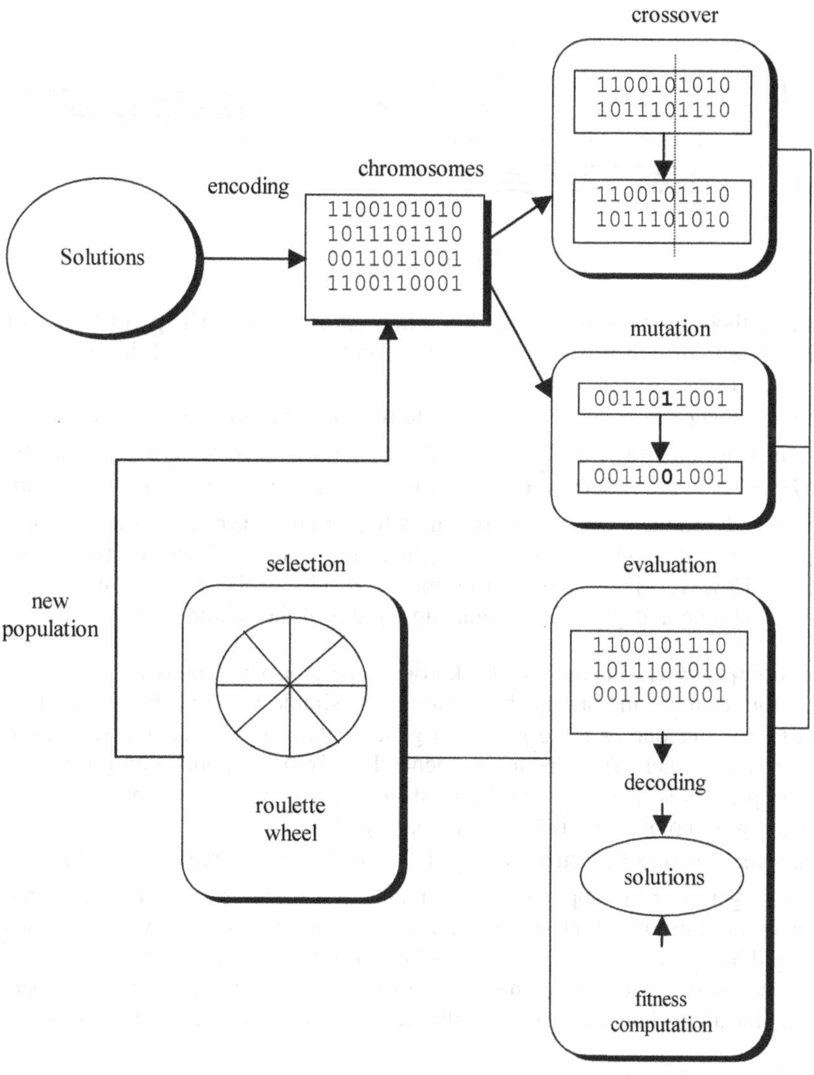

Fig. 1.6. The general structure of genetic algorithms

1.2.2 Genetic Operators

Crossover operator. Crossover is the main genetic operator. It operates on two chromosomes at a time and generates offspring by combining both chromosomes' features. A simple way to achieve crossover would be to choose a random cut-point and generate the offspring by combining the segment of one parent to the left of the cut-point with the segment of the other parent to the right of the cut-point (Fig. 1.7).

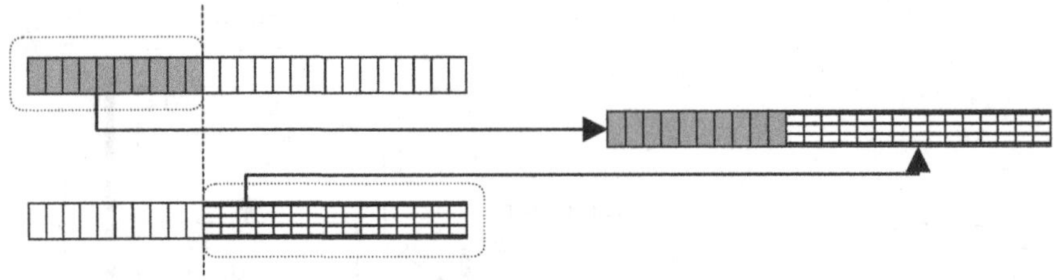

Fig. 1.7. Crossover operator

This method works well with the bit string representation. The performance of genetic algorithms depends, to a great extent, on the performance of the crossover operator used.

The *crossover rate* (denoted by p_c) is defined as the ratio of the number of offspring produced in each generation to the population size (usually denoted by *pop_size*). This ratio controls the expected number $p_c \times$ *pop_size* of chromosomes to undergo the crossover operation. A higher crossover rate allows exploration of more of the solution space and reduces the chances of settling for a false optimum. However, if this rate is too high, it results in the wastage of a lot of computation time in exploring unpromising regions of the solution space.

Mutation operator. Mutation is a background operator which produces spontaneous random changes in various chromosomes. A simple way to achieve mutation would be to alter one or more genes. In genetic algorithms, mutation serves the crucial role of either (a) replacing the genes lost from the population during the selection process so that they can be tried in a new context or (b) providing the genes that were not present in the initial population.

The *mutation rate* (denoted by p_m) is defined as the percentage of the total number of genes in the population. The mutation rate controls the rate at which new genes are introduced into the population for trial. If it is too low, many genes that would have been useful are never tried out; if it is too high, there will be much random perturbation, the offspring will start losing their resemblance to the parents, and the algorithm will lose the ability to learn from the history of the search.

1.2.3 Search Techniques

Search is one of the more universal problem-solving methods for such problems where one cannot determine *a priori* the sequence of steps leading to a solution. Search can be performed with either blind strategies or heuristic strategies. Blind search strategies do not use information about the problem domain. Heuristic search strategies use additional information to guide the search along with the best search directions. There are two important issues in search strategies: exploiting the best solution and exploring the search space. Hill-climbing is an example of a strategy which exploits the best solution for possible improvement while ignoring the exploration of the search space. Random search is an example of a strategy which explores the search space while ignoring the exploitation of the promising regions of the search space. Genetic algorithms are a class of general-purpose search methods combining elements of directed and stochastic search which can make a remarkable balance between exploration and exploitation of the search space. At the beginning of genetic search, there is a widely random and diverse population and the crossover operator tends to perform a widespread search for exploring the complete solution space. As the high fitness solutions develop, the crossover operator provides exploration in the neighbourhood of each of them. In other words, the types of searches (exploration or exploitation) a crossover performs would be determined by the environment of the genetic system (the diversity of population), but not by the operator itself. In addition, simple genetic operators are designed as general-purpose search methods (the domain-independent search methods); they perform essentially a blind search and could not guarantee to yield an improved offspring.

1.2.4 Comparison of Conventional and Genetic Approaches

Generally, the algorithm for solving optimization problems is a sequence of computational steps which asymptotically converge to an optimal solution. Most classical optimization methods generate a deterministic sequence of computation based on the gradient or higher-order derivatives of the objective function. The methods are applied to a single point in the search space. The point is then improved along the deepest descending/ascending direction gradually through iterations. This point-to-point approach has the danger of falling in local optima. Genetic algorithms perform a multiple directional search by maintaining a population of potential solutions. The population-to-population approach attempts to make the search escape from local optima. Population undergoes a simulated evolution: at each generation the relatively good solutions are reproduced, while the relatively bad solutions die. Genetic algorithms use probabilistic transition rules to select someone to be reproduced and someone to die so as to guide their search toward regions of the search space with likely improvement.

1.2.5 Advantages of Genetic Algorithms

Genetic algorithms have received considerable attention regarding their potential as a novel optimization technique. There are three major advantages when applying genetic algorithms to optimization problems:

1. Genetic algorithms do not have much mathematical requirements about the optimization problems. Due to their evolutionary nature, genetic algorithms will search for solutions without regard to the specific inner workings of the problem. Genetic algorithms can handle any kind of objective functions and any kind of constraints (i.e., linear or nonlinear) defined on discrete, continuous, or mixed search spaces.

2. The ergodicity of evolution operators makes genetic algorithms very effective at performing a global search (in probability). The traditional approaches perform a local search by a convergent stepwise procedure, which compares the values of nearby points and moves to the relative optimal points. Global optima can be found only if the problem possesses certain convexity properties that essentially guarantee that any local optima is a global optima.

3. Genetic algorithms provide us with a great flexibility to hybridize with domain-dependent heuristics to make an efficient implementation for a specific problem.

1.2.6 Genetic Algorithm Vocabulary

Because genetic algorithms are rooted in both natural genetics and computer sciences, the terminology used in genetic algorithm literature is a mixture of the natural and the artificial.

In a biological organism, the structure that encodes the prescription specifying how the organism is to be constructed is called a *chromosome*. One or more chromosomes may be required to specify the complete organism. The complete set of chromosomes is called a *genotype*, and the resulting organism is called a *phenotype*. Each chromosome comprises a number of individual structures called *genes*. Each gene encodes a particular feature of the organism, and the location, or *locus*, of the gene within the chromosome structure determines what particular characteristic the gene represents. At a particular locus, a gene may encode any of several different values of the particular characteristic it represents. The different values of a gene are called *alleles*.

The correspondence of genetic algorithm terms and optimization terms is summarized in Table. 1.1.

Table 1.1. Explanation of genetic algorithm terms

Genetic algorithms	Explanation
1. Chromosome	Solution (coding)
2. Gene (bits)	Part of solution
3. Locus	Position of gene
4. Alleles	Values of gene
5. Phenotype	Decoded solution
6. Genotype	Encoded solution

1.2.7 Examples with Genetic Algorithms

In this section we explain in detail about how a genetic algorithm actually works, using two simple examples.

Example 1.9. Optimization problem. The numerical example of optimization problem is given as follows:

$$f(x_1,x_2) = (-2x_2^3 + 6x_2^2 + 6x_2 + 10) \cdot \sin(\ln(x_1) \cdot e^{x_2})$$
$$0.5 \le x_1 \le 1.1, \qquad 1.0 \le x_2 \le 4.6$$

It is necessary to find: $\max_{x_1,x_2} f(x_1,x_2)$.

A three-dimensional plot of the objective function is shown in Fig. 1.8.

Representation. First, we need to encode decision variables into binary strings. The length of the string depends on the required precision. For example, the domain of variable x_j is $\left[a_j,b_j\right]$ and the required precision is five places after the decimal point. The precision requirements imply that the range of the domain of each variable should be divided into at least $\left(b_j - a_j\right)\times 10^5$ size ranges. The required bits (denoted with m_j) for a variable is calculated as follows:

$$2^{m_j-1} < \left(b_j - a_j\right)\times 10^5 \le 2^{m_j} - 1$$

The mapping from a binary string to a real number for variable x_j is straightforward and completed as follows:

$$x_j = a_j + decimal(\ substring_j\ ) \times \frac{b_j - a_j}{2^{m_j} - 1} ,$$

where $decimal(\ substring_j\ )$ represents the decimal value of $substring_j$ for decision variable x_j .

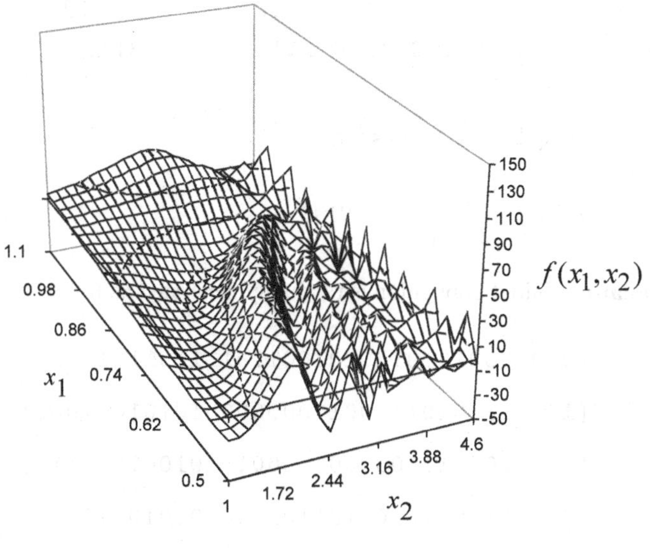

Fig. 1.8. Objective function

Suppose that the precision is set as five places after the decimal point. The required bits for variables x_1 and x_2 is calculated as follows:

$$(1.1 - 0.5) \times 100{,}000 = 60{,}000$$

$$2^{15} < 60{,}000 \leq 2^{16} - 1, \qquad m_1 = 16$$

$$(4.6 - 1.0) \times 100{,}000 = 360{,}000$$

$$2^{18} < 360{,}000 \leq 2^{19} - 1, \qquad m_2 = 19$$

$$m = m_1 + m_2 = 16 + 19 = 35 \ .$$

The total length of a chromosome is 35 bits which can be represented as follows:

|← 35 bits →|

v_j: 0100000101010010 1001101111011111110

|← 16 bits →| |← 19 bits →|

The corresponding values for variables x_1 and x_2 are given below:

	Binary number	Decimal number
x_1	0100000101010010	16722
x_2	1001101111011111110	319230

$$x_1 = 0.5 + 16722 \times \frac{1.1 - 0.6}{2^{16} - 1} = 0.65310 \ ,$$

$$x_2 = 1.0 + 319230 \times \frac{4.6 - 1.0}{2^{19} - 1} = 3.19198 \ .$$

Initial population. Initial population is randomly generated as follows:

v_1 = [01000001010100101001101111011111110]

v_2 = [10001110101110011000000010101001000]

v_3 = [11111000111000001000010101001000110]

v_4 = [01100110110100101101000000010111001]

v_5 = [00000010111101100010001110001101000]

v_6 = [10111110101011011000000010110011001]

v_7 = [00110100010011111000100110011101101]

v_8 = [11001011010100001100010110011001100]

v_9 = [01111110001011101100011101000111101]

v_{10} = [01111101001110101010000010101101010]

The corresponding decimal values are:

$v_1 = [x_1, x_2] =$ [0.653097, 3.191983]

$v_2 = [x_1, x_2] =$ [0.834511,2.809287]

$v_3 = [x_1, x_2] =$ [1.083310,2.874312]

$v_4 = [x_1, x_2] =$ [0.740989,3.926276]

$v_5 = [x_1, x_2] =$ [0.506940,1.499934]

$v_6 = [x_1, x_2] =$ [0.946903,2.809843]

$v_7 = [x_1, x_2] =$ [0.622600,2.935225]

$v_8 = [x_1, x_2] =$ [0.976521,3.778750]

$v_9 = [x_1, x_2] =$ [0.795738,3.802377]

$v_{10} = [x_1, x_2] =$ [0.793504,3.259521]

Evaluation. The process of evaluating the fitness of a chromosome consists of the following three steps:

1°. Convert the chromosome's genotype to its phenotype. Here, this means converting the binary string into relative real values $\mathbf{x}^k = (x_1^k, x_2^k)$, $k = 1, 2, ..., pop\_size$.

2°. Evaluate the objective function $f(\mathbf{x}^k)$.

3°. Convert the value of the objective function into fitness. For the maximization problem, the fitness is simply equal to the value of the objective function $eval(v_k) = f(\mathbf{x}^k)$, $k = 1, 2, ..., pop\_size$.

An evaluation function plays the role of the environment, and it rates chromosomes in terms of their fitness.

The fitness function values of the above chromosomes are as follows:

$eval(v_1)$ = $f(0.653097, 3.191983) = 20.432394$

$eval(v_2)$ = $f(0.834511, 2.809287) = -4.133627$

$eval(v_3)$ = $f(1.083310, 2.874312) = 28.978472$

$eval(v_4)$ = $f(0.740989, 3.926276) = -2.415740$

$eval(v_5)$ = $f(0.506940, 1.499934) = -2.496340$

$eval(v_6)$ = $f(0.946903, 2.809843) = -23.503709$

$eval(v_7)$ = $f(0.622600, 2.935225) = -13.878172$

$eval(v_8)$ = $f(0.976521, 3.778750) = -8.996062$

$eval(v_9)$ = $f(0.795738, 3.802377) = 6.982708$

$eval(v_{10})$ = $f(0.793504, 3.259521) = 6.201905$

It is clear that chromosome v_3 is the strongest one and that chromosome v_6 is the weakest one.

Selection. In most practices, a *roulette wheel* approach is adopted as the selection procedure [22]; it belongs to the fitness-proportional selection and can select a new population with respect to the probability distribution based on fitness values. The roulette wheel can be constructed as follows:

1. Calculate the fitness value $eval(v_k)$ for each chromosome v_k:

$$eval\ (v_k) = f\ (\mathbf{x}^k)\ ,\ k = 1, 2, ..., pop\_size\ .$$

2. Calculate the total fitness for the population:

$$F = \sum_{k=1}^{pop\_size} \left(eval\ (v_k) - \min_{j=1,pop\_size} \{eval\ (v_j)\} \right)$$

3. Calculate selection probability p_k for each chromosome v_k:

$$p_k = \frac{eval(v_k) - \min_{j=1,pop\_size} \{eval\ (v_j)\}}{F}\ ,\quad k = 1, 2, ..., pop\_size\ .$$

4. Calculate cumulative probability q_k for each chromosome v_k:

$$q_k = \sum_{j=1}^{k} p_j \;, \qquad k = 1, 2, \ldots, pop\_size \;.$$

The selection process begins by spinning the roulette wheel pop_size times; each time, a single chromosome is selected for a new population in the following way:

1°. Generate a random number r from the range $[0,1]$.

2°. If $r \le q_1$, then select the first chromosome v_1; otherwise, select the k th chromosome v_k ($2 \le k \le pop\_size$) such that $q_{k-1} < r \le q_k$.

The total fitness F of the population is:

$$F = \sum_{k=1}^{10} \left(eval\ (v_k) - \min_{j=1,10} \{ eval\ (v_j) \} \right) = 242.208919 \;.$$

The probability of a selection p_k for each chromosome v_k ($k = 1, 2, \ldots, 10$) is as follows:

$p_1 = 0.181398,$ $p_2 = 0.079973,$ $p_3 = 0.216681,$

$p_4 = 0.087065,$ $p_5 = 0.086732,$ $p_6 = 0.000000,$

$p_7 = 0.039741,$ $p_8 = 0.059897,$ $p_9 = 0.125868,$

$p_{10} = 0.122645$.

The cumulative probability q_k for each chromosome v_k ($k = 1, 2, \ldots, 10$) is as follows:

$q_1 = 0.181398,$ $q_2 = 0.261370,$ $q_3 = 0.478052,$

$q_4 = 0.565117,$ $q_5 = 0.651849,$ $q_6 = 0.651849,$

$q_7 = 0.691590,$ $q_8 = 0.751487,$ $q_9 = 0.877355,$

$q_{10} = 1.000000$.

Now we are ready to spin the roulette wheel 10 times, and each time we select a single chromosome for a new population. Let us assume that a random sequence of 10 numbers from the range $[0,1]$ is as follows:

0.301431	0.322062	0.766503	0.881893
0.350871	0.583392	0.177618	0.343242
0.032685	0.197577.		

The first number $r_1 = 0.301431$ is greater than q_2 and smaller than q_3, meaning that the chromosome v_3 is selected for the new population; the second number $r_2 = 0.322062$ is greater than q_2 and smaller than q_3, meaning that the chromosome v_3 is again selected for the new population; and so on. Finally, the new population consists of the following chromosomes:

$v_1' =$ [11111000111000001000010101001000110] (v_3)

$v_2' =$ [11111000111000001000010101001000110] (v_3)

$v_3' =$ [11001011010100001100010110011001100] (v_8)

$v_4' =$ [01111110001011101100011101000111101] (v_9)

$v_5' =$ [11111000111000001000010101001000110] (v_3)

$v_6' =$ [01100110110100101101000000010111001] (v_4)

$v_7' =$ [01000001010100101001101111011111110] (v_1)

$v_8' =$ [11111000111000001000010101001000110] (v_3)

$v_9' =$ [01000001010100101001101111011111110] (v_1)

$v_{10}' =$ [10001110101110011000000010101001000] (v_2)

Crossover. Crossover used here is one-cut-point method, which randomly selects one cut-point and exchanges the right parts of two parents to generate offspring. Consider two chromosomes as follows, and the cut-point is randomly selected after the 17th gene:

↓

$v_1 =$ [11111000111000001000010101001000110]
$v_2 =$ [10001110101110011000000010101001000]

The resulting offspring by exchanging the right parts of their parents would be as follows:

$v_1 =$ [11111000111000001 000000010101001000]
$v_2 =$ [10001110101110011 000010101001000110]

The probability of crossover is set as $p_c = 0.25$, so we expect that, on average, 25% of chromosomes undergo crossover. Crossover is performed in the following way:

Procedure: Crossover
begin
  $k := 0$;
  while ($k \leq 10$) **do**
    r_k := random number from [0,1] ;
    if ($r_k < 0.25$) **then**
      select v_k as one parent for crossover;
    end ;
    $k := k + 1$;
  end ;
end.

Assume that the sequence of random numbers is:

0.625721	0.266823	0.288644	0.295114
0.163274	0.567461	0.085940	0.392865
0.770714	0.548656 .		

This means that the chromosomes v_5' and v_7' were selected for crossover. We generate a random integer number *pos* from the range [1, 34] (because 35 is the total length of a chromosome) as cutting point or in other words, the position of the crossover point. Assume that the generated number *pos* equals 1, the two chromosomes are cut after the first bit, and offspring are generated by exchanging the right parts of them as follows:

$$\downarrow$$

v_5' = [11111000111000001000010101001000110]
v_7' = [01000001010100101001101111011111110]

$$\Downarrow$$

v_5' = [1 1000001010100101001101111011111110]
v_7' = [0 1111000111000001000010101001000110]

Mutation. Mutation alters one or more genes with a probability equal to the mutation rate. Assume that the 18th gene of the chromosome v_1' is selected for a mutation. Since the gene is 1, it would be flipped into 0. Thus the chromosome after mutation would be:

v_1' = [11111000111000001**0**00010101001000110]

$$\Downarrow$$

v_1' = [11111000111000001**1**00010101001000110]

The probability of mutation is set as $p_m = 0.01$, so we expect that, on average, 1% of the total bit of the population would undergo mutation. There are $m \times pop\_size = \; = 35 \times 10 = 350$ bits in the whole population; we expect 3.5 mutations per generation. Every bit has an equal chance to be mutated. Thus we need to generate a sequence of random numbers r_k $(k = 1..350)$ from the range [0,1]. Suppose that the following genes will go through mutation:

Position of gene in population	Number of chromosome	Position of gene in population	Random number I r_k
111	4	6	0.009857
172	5	32	0.003113
211	7	1	0.000946
347	10	32	0.001282

After mutation, we get the final population as follows:

v_1' = [11111000111000001000010101001000110]
v_2' = [11111000111000001000010101001000110]
v_3' = [11001011010100001100010110011001100]
v_4' = [01111010001011101100011101000111101]
v_5' = [11000001010100101001101111011110110]
v_6' = [01100110110100101101000000010111001]
v_7' = [11111000111000001000010101001000110]
v_8' = [11111000111000001000010101001000110]
v_9' = [01000001010100101001101111011111110]
v_{10}' = [10001110101110011000000010101000000] .

The corresponding decimal values of variables x_1 and x_2 and fitness are as follows:

$f(1.083310,2.874312)=28.978472$
$f(1.083310,2.874312)=28.978472$
$f(0.976521,3.778750)=-8.996062$
$f(0.786363,3.802377)=9.366723$
$f(0.953101,3.191928)=-23.229745$
$f(0.740989,3.926276)=-2.415740$
$f(1.083310,2.874312)=28.978472$
$f(1.083310,2.874312)=28.978472$
$f(0.653097,3.191983)=20.432394$
$f(0.834511,2.809232)=-4.138564$

Now we just completed one iteration of the genetic algorithm. The test run is terminated after 1000 generations. We have obtained the best chromosome in the 419th generation:

v^* = [01000011000100110110010011011101001]

$eval\ (v^*) = f(0.657208,2.418399) = 31.313555$
$x_1^* = 0.657208 \qquad x_2^* = 2.418399$
$f(x_1^*, x_2^*) = 31.313555.$

Example 1.10. Word matching problem. Another nice example to show the power of genetic algorithms, the *word matching problem* tries to evolve an expression of «live and learn» from the randomly-generated lists of letters with a genetic algorithm. Since there are 26 possible letters plus space character for each of 14 locations in the list, the probability that we get the correct phrase in a pure random way is $(1/27)^{14} = 9.14\times 10^{-22}$, which is almost equal to zero.

We use a list of ASCII integers to encode the string of letters. The lowercase letters in ASCII are represented by numbers in the range [97,122] and the space character is 32 in the decimal number system. For example, the string «live and learn» is converted into the following chromosome represented with ASCII integers:

```
[108,105,118,101, 32, 97,110,100, 32,108,101, 97,114,110]
```

Generate an initial population of 10 random phrases as follows:

```
[115,111,113,114,100,109,119,115,118,106,108,116,112,106]
[116,111,112,122,122,119,103,106,122,100,114, 99,115,103]
[117,106,111,102,113, 97, 32,114,114,112,117,117,103,115]
[ 32, 97,114,118,104, 99,117,105,100,118, 98,114,102, 32]
[119, 99,117,103,102,122,112, 32,114,122,101,107,101,106]
```

```
[116,117,100,120, 32, 32, 97,122,118,121,104,103, 97,113]
[118,100,104,122,101,102,114,113,113, 98,111,114, 98,116]
[120,106,105,101, 98,110,108,116, 97,118,104,116,103,118]
[102,117,115,100,122,107,118,104,107,112, 99,109,120,109]
[100,110,100,102,115, 32,107,104,104, 32,121,109, 99,120]
```

Now, we convert this population to string to see what they look like:

```
«soqrdmwsvjltpj»
«topzzwgjzdrcsg»
«ujofqa rrpuugs»
« arvhcuidvbrf »
«wcugfzp rzekej»
«tudx  azvyhgaq»
«vdhzefrqqborbt»
«xjiebnltavhtgv»
«fusdzkvhkpcmxm»
«dndfs khh ymcx»
```

Fitness is calculated as the number of matched letters. For example, the fitness for string «ujofqa rrpuugs» is 1. Only mutation is used which results in a change to a given letter with a given probability. Now, we run our genetic algorithm with 32 generations to see how well it works. The best one of each generation is listed in Table 1.2.

Table 1.2. The best string for each generation

Gen.	String	Fitness function	Gen.	String	Fitness function
1	ujofqa rrpuugs	1	17	liie xnd leaez	10
2	wfugfzpnrzewen	2	18	liye xnt learn	11
3	wiipvap ozekej	3	19	liye xnt learn	11
4	wi gvahdlzerej	4	20	liye xnt learn	11
5	liigvapt yekej	5	21	live xnd nearn	12
6	liigvapt yekej	5	22	live xnd nearn	12
7	lqie zp zekrj	6	23	live xnd nearn	12
8	lqie zp zekrj	6	24	live xnd nearn	12
9	lqie zp zekrj	6	25	live gnd learn	13
10	ljie zni yeaez	7	26	live gnd learn	13
11	ljie zni yeaez	7	27	live gnd learn	13
12	liie xnt beaez	8	28	live gnd learn	13
13	liye nd yeaez	9	29	live and learn	14
14	liye nd yeaez	9	30	live and learn	14
15	liye nd yeaez	9	31	live and learn	14
16	liie xnd leaez	10	32	live and learn	14

After 29 generations, the population produced the desired phrase. The total examined chromosomes are 290. If we use pure random method to produce 290 random phrases, could we have a match?

1.3 Neural Networks

This chapter is written on the basis of the works [3, 4, 24, 25]. The additional information relative to artificial neural networks can be found in the works [26 – 31].

1.3.1 Neural Net Basics

The imitation of human minds in machines has inspired scientists for the last century. About 50 years ago, researchers created the first electronic hardware models of nerve cells. Since then, the greater scientific community has been working on new mathematical models and training algorithms. Today, so-called neural nets absorb most of the interest in this domain. Neural nets use a number of simple computational units called "neurons", of which each tries to imitate the behavior of a single human brain cell. The brain is considered as a "biological neural net" and implementations on computers are considered as "neural nets". Fig. 1.9 shows the basic structure of such a neural net.

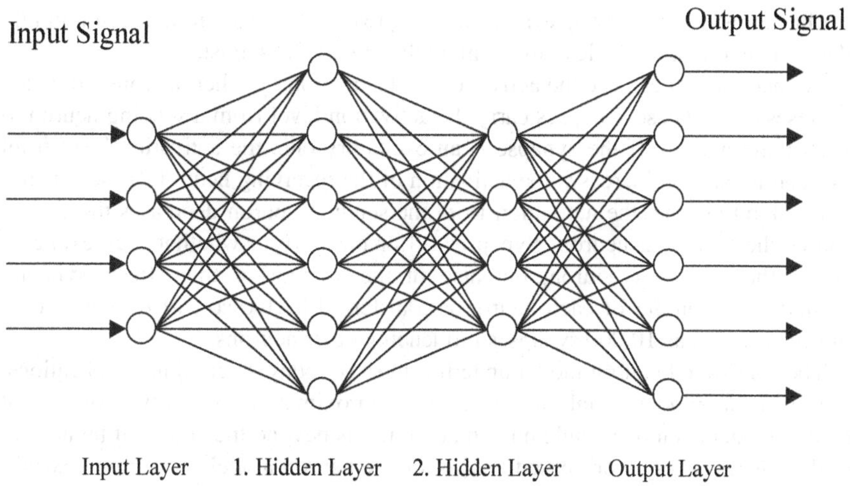

Fig 1.9. Basic structure of an artificial neural net

Each neuron in a neural net processes the incoming inputs to an output. The output is then linked to other neurons. Some of the neurons form the interface of the neural net. The neural net shown in Fig. 1.9 has a layer for the input signals and one for the output signals. The information enters the neural net at the input layer. All layers of the neural net process these signals through the net until they reach the output layer.

The objective of a neural net is to process the information in a way that it is previously trained. Training uses either sample data sets of inputs and corresponding outputs or a teacher who rates the performance of the neural net. For this training, neural nets use so-called learning algorithms. Upon creation, a neural net is dumb and does not exhibit any behavior at all. The learning algorithms then modify the individual neurons of the net and the weight of their connections in such a way that the behavior of the net reflects the desired one.

1.3.2 Mimic of Human Nerve Cells

Researchers in the area of neural nets have analyzed various models of human brain cells. In the following, we only describe the one most commonly used in industrial applications.

The human brain contains about 10^{11} nerve cells with about 10^{14} connections to each other. Fig. 1.10 shows the simplified scheme of such a human neuron. The cell itself contains a kernel, and the outside is an electrical membrane. Each neuron has an activation level, which ranges between a maximum and a minimum. Hence, in contrast to Boolean logic, more then two values exist.

To increase or decrease the activation of this neuron by other neurons, so-called synapses exist. These synapses carry the activation level from a sending neuron to a receiving neuron. If the synapse is an excitatory one, the activation level from the sending neuron increases the activation of the receiving neuron. If the synapse is an inhibiting one, the activation from the sending neuron decreases the activation of the receiving neuron. Synapses differ not only in whether they excite or inhibit the receiving neuron, but also in the amount of this effect (synaptic strength). The output of each neuron is transferred by the so-called axon, which ends in as much as 10,000 synapses influencing other neurons.

The considered neuron model underlies most of today's neural net applications. Note that this model is only a very coarse approximation of reality. You cannot exactly model even one single human neuron; it is beyond the ability of humans to model. Hence, every work based on this simple neuron model is unable to exactly copy the human brain. However, many successful applications using this technique prove the benefit of neural nets based on the simple neuron model.

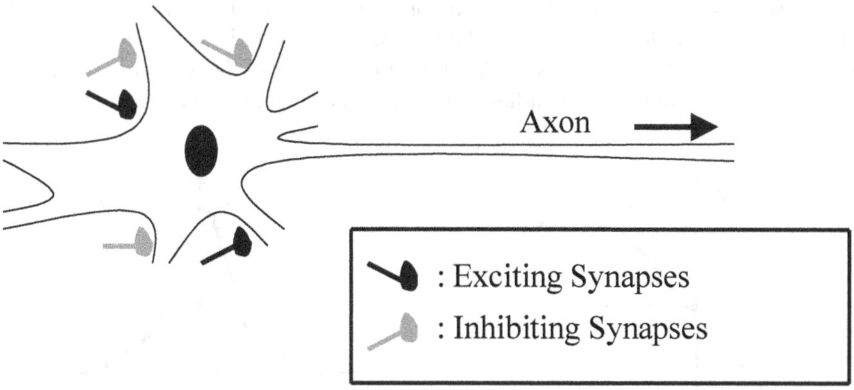

Fig. 1.10. Simplified scheme of a human neuron

1.3.3 Mathematical Model of a Neuron

Various mathematical models are based on the simple neuron concept. Fig. 1.11 shows the most common one. First, the so-called propagation function combines all inputs X_i that stem from the sending neurons. The means of combination is a weighted sum, where the weights w_i represent the synaptic strength. Exciting synapses have positive weights, inhibiting synapses have negative weights. To express a background activation level of the neuron, an offset (bias) Θ is added to the weighted sum.

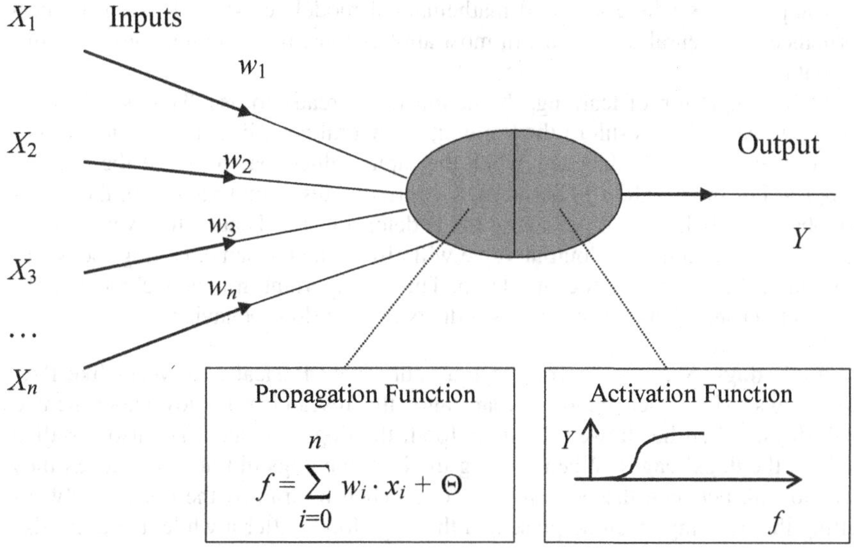

Fig. 1.11. Simple mathematical model of a neuron.

The so-called activation function computes the output signal Y of the neuron from the activation level f. For this, the activation function is of the sigmoid type as plotted in the lower right box of Fig. 1.11. Other types of the activation function are the linear function and the radial-symmetric function showed in Fig. 1.12.

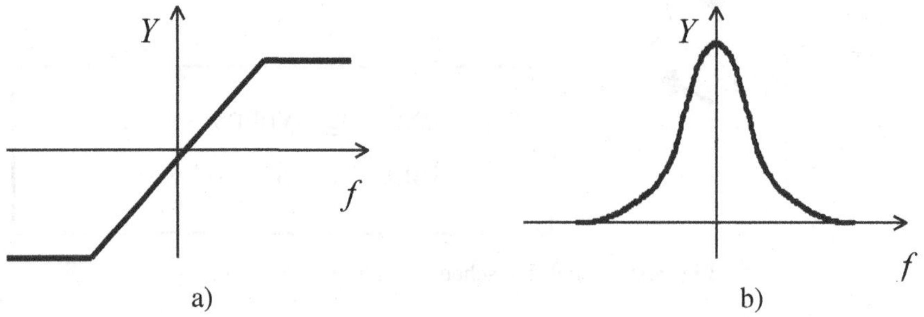

Fig. 1.12. Activation functions of a neuron:a) linear; b) radial-symmetric

1.3.4 Training Neural Nets

There are multiple ways to build a neural net. They differ in their topology and the learning methods they employ.

The first step in designing a neural net solution is teaching the desired behavior. This is called the learning phase. Here, you can either use sample data sets or a "teacher". A teacher is either a mathematical function or a person who rates the quality of the neural net performance. Since neural nets are mostly used for complex applications where no good mathematical models exist, and rating the performance of a neural net is hard in most applications, most applications use sample data training.

After completion of learning, the neural net is ready to use. This is called the working phase. As a result of the training, the neural net will output values similar to those in the sample data sets when the input values match one of the training samples. For input values in between, it approximates output values. In the working phase, the behavior of the neural net is deterministic. That is, for every combination of input values, the output value will always be the same. During the working phase, the neural net does not learn. This is important in most technical applications to ensure that the system never drifts to hazardous behavior.

Pavlov's dogs. So, how do you teach a neural net? Basically, it works like Pavlov's dogs. More then hundred years ago, the researcher Pavlov experimented with dogs. When he showed the dogs food, the dogs salivated. He also installed bells in the dogs' cages. When he rang the bell, the dogs did not salivate, as they saw no link between the bell and the food. Then he trained the dogs by always letting the bell ring when he presented the dogs food. After a while, the dogs also salivated when just the bell rang and he showed no food.

Fig. 1.13 shows how the simple neuron model can represent Pavlov's experiment. There are two input neurons: one represents the fact that the dog sees food, the other one the fact that the bell rings. Both input neurons have links to the output neuron. These links are the synapses. The thickness of the lines represents synapse weights. Before learning, the dog only reacts to the food and not the bell. Hence, the line from the left input neuron to the output neuron is thick, while the line from the right input neuron to the output neuron is very thin.

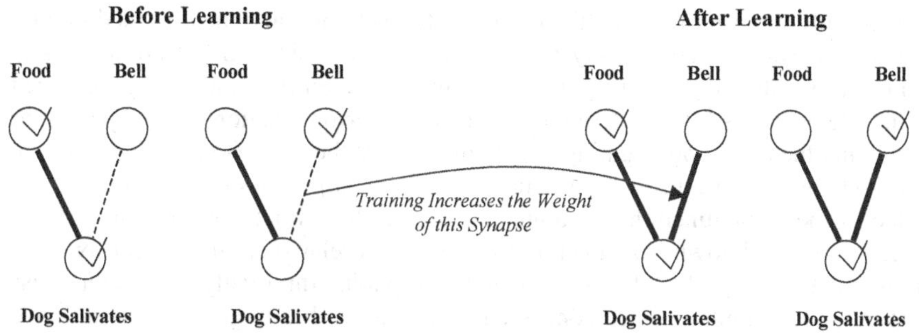

Fig. 1.13. Principle of the Pavlov dog experiment

The Hebbian learning rule. Constantly letting the bell ring when food is presented creates an association between the bell and the food. Hence, the right line also becomes thicker - the synapse weight increases. From these experiments, in 1949 a researcher by the name of Hebb deduced the following learning rule:

Increase weight* to active input neuron, if the output of this neuron *should be active.
Decrease weight* to active input neuron, if the output of this neuron *should be inactive.

This rule, called the Hebbian rule, is the forerunner of all learning rules, including today's most used neural net learning algorithm, the so-called error back propagation algorithm.

1.3.5 Error Back Propagation Algorithm

The learning rule for multilayer neural nets is called the "generalized delta rule", or the "back propagation rule", and was suggested in 1986 by Rumelhart, McClelland, and Williams. It signaled the renaissance of the entire subject. It was later found that Parker had published similar results in 1982, and then Werbos was shown to have done the work in 1984. Such is the nature of science; groups working in diverse fields cannot keep up with all the advances in other areas, and there is often duplication of effort. However, the paper of Rumelhart et al. published in "Nature" (1986) is still one of the most important works in this field.

Learning of the net is begun by the net being shown a pattern and calculating its response. Comparison with the desired response enables the weights to be altered so that the network can produce a more accurate output the next time. The learning rule provides the method for adjusting the weights in the network. Information about the output is available to units in earlier layers, so that these units can have their weights adjusted so as to decrease the error the next time.

When we show the untrained network an input pattern, it will produce any random output. An error function represents the difference between the network's current output and the correct output that we want it to produce. In order to learn successfully we want to make the output of the net approach the designed output, that is, we want to continually reduce the value of this error function. This is achieved by adjusting the weights on the links between the units; the generalized delta rule does this by calculating the value of the error function for that particular input, and then back-propagating (hence the name!) the error from one layer to the previous one. Each unit in the net has its weights adjusted so that it reduces the value of the error function; for units actually on the output, their output and desired output are known, so adjusting the weights is relatively simple, but for units in the middle layer, the adjustment is not so obvious. Intuitively, we might guess that the hidden units that are connected to outputs with a large error should have their weights adjusted a lot, while units that feed almost correct outputs should not be altered much. In other words, the weights for a particular node should be adjusted in direct proportion to the error in the units to which it is connected; that is why back-propagating these errors through the net allows the weights between all the layers to be correctly adjusted. In this way the error function is reduced and the network learns.

The main formulae for the error back propagation method have been obtained in [3, 4].

The notation used is as follows:

E_p is the error function for pattern p;

t_{pj} is the target output for pattern p on node j;

o_{pj} is the actual output for pattern p on node j;

w_{ij} is the weight from node i to node j.

Let us define the error function to be proportional to the square of the difference between the actual and desired output, for all the patterns to be learnt:

$$E_p = \frac{1}{2}\sum_j \left(t_{pj} - o_{pj}\right)^2 . \tag{1.2}$$

The $\frac{1}{2}$ makes the math a bit simpler, and brings this specific error function into line with other similar measures.

The activation of each unit j, for pattern p, can be written simply as the weighted sum:

$$net_{pj} = \sum_i w_{ij} o_{pi} \; . \tag{1.3}$$

The output of each unit j is the threshold function f_j activated on the weighted sum. In the multilayer networks, it is usually the sigmoid function, although any continuously differentiable monotonic function can be used:

$$o_{pj} = f_j\left(net_{pj}\right) \; . \tag{1.4}$$

We can write by the chain rule:

$$\frac{\partial E_p}{\partial w_{ij}} = \frac{\partial E_p}{\partial\, net_{pj}} \frac{\partial\, net_{pj}}{\partial w_{ij}} \; . \tag{1.5}$$

Looking at the second term in (1.5), and substituting in (1.3)

$$\frac{\partial net_{pj}}{\partial w_{ij}} = \frac{\partial}{\partial w_{ij}} \sum_k w_{kj} o_{pk} = \sum_k \frac{\partial w_{jk}}{\partial w_{ij}} o_{pk} = o_{pi} \; , \tag{1.6}$$

since $\frac{\partial w_{kj}}{\partial w_{ij}} = 0$, except when $k = i$, and this derivative is equal to unity.

We can define the change in error as a function of the change in the net inputs to a unit as

$$-\frac{\partial E_p}{\partial net_{pj}} = \delta_{pj} \; , \tag{1.7}$$

and so (1.5) becomes

$$-\frac{\partial E_p}{\partial w_{ij}} = \delta_{pj} o_{pi} \; . \tag{1.8}$$

Decreasing the value E_p therefore means making the weight changes proportional to $\delta_{pj} o_{pj}$, i.e.,

$$\Delta_p w_{ij} = \eta \delta_{pj} o_{pi} \; , \tag{1.9}$$

where η is a learning rate.

We now need to know what δ_{pj} is for each of the units. Using (1.7) and the chain rule, we can write:

$$\delta_{pj} = -\frac{\partial E_p}{\partial net_{pj}} = -\frac{\partial E_p}{\partial o_{pj}} \frac{\partial o_{pj}}{\partial net_{pj}} \; . \tag{1.10}$$

Consider the second term, and from (1.4):

$$\frac{\partial o_{pj}}{\partial net_{pj}} = f_j'\left(net_{pj}\right) \; . \tag{1.11}$$

Consider now the first term in (1.10). From (1.2), we can easy obtain

$$\frac{\partial E_p}{\partial o_{pj}} = -\left(t_{pj} - o_{pj}\right) . \tag{1.12}$$

Thus

$$\delta_{pj} = f_j'\left(net_{pj}\right)\left(t_{pj} - o_{pj}\right) . \tag{1.13}$$

This is useful for the output units, since the target and output are both available, but not for the hidden units, since their targets are not known.

Therefore, if unit j is not an output unit, we can write, by the chain rule again, that

$$\frac{\partial E_p}{\partial o_{pj}} = \sum_k \frac{\partial E_p}{\partial net_{pk}} \frac{\partial net_{pk}}{\partial o_{pj}} = \sum_k \frac{\partial E_p}{\partial net_{pk}} \frac{\partial}{\partial o_{pj}} \sum_i w_{ik} o_{pi} , \tag{1.14}$$

$$\sum_k \frac{\partial E_p}{\partial net_{pk}} \frac{\partial}{\partial o_{pj}} \sum_i w_{ik} o_{pi} = -\sum_k \delta_{pk} w_{jk} , \tag{1.15}$$

using (1.3) and (1.7), noticing that the sum drops out since the partial differential is non-zero for only one value, just as in (1.6). Substituting (1.15) in (1.10), we get finally

$$\delta_{pj} = f_j'\left(net_{pj}\right) \sum_k \delta_{pk} w_{jk} . \tag{1.16}$$

Equations (1.13) and (1.16) are the basis of the multilayer network learning method.

One advantage of using the sigmoid function as the nonlinear threshold function is that it is quite like the step function, and so should demonstrate behavior of a similar nature. The sigmoid function is defined as

$$f(net) = \frac{1}{1 + e^{-k \cdot net}}$$

and has the range $0 < f(net) < 1$. k is a positive constant that controls the "spread" of the function - large values of k squash the function until as $k \to \infty$ when $f(net) \to$ Heaviside function. It also acts as an automatic gain control, since for small input signals the slope is quite steep and so the function is changing quite rapidly, producing a large gain. For large inputs, the slope and thus the gain is much less. This means that the network can accept large inputs and still remain sensitive to small changes.

A major reason for its use is that it has a simple derivative, however, and this makes the implementation of the back-propagation system much easier. Given that the output of unit, o_{pj} is given by

$$o_{pj} = f(net) = \frac{1}{1+e^{-k \cdot net}},$$

the derivative with respect to that unit, $f'(net)$, is give by

$$f'(net) = \frac{ke^{-k \cdot net}}{(1+e^{-k \cdot net})^2} = \frac{kf(net)}{1-f(net)} = ko_{pj}(1-o_{pj}).$$

The derivative is therefore a simple function of the outputs.

1.3.6 The Multilayer Neural Network Learning Algorithm

The algorithm for the multilayer neural network learning that implements the back-propagation training rule is shown below. It requires the units to have thresholding nonlinear functions that are continuously differentiable, i.e. smooth everywhere. We have assumed the use of the sigmoid function, $f(net) = \frac{1}{1+e^{-k \cdot net}}$, since it has a simple derivative.

The multilayer neural network learning algorithm includes the following steps.

1°. Initialize weights and thresholds. Set all weights and thresholds to small random values.

2°. Present input and desired output.

Present input $X_p = \{x_0, x_1, ..., x_{n-1}\}$ and target output $T_p = \{t_0, t_1, ..., t_{m-1}\}$, where n is a number of input nodes and m is a number of output nodes. Set $w_0 = -\Theta$ the bias, and $x_0 = 1$.

For classification, T_p is set to zero except for one element set to 1 that corresponds to the class that X_p is in.

3°. Calculate actual output.

Each layer calculates

$$y_{pj} = f\left[\sum_{i=0}^{n-1} w_i x_i\right]$$

and passes that as input to the next layer. The final layer output values are o_{pj}.

4°. Adapt weights.

Start from the output layer, and work backwards

$$w_{ij}(t+1) = w_{ij}(t) + \eta\delta_{pj}o_{pj},$$

where $w_{ij}(t)$ represents the weights from node i to node j at time t, η is a learning rate, and δ_{pj} is an error term for pattern p on node j.

For output units

$$\delta_{pj} = ko_{pj}\left(1-o_{pj}\right)\left(t_{pj}-o_{pj}\right) .$$

For hidden units

$$\delta_{pj} = ko_{pj}\left(1-o_{pj}\right)\sum_{k}\delta_{pk}w_{jk} ,$$

where the sum is over the k nodes in the layer above node j.

References

1. Zadeh, L.A.: The Concept of a Linguistic Variable and its Application to Approximate Reasoning, Part 1-3. Information Sciences 8, 199–251 (1975); 9, 301 - 357, 43 - 80 (1976)
2. Zadeh, L.A.: Fuzzy Sets as a Basic for a Theory of Possibility. Fuzzy Sets and Systems 1, 3–28 (1978)
3. Rummelhart, D.E., McClelland, J.L.: Parallel Distributed Processing: Explorations in the Microstructure of Cognition, vol. 1,2, p. 320. The MIT Press (1986)
4. Rumelhart, D.E., Hinton, G.E., Williams, R.J.: Learning Internal Representation by Back - Propagation Errors. Nature 323, 533–536 (1986)
5. Eickhoff, P.: System Identification: Parameter and State Estimation. Wiley, London (1974)
6. Tsypkin, Y.Z.: Information Theory of Identification, p. 320. Nauka, Moscow (1984) (in Russian)
7. Shteinberg, S.E.: Identification in Control Systems, p. 81. Energoatomizdat, Moscow (1987) (in Russian)
8. Reklaitis, G.V., Ravindran, A., Ragsdell, K.M.: Engineering Optimization. In: Methods and Applications. John Wiley & Sons, New York (1983)
9. Goldberg, D.: Genetic Algorithms in Search, Optimization and Machine Learning. Addison Wesley (1989)
10. Tang, K.S., Man, K.F., Kwong, S., He, Q.: Genetic Algorithms and Their Applications. IEEE Signal Processing Magazine, 22–36 (November 1996)
11. Kaufmann, A., Gupta, M.M.: Introduction to Fuzzy Arithmetic: Theory and Applications. Van Nostrand Reinhold, New York (1985)
12. Pospelov, D.A. (ed.): Fuzzy Sets in Management Models and Artificial Intelligence. Nauka, Moscow (1986) (in Russian)
13. Bellman, R.E., Zadeh, L.A.: Decision-Making in a Fuzzy Environment. Management Science 17(4), 141–164 (1970)
14. Yager, R.R.: Fuzzy Set and Possibility Theory: Recent Developments. Pergamon Press, New York (1982) (Russian Translation, 1986)
15. Dubois, D., Prade, H.: Possibility Theory: An Approach to Computerized Processing of Uncertainty, p. 263. Plenum Press, New York (1988)

16. Borisov, A.N., Krumberg, O.A., Fedorov, I.P.: Decision Making based on Fuzzy Models: Examples Use, p. 184. Zinatne, Riga (1990) (in Russian)
17. Zimmermann, H.-J.: Fuzzy Set Theory and Its Applications, p. 315. Kluwer, Dordrecht (1991)
18. Zadeh, L., Kacprzyk, J.: Fuzzy Logic for the Management of Uncertainty, p. 676. John Wiley & Sons, Chichester (1992)
19. Klir, G., Yuan, B.: Fuzzy Sets and Fuzzy Logic: Theory and Applications, p. 592. Prentice Hall PTR, New York (1995)
20. Ross, T.J.: Fuzzy Logic with Engineering Applications, p. 593. Wiley, Chichester (1995)
21. Pedrycz, W., Gomide, F.: An introduction to fuzzy sets: Analysis and Design. A Bradford Book, p. 465. The MIT Press (1998)
22. Gen, M., Cheng, R.: Genetic Algorithms and Engineering Design, p. 352. John Wiley & Sons, New York (1997)
23. Haupt, R., Haupt, S.: Practical Genetic Algorithms, p. 177. John Willey & Sons, New York (1998)
24. Hinton, G.E.: How Neural Networks Learn from Experience. Scientific American, 145–151 (September 1992)
25. Hung, S.L., Adeli, H.: Machine Learning, p. 211. John Willey & Sons, New York (1995)
26. Mkrtchan, S.O.: Neurons and Neural Networks, p. 272. Energia, Moscow (1971) (in Russian)
27. Amosov, N.M. (ed.): Neurocomputers and Intelligent Robots, Kiev, Naukova Dumka, p. 272 (1991) (in Russian)
28. von Altrock, C.: Fuzzy Logic & NeuroFuzzy Applications Explained, p. 350. Prentice Hall PTR, New Jersey (1995)
29. Lin, C.T., Lee, C.S.: Neural Fuzzy Systems. Prentice Hall PTR, New York (1996)
30. Nauck, D., Klawonn, F., Kruse, R.: Foundation of Neuro–Fuzzy Systems, p. 305. John Willey & Sons, New York (1997)
31. Bishop, C.M.: Neural Networks for Pattern Recognition, p. 482. Oxford University Press (2002)

Chapter 2
Direct Inference Based on Fuzzy Rules

This chapter is devoted to the methodology aspects of identification and decision making on the basis of intellectual technologies. The essence of intellectuality consists of representation of the structure of the object in the form of linguistic IF-THEN rules, reflecting human reasoning on the common sense and practical knowledge level. The linguistic approach to designing complex systems based on linguistically described models was originally initiated by Zadeh [1] and developed further by Tong [2], Gupta [3], Pedrych [4 – 6], Sugeno [7], Yager [8], Zimmermann [9], Kacprzyk [10], Kandel [11]. The main principles of fuzzy modeling were formulated by Yager [8]. The linguistic model is a knowledge-based system. The set of fuzzy IF-THEN rules takes the place of the usual set of equations used to characterize a system [12 – 14]. The fuzzy sets associated with input and output variables are the parameters of the linguistic model [15]; the number of the rules determines its structure. Different interpretations of the knowledge contained in these rules, which are due to different reasoning mechanisms, result in different types of models.

This monograph can be regarded as one of the possible approaches to modeling intellectual activity on the basis of knowledge engineering. The herein proposed intellectual technique of identification, which supports the human-system approach to the solution of the simulation tasks [16], represents some general framework for design of fuzzy expert systems. The aim of this chapter is to introduce the main formalisms necessary for the definition of fuzzy knowledge bases being the medium of expert information. All intellectual tasks discussed above can be considered to be the tasks of identification having the following common properties [17]:

1) the output variable is associated with the object of identification, that is with the type of the decision made,
2) the input variables are associated with the parameters of the identification object state,
3) output and input variables can have quantitative and qualitative estimations,
4) the structure of the interconnection between output and input variables is described by IF <inputs> THEN <outputs> rules using qualitative estimations of variables and representing fuzzy knowledge bases.

A.P. Rotshtein et al.: Fuzzy Evidence in Identif., Forecast. and Diagn., STUDFUZZ 275, pp. 39–53.
springerlink.com

A fuzzy knowledge base represents some combination of IF <inputs>, THEN <output> rules, which reflect expert experience and the understanding of cause-effect connections in the decision making task considered (control, diagnosis, prediction and other ones). Peculiarity of the similar expressions consists in the fact that their adequacy doesn't change with the insignificant deviations of experiment conditions. Therefore, formation of the fuzzy knowledge base can be treated as an analog of the structural identification [12 – 14] stage, which involves simulation of the rough object model. In this case, the results of fuzzy evidence depend on the forms of fuzzy terms membership functions, which are used to estimate object inputs and outputs. In addition, the combination of IF-THEN rules can be considered as a set of expert points in input-output space. Application of the fuzzy logic evidence apparatus allows us to restore and identify the multidimensional surface according to these points, which allows us to receive output values with various combinations of input variables values available.

Work [17] is the basis of this chapter.

2.1 Formalization of Source Information

2.1.1 Inputs and Outputs of an Object

Here we consider an object with one output and n inputs of the form:

$$y = f_y(x_1, x_2, ..., x_n) \quad , \tag{2.1}$$

where y is the output variable; $x_1, x_2, ..., x_n$ are the input variables.

Variables $x_1, x_2, ..., x_n$ and y can be quantitative and qualitative. The examples of quantitative variables are: VEHICLE SPEED = [0, 160] km/h, PATIENT TEMPERATURE = [36, 41] °C, REACTOR LOAD DOZE = [6, 20]%, and other variables, easily measured using accepted for them quantitative scales.

The example of a variable for which there is no natural scale is the LEVEL OF OPERATOR STRESS, which can be estimated by qualitative terms (low, average, high) or measured by artificial scales, for example, using 5-, 10- or 100- points systems.

For quantitative variables some known intervals of change are suggested:

$$U_i = [\, \underline{x}_i, \overline{x}_i \,] \ , \ i = \overline{1,n} \quad , \tag{2.2}$$

$$Y = [\, \underline{y}, \overline{y} \,] \quad , \tag{2.3}$$

where $\underline{x}_i$ $(\overline{x}_i)$ is the lower (upper) value of input variable x_i , $i = \overline{1,n}$;

$\underline{y}$ ($\overline{y}$) is the lower (upper) value of output variable y.

It is suggested that the sets of all possible values for qualitative variables $x_1 \div x_n$ and y are known:

$$U_i = \{v_i^1, v_i^2, ..., v_i^{q_i}\}, \; i = \overline{1,n}, \tag{2.4}$$

$$Y = \{y^1, y^2, ..., y^{q_m}\}, \tag{2.5}$$

where v_i^1 $\left(v_i^{q_i}\right)$ is the point estimation corresponding to the smallest (largest) value of input variable x_i;

y^1 (y^{q_m}) is the point estimation corresponding to the smallest (largest) value of output variable y;

q_i, $i = \overline{1,n}$ and q_m are the cardinalities of sets (2.4) and (2.5), where in the general case $q_1 \neq q_2 \neq \ldots \neq q_n \neq q_m$.

2.1.2 Linguistic Variables

Let $\mathbf{X}^* = \langle x_1^*, x_2^*, ..., x_n^* \rangle$ be some vector of the input variables fixed values of the considered object, where $x_i^* \in U_i$, $i = \overline{1,n}$. The task of decision making consists of defining the output $y^* \in Y$ on the basis of the information about the vector of inputs $\mathbf{X}^*$. The necessary condition for a formal solution of this task is the availability of dependence (2.1). To define this dependence we consider input variables x_i, $i = \overline{1,n}$, and output variable y as linguistic variables [15], given on universal sets (2.2), (2.3) or (2.4), (2.5).

To make an estimation of the linguistic variables x_i, $i = \overline{1,n}$, and y we use qualitative terms from the following term-sets:

$A_i = \{a_i^1, a_i^2, ..., a_i^{l_i}\}$ is the term-set of variable x_i, $i = \overline{1,n}$,

$D = \{d_1, d_2, ..., d_m\}$ is the term-set of variable y,

where a_i^p is the p-th linguistic term of variable x_i, $p = \overline{1,l_i}$, $i = \overline{1,n}$;

d_j is the j-th linguistic term of variable y,

m is the number of various solutions in the considered region.

Cardinalities of term-sets A_i, $i = \overline{1,n}$, in the general case can be various, that is $l_1 \neq l_2 \neq \ldots \neq l_n$.

The names of separate terms $a_i^1, a_i^2, ..., a_i^{l_i}$ can also differ for various linguistic variables x_i, $i = \overline{1,n}$.

For example, VEHICLE SPEED { low, average, high, very high }, CONVERSION TEMPERATURE { psychrophilic, mesophilic, thermophilic }, PULSE FREQUENCE { delayed, normal, increased }.

Linguistic terms $a_i^p \in A_i$ and $d_j \in D$, $p = \overline{1,l_i}$, $i = \overline{1,n}$, $j = \overline{1,m}$, are considered as fuzzy sets given on universal sets U_i and Y defined by relations (2.2)$\div$(2.5).

In the case of quantitative variables x_i, $i = \overline{1,n}$, and y fuzzy sets a_i^p and d_j are defined by relations:

$$a_i^p = \int_{\underline{x_i}}^{\overline{x_i}} \mu^{a_i^p}(x_i) / x_i \ , \tag{2.6}$$

$$d_j = \int_{\underline{d}}^{\overline{d}} \mu^{d_j}(d) / d \ , \tag{2.7}$$

where $\mu^{a_i^p}(x_i)$ is the membership function of the input variable $x_i \in [\underline{x_i}, \overline{x_i}]$ value to the term $a_i^p \in A_i$, $p = \overline{1,l_i}$, $i = \overline{1,n}$;

$\mu^{d_j}(d)$ is the membership function of the output variable $y \in [\underline{y}, \overline{y}]$ to the term - solution $d_j \in D$, $j = \overline{1,m}$.

In the case of qualitative variables x_i, $i = \overline{1,n}$ and y fuzzy sets a_i^p and d_j are defined as:

$$a_i^p = \sum_{k=1}^{q_i} \mu^{a_i^p}(v_i^k) / v_i^k \ , \tag{2.8}$$

$$d_j = \sum_{r=1}^{q_m} \mu^{d_j}(y^r) / y^r \ , \tag{2.9}$$

where $\mu^{a_i^p}(v_i^k)$ is the membership degree of the element $v_i^k \in U_i$ to the term $a_i^p \in A_i$, $p = \overline{1,l_i}$, $i = \overline{1,n}$, $k = \overline{1,q_i}$;

$\mu^{d_j}(y^r)$ is the membership degree of the element $y^r \in Y$ to the term - solution $d_j \in D$, $j = \overline{1,m}$;

U_i and Y are defined by relations (2.4) and (2.5).

Note that integral and summation signs in relations (2.6) – (2.9) designate joining of pairs $\mu(u)/u$.

This stage of fuzzy model construction is named *fuzzification* of variables in fuzzy logic literature [9]. At this stage the linguistic estimations of variables and the membership functions necessary for their formalization are defined.

2.1.3 Fuzzy Knowledge Base

Let us take N experimental data connecting inputs and output of the identification object, and distribute it in the following way:

$$N = k_1 + k_2 + ... + k_m,$$

where k_j is the number of experimental data corresponding to output solution d_j, $j = \overline{1,m}$, m is the number of output decisions where in the general case $k_1 \neq k_2 \neq ... \neq k_m$.

It is supposed that $N < l_1 \cdot l_2 \cdot ... \cdot l_n$, that is, the number of the selected experimental data is smaller than the complete set of various combinations of object input variables change levels $(l_i, i = \overline{1,n})$.

Let us number N experimental data in the following way:

11, 12, ..., 1 k_1 – numbers of input variables combinations for solution d_1;

...

$j\,1$, $j\,2$, ..., $j\ k_j$ – numbers of input variables combinations for solution d_j;

...

$m\,1, m\,2$, ..., $m\ k_m$ – numbers of input variables combinations for solution d_m.

Let us designate Table 2.1 as a knowledge matrix formed according to such rules:

1) Dimension of this matrix is equal to $(n+1) \times N$, where $(n+1)$ is the number of columns and $N = k_1 + k_2 + ... + k_m$ is the number of rows.

2) The first n columns of the matrix correspond to input variables x_i, $i = \overline{1,n}$, and the $(n+1)$-th column corresponds to values d_j of output variable y ($j = \overline{1,m}$).

3) Each row of the matrix represents some combination of input variables values referred to one of possible output variable y values. In this connection: the first k_1 rows correspond to output variable $y = d_1$ value, the second k_2 rows correspond to $y = d_2$ value, . . . , the last k_m rows correspond to value $y = d_m$.

4) Element a_i^{jp}, placed at the crossing of i-th column and jp-th row, corresponds to the linguistic estimation of parameter x_i in row number jp of the fuzzy

knowledge base, where linguistic estimation a_i^{jp} is selected from a term-set corresponding to variable x_i, that is $a_i^{jp} \in A_i$, $i=\overline{1,n}$, $j=\overline{1,m}$, $p=\overline{1,k_j}$.

Thus introduced knowledge base defines some system of logical expressions of the type «IF - THEN, OTHERWISE», interconnecting input variables values $x_1 \div x_n$ with one of the possible types of solution d_j, $j=\overline{1,m}$:

Table 2.1. Knowledge base

Number of the input combination of values	Input variables				Output variable
	x_1	x_2	$\dots x_i \dots$	x_n	y
11	a_1^{11}	a_2^{11}	$\dots a_i^{11} \dots$	a_n^{11}	
12	a_1^{12}	a_2^{12}	$\dots a_i^{12} \dots$	a_n^{12}	d_1
...					
$1k_1$	$a_1^{1k_1}$	$a_2^{1k_1}$	$\dots a_i^{1k_1} \dots$	$a_n^{1k_1}$	
...					
$j1$	a_1^{j1}	a_2^{j1}	$\dots a_i^{j1} \dots$	a_n^{j1}	
$j2$	a_1^{j2}	a_2^{j2}	$\dots a_i^{j2} \dots$	a_n^{j2}	d_j
...					
jk_j	$a_1^{jk_j}$	$a_2^{jk_j}$	$\dots a_i^{jk_j} \dots$	$a_n^{jk_j}$	
...					
$m1$	a_1^{m1}	a_2^{m1}	$\dots a_i^{m1} \dots$	a_n^{m1}	
$m2$	a_1^{m2}	a_2^{m2}	$\dots a_i^{m2} \dots$	a_n^{m2}	d_m
...					
mk_m	$a_1^{mk_m}$	$a_2^{mk_m}$	$\dots a_i^{mk_m} \dots$	$a_n^{mk_m}$	

IF $(x_1 = a_1^{11})$ AND $(x_2 = a_2^{11})$ AND ... AND $(x_n = a_n^{11})$ OR
$(x_1 = a_1^{12})$ AND $(x_2 = a_2^{12})$ AND ... AND $(x_n = a_n^{12})$ OR ...
$(x_1 = a_1^{1k_1})$ AND $(x_2 = a_2^{1k_1})$ AND ... AND $(x_n = a_n^{1k_1})$,
THEN $y = d_1$, OTHERWISE
IF $(x_1 = a_1^{21})$ AND $(x_2 = a_2^{21})$ AND ... AND $(x_n = a_n^{21})$ OR
$(x_1 = a_1^{22})$ AND $(x_2 = a_2^{22})$ AND ... AND $(x_n = a_n^{22})$ OR ...
$(x_1 = a_1^{2k_2})$ AND $(x_2 = a_2^{2k_2})$ AND ... AND $(x_n = a_n^{2k_2})$,
THEN $y = d_2$, OTHERWISE ...

IF $(x_1 = a_1^{m1})$ AND $(x_2 = a_2^{m1})$ AND ... AND $(x_n = a_n^{m1})$ OR
$(x_1 = a_1^{m2})$ AND $(x_2 = a_2^{m2})$ AND ... AND $(x_n = a_n^{m2})$ OR ...
$(x_1 = a_1^{mk_m})$ AND $(x_2 = a_2^{mk_m})$ AND ... AND $(x_n = a_n^{mk_m})$,
THEN $y = d_m$, (2.10)

where $d_j (j = \overline{1,m})$ is a linguistic estimation of output variable y defined from term-set D;

a_i^{jp} is a linguistic estimation of input variable x_i in p-th row of j-th disjunction selected from the corresponding term-set A_i, $i = \overline{1,n}$, $j = \overline{1,m}$, $p = \overline{1,k_j}$;

k_j is the number of rules defining output variable value $y = d_j$.

Let us call the system of logic statements like this one *the fuzzy knowledge base* system.

Using operations $\bigcup$ (OR) and $\bigcap$ (AND) the system of logical statements (2.10) can be rewritten in a more compact form:

$$\bigcup_{p=1}^{k_j}\left[\bigcap_{i=1}^{n}(x_i = a_i^{jp})\right] \longrightarrow y = d_j,\ j = \overline{1,m}. \quad (2.11)$$

Thus, the required relation (2.1) defining interconnection between input parameters x_i and output variable y, is formalized in the form of fuzzy logical statements (2.11) system, which is based on the above introduced knowledge matrix.

2.1.4 Membership Functions

According to definition [15], membership function $\mu^T(x)$ characterizes some subjective measure (in the range of [0, 1]) of expert certainty in the fact that crisp value x corresponds to fuzzy term T. The most spread in practical applications [9] are triangle, trapezoidal and bell shape Gaussian membership functions, parameters of which allow us to change function shapes.

We suggest an analytical model of a variable x membership function to an arbitrary fuzzy term T in the form of:

$$\mu^T(x) = \frac{1}{1+\left(\dfrac{x-b}{c}\right)^2}, \quad (2.12)$$

which is simple and convenient for tuning, where b and c are tuning parameters: b is the function maximum coordinate, $\mu^T(b) = 1$; c is the function concentration-extension ratio (Fig. 2.1). For fuzzy term T number b represents the most possible value of variable x.

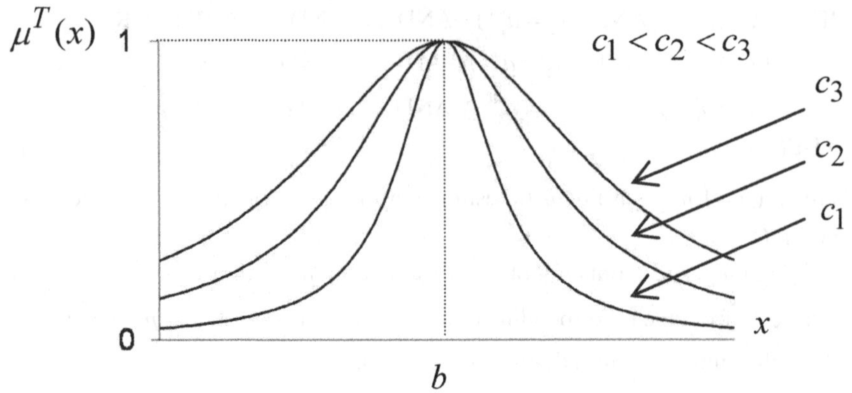

Fig. 2.1. Membership function model

2.2 Fuzzy Approximator for System with Discrete Output

2.2.1 Problem Statement

Let us consider the following as known:

* the set of decisions $D = \{d_1, d_2, ..., d_m\}$, corresponding to output variable y,
* the set of input variables $\mathbf{X} = (x_1, x_2, ..., x_n)$,
* the ranges of quantitative change of each input variable $x_i \in [\underline{x_i}, \overline{x_i}]$, $i = \overline{1,n}$,
* the membership functions allowing to represent variables x_i, $i = \overline{1,n}$, in the form of fuzzy sets (2.6) or (2.8),
* the knowledge matrix defined according to the rules introduced in Section 2.1.3.

It is thus required to design such an algorithm of decision making which allows us to bring the fixed vector of input variables $\mathbf{X}^* = \left\langle x_1^*, x_2^*, ..., x_n^* \right\rangle$, $x_i^* \in [\underline{x_i}, \overline{x_i}]$, into correspondence with decision $y \in D$.

The task of object approximation with a discrete output is shown in the form of a diagram in Fig. 2.2, where it is emphasized that the object inputs are given by three methods: 1- by number, 2- by linguistic term, 3- by thermometer principle.

The idea behind the method suggested below for the solution of this task consists of using fuzzy logic equations. These equations are constructed on the basis of a knowledge matrix or of some system of logical statements (2.10) which is isomorphic to this matrix and allow us to calculate the values of membership functions of various decisions (solutions) for fixed values of object input variables. The solution with the greatest value of membership function is chosen as the required one.

2.2.2 Fuzzy Logical Equations

Linguistic estimations a_i^{jp} of variables $x_1 \div x_n$, contained in logic statements about decisions d_j (2.10), are considered as fuzzy sets defined on universal sets $U_i = [\underline{x}_i, \overline{x}_i]$, $i = \overline{1,n}$, $j = \overline{1,m}$.

Let $\mu^{a_i^{jp}}(x_i)$ be the membership function of parameter $x_i \in [\underline{x}_i, \overline{x}_i]$ to fuzzy term a_i^{jp}, $i = \overline{1,n}$, $j = \overline{1,m}$, $p = \overline{1,k_j}$;

$\mu^{d_j}(x_1, x_2, ..., x_n)$ is the membership function of input variables $X = (x_1, x_2, ..., x_n)$ vector to the value of output variable $y = d_j$, $j = \overline{1,m}$.

Interconnection between these functions is defined by fuzzy knowledge base (2.11) and can be represented in the form of the following equations:

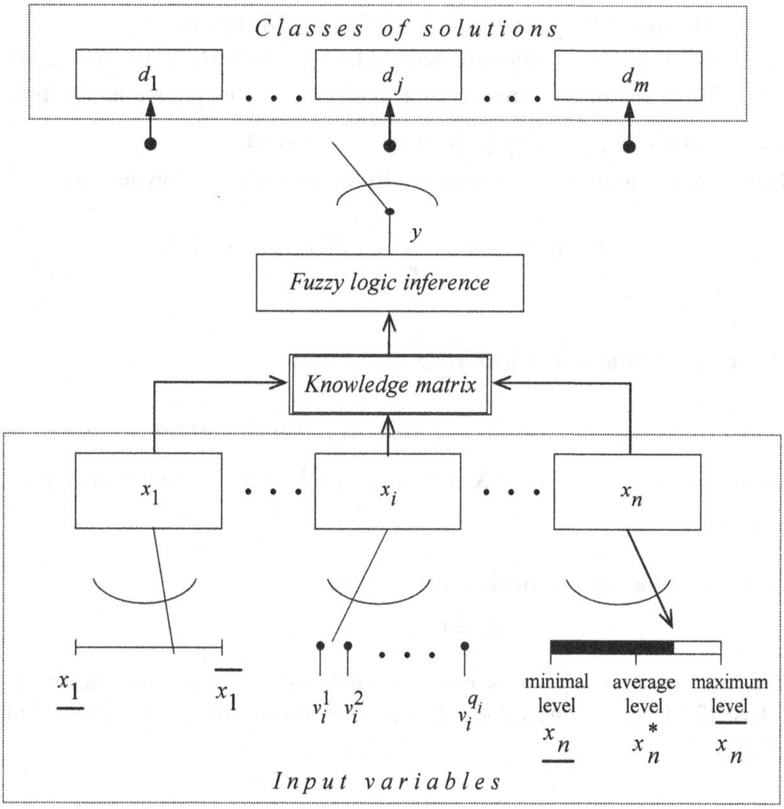

Fig. 2.2. Approximation of a nonlinear object with discrete output

$$\mu^{d_1}(x_1,x_2,...,x_n) = \mu^{a_1^{11}}(x_1) \wedge \mu^{a_2^{11}}(x_2) \wedge ... \wedge \mu^{a_n^{11}}(x_n) \vee$$
$$\vee\, \mu^{a_1^{12}}(x_1) \wedge \mu^{a_2^{12}}(x_2) \wedge ... \wedge \mu^{a_n^{12}}(x_n) \vee ...$$
$$... \vee\, \mu^{a_1^{1k_1}}(x_1) \wedge \mu^{a_2^{1k_1}}(x_2) \wedge ... \wedge \mu^{a_n^{1k_1}}(x_n) \;,$$

$$\mu^{d_2}(x_1,x_2,...,x_n) = \mu^{a_1^{21}}(x_1) \wedge \mu^{a_2^{21}}(x_2) \wedge ... \wedge \mu^{a_n^{21}}(x_n) \vee$$
$$\vee\, \mu^{a_1^{22}}(x_1) \wedge \mu^{a_2^{22}}(x_2) \wedge ... \wedge \mu^{a_n^{22}}(x_n) \vee ...$$
$$... \vee\, \mu^{a_1^{2k_2}}(x_1) \wedge \mu^{a_2^{2k_2}}(x_2) \wedge ... \wedge \mu^{a_n^{2k_2}}(x_n) \;,$$

...

$$\mu^{d_m}(x_1,x_2,...,x_n) = \mu^{a_1^{m1}}(x_1) \wedge \mu^{a_2^{m1}}(x_2) \wedge ... \wedge \mu^{a_n^{m1}}(x_n) \vee$$
$$\vee\, \mu^{a_1^{m2}}(x_1) \wedge \mu^{a_2^{m2}}(x_2) \wedge ... \wedge \mu^{a_n^{m2}}(x_n) \vee ...$$
$$... \vee\, \mu^{a_1^{mk_m}}(x_1) \wedge \mu^{a_2^{mk_m}}(x_2) \wedge ... \wedge \mu^{a_n^{mk_m}}(x_n) \;,$$

where $\vee$ is the logic OR operation, $\wedge$ is the logic AND operation.

These fuzzy logical equations are derived from fuzzy knowledge base (2.11) by way of replacing linguistic terms a_i^{jp} and d_j by corresponding membership functions, and operations $\bigcup$ and $\bigcap$ by operations $\vee$ and $\wedge$.

The logical equation system can be briefly written in the following way:

$$\mu^{d_j}(x_1,x_2,...,x_n) = \bigvee_{p=1}^{k_j}\left[\bigwedge_{i=1}^{n} \mu^{a_i^{jp}}(x_i)\right], \; j=\overline{1,m} \;. \tag{2.13}$$

2.2.3 Approximation Algorithm

The making of decision $d^* \in D = \{d_1, d_2, ..., d_m\}$, which corresponds to the fixed values vector of input variables $\mathbf{X}^* = \left\langle x_1^*, x_2^*, ..., x_n^* \right\rangle$, is performed in the following sequence.

1°. Let us fix the input variables values vector

$$\mathbf{X}^* = (x_1^*, x_2^*, ..., x_n^*) \;.$$

2°. Let us assign fuzzy terms membership functions used in the fuzzy knowledge base (2.11) and define values of these functions for the given values of input variables $x_1^* \div x_n^*$.

3°. Using logical equations (2.13) we calculate multidimensional membership functions $\mu^{d_j}(x_1^*, x_2^*, ..., x_n^*)$ of vector $\mathbf{X}^*$ for all the values d_j, $j = \overline{1,m}$ of output variable y. Logic operations AND ($\wedge$) and OR ($\vee$) performed on membership functions are replaced by the operations *min* and *max*.

$$\mu(a) \wedge \mu(b) = \min[\mu(a), \mu(b)] \ ,$$

$$\mu(a) \vee \mu(b) = \max[\mu(a), \mu(b)] \ .$$

4°. Let us define value d_j^*, the membership function of which is maximal:

$$\mu^{d_j}(x_1^*, x_2^*, ..., x_n^*) = \max_{j=\overline{1,m}} \left(\mu^{d_j}(x_1^*, x_2^*, ..., x_n^*) \right).$$

It is this solution that is required for the input variables values vector $\mathbf{X}^* = (x_1^*, x_2^*, ..., x_n^*)$.

Thus, the suggested algorithm uses the idea of linguistic term identification by membership function maximum and generalizes this idea over the entire knowledge base.

The computational part of the suggested algorithm is easily realized with the membership functions values matrix derived from the knowledge matrix by way of doing *min* and *max* operations (Fig. 2.3).

The suggested algorithm of finding discrete values $\{d_1, d_2, ..., d_m\}$ of output variable y by the given input variables fixed values vector $\mathbf{X}^* = \langle x_1^*, x_2^*, ..., x_n^* \rangle$ and by the knowledge matrix allows to approximate the object $y = f_y(x_1, x_2, ..., x_n)$ with a discrete output.

2.3 Fuzzy Approximator for System with Continuous Output

Let us break interval $[\underline{y}, \overline{y}]$, with which object output y changes, into m parts:

$$[\underline{y}, \overline{y}] = \underbrace{[\underline{y}, y_1)}_{d_1} \cup \underbrace{[y_1, y_2)}_{d_2} \cup ... \cup \underbrace{[y_{j-1}, y_j)}_{d_j} \cup ... \cup \underbrace{[y_{m-1}, \overline{y}]}_{d_m} \ . \qquad (2.14)$$

Known expert information about the object with continuous output we give in the form of fuzzy logical expressions system:

$$\begin{array}{ll} \text{IF} & [(x_1 = a_1^{j1}) \text{ AND } (x_2 = a_2^{j1}) \text{ AND } ... \ (x_n = a_n^{j1})] \\ \text{OR} & [(x_1 = a_1^{j2}) \text{ AND } (x_2 = a_2^{j2}) \text{ AND } ... \ (x_n = a_n^{j2})] \quad ... \\ ... \text{ OR} & \left[(x_1 = a_1^{jk_j}) \text{ AND } (x_2 = a_2^{jk_j}) \text{ AND } ... \ (x_n = a_n^{jk_j})\right], \\ \text{THEN} & y \in d_j = [y_{j-1}, y_j) \ , \text{ for all } j = \overline{1,m}, \end{array} \qquad (2.15)$$

where a_j^p is the linguistic term by which variable x_i in the row with number $p = k_j$ is estimated;

k_j is the number of rows-conjunctions corresponding to interval d_j, $j = \overline{1,m}$.

$\mu^{11}(x_1)$	$\mu^{11}(x_2)$	...	$\mu^{11}(x_n)$	min	max	max
$\mu^{12}(x_1)$	$\mu^{12}(x_2)$	...	$\mu^{12}(x_n)$	min		
...	...	...	...			
$\mu^{1k_1}(x_1)$	$\mu^{1k_1}(x_2)$	...	$\mu^{1k_1}(x_n)$	min		
...	...	...	...			
$\mu^{21}(x_1)$	$\mu^{21}(x_2)$	...	$\mu^{21}(x_n)$	min	max	
$\mu^{22}(x_1)$	$\mu^{22}(x_2)$	...	$\mu^{22}(x_n)$	min		
...	...	...	...			
$\mu^{2k_2}(x_1)$	$\mu^{2k_2}(x_2)$	...	$\mu^{2k_2}(x_n)$	min		
...	...	...	...			
$\mu^{m1}(x_1)$	$\mu^{m1}(x_2)$	...	$\mu^{m1}(x_n)$	min	max	
$\mu^{m2}(x_1)$	$\mu^{m2}(x_2)$	...	$\mu^{m2}(x_n)$	min		
...	...	...	...			
$\mu^{mk_m}(x_1)$	$\mu^{mk_m}(x_2)$	...	$\mu^{mk_m}(x_n)$	min		

(On the printed figure, each inner "max" is a bracket over the "min" values of one group of rows, from 1 to k_j. The outer "max" is a bracket over all the groups.)

Fig. 2.3. Matrix realization of decision making algorithm

2.3.1 Problem Statement

Let us consider the following as known:

* the interval of change $[\underline{y}, \overline{y}]$ of output variable y,
* the input variables set $\mathbf{X} = (x_1, x_2, ..., x_n)$,
* the ranges of quantitative change of each input variable $x_i \in [\underline{x_i}, \overline{x_i}]$, $i = \overline{1,n}$,
* the membership functions allowing to represent variables x_i, $i = \overline{1,n}$, in the form of fuzzy sets (2.6) or (2.8),
* the system of logical expressions of form (2.15), which can be represented in the form of the knowledge base from Section 2.1.3.

It is thus required to design such a decision making algorithm that allows to bring the fixed vector of input variables $\mathbf{X}^* = \langle x_1^*, x_2^*, ..., x_n^* \rangle$, $x_i^* \in [\underline{x_i}, \overline{x_i}]$ into correspondence with decision $y \in [\underline{y}, \overline{y}]$.

The fuzzy logic evidence algorithm presented in Section 2.2.3 allows us to calculate the output value y in the form of a fuzzy set:

$$\tilde{y} = \left\{ \frac{\mu^{d_1}(y)}{[\underline{y}, y_1)}, \frac{\mu^{d_2}(y)}{[y_1, y_2)}, \dots, \frac{\mu^{d_m}(y)}{[y_{m-1}, \overline{y}]} \right\}. \tag{2.16}$$

To obtain a crisp number corresponding to the fuzzy value (2.16) from interval $[\underline{y}, \overline{y}]$ it is necessary to use the defuzzification operation [9]. Defuzzification is the operation of transforming fuzzy information into its crisp form. Let us define a crisp number y^* which corresponds to fuzzy set (2.16) such that:

$$y^* = \frac{\underline{y}\mu^{d_1}(y) + y_1\mu^{d_2}(y) + \dots + y_{m-1}\mu^{d_m}(y)}{\mu^{d_1}(y) + \mu^{d_2}(y) + \dots + \mu^{d_m}(y)}. \tag{2.17}$$

Where there is probability interpretation of membership degrees, formula (2.17) can be considered as an analog to mathematical expectation of a discrete random value.

If we break interval $[\underline{y}, \overline{y}]$ into m equal parts, that is,

$$y_1 = \underline{y} + \Delta,\ y_2 = \underline{y} + 2\Delta,\ \dots,\ y_{m-1} = \overline{y} - \Delta,\ \Delta = \frac{\overline{y} - \underline{y}}{m-1},$$

then formula (2.17) is simplified and takes the form which is convenient for calculations:

$$y^* = \frac{\sum_{j=1}^{m} [\underline{y} + (j-1)\Delta]\, \mu^{d_j}(y)}{\sum_{j=1}^{m} \mu^{d_j}(y)}. \tag{2.18}$$

2.3.2 Approximation Algorithm

To solve the stated problem of the approximation of a nonlinear object with continuous output we use the fuzzy logic evidence algorithm from Section 2.2.3 and the defuzzification operation (2.18). Then the value of the output variable $y^* \in [\underline{y}, \overline{y}]$, which corresponds to the vector of input variables fixed values $\mathbf{X}^* = \langle x_1^*, x_2^*, ..., x_n^* \rangle$, is found in such a sequence.

1°. Using the fuzzy logic evidence algorithm from Section 2.2.3 we calculate multi-dimensional membership functions $\mu^{d_j}(x_1^*, x_2^*, ..., x_n^*)$ of vector $\mathbf{X}^*$ for all the subintervals $d_j = [y_{j-1}, y_j)$, $j = \overline{1, m}$, into which interval $[\underline{y}, \overline{y}]$ of output variable y is broken.

2°. Using defuzzification operation (2.18) we obtain the required value y^*.

Approximation of a nonlinear object with continuous output is shown in Fig. 2.4.

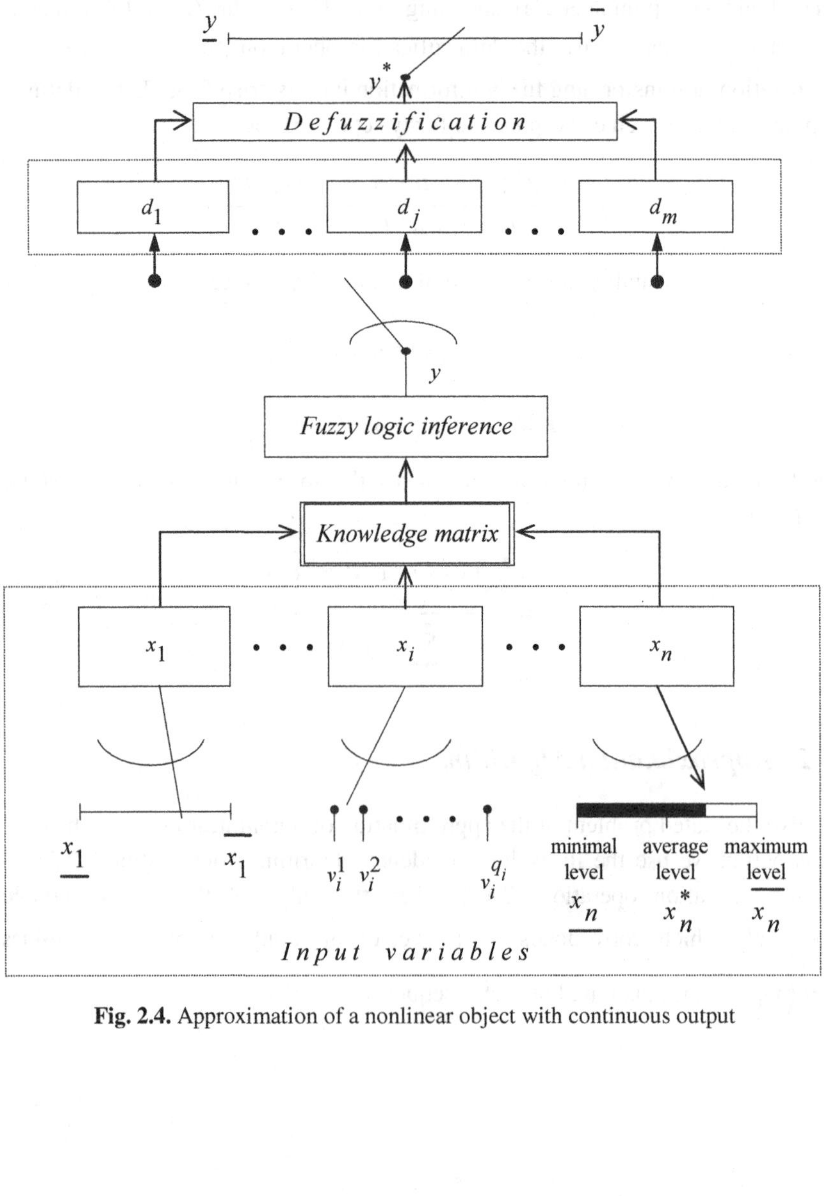

Fig. 2.4. Approximation of a nonlinear object with continuous output

References

1. Zadeh, L.: Outline of a New Approach to the Analysis of Complex Systems and Decision Processes. IEEE Transactions on Systems, Man, and Cybernetics SMC-3, 28–44 (1973)
2. Tong, R.M.: The construction and evaluation of fuzzy models. In: Gupta, M.M., Ragade, R.K., Yager, R.R. (eds.) Advances in Fuzzy Set Theory and Applications, pp. 559–576. Amsterdam, North-Holland (1979)
3. Gupta, M.M., Kiszka, J.B., Trojan, G.J.: Multivariable structure of fuzzy control systems. IEEE Transactions on Systems, Man, and Cybernetics SMC -16, 638–656 (1986)
4. Czogala, E., Pedrycz, W.: Control problems in fuzzy systems. Fuzzy Sets and Systems 7, 257–274 (1982)
5. Pedrycz, W.: Identification in fuzzy systems. IEEE Transactions on Systems, Man and Cybernetics 14, 361–366 (1984)
6. Pedrycz, W.: Fuzzy Control and Fuzzy Systems. Wiley, New York (1989)
7. Sugeno, M., Yasukawa, T.: A fuzzy logic based approach to qualitative modeling. IEEE Transactions on Fuzzy Systems 1, 7–31 (1993)
8. Yager, R.R., Filev, D.P.: Essentials of Fuzzy Modeling and Control, p. 388. John Willey & Sons, New York (1994)
9. Zimmermann, H.-J.: Fuzzy Sets, Decision Making and Expert Systems, p. 335. Kluwer, Dordrecht (1987)
10. Kacprzyk, J.: Multistage Fuzzy Control: A Model-based Approach to Fuzzy Control and Decision Making, p. 327. John Willey & Sons (1997)
11. Schneider, M., Kandel, A., Langholz, G., Chew, G.: Fuzzy Expert System Tools, p. 198. John Willey & Sons, New York (1996)
12. Eickhoff, P.: System Identification: Parameter and State Estimation. Wiley, London (1974)
13. Tsypkin, Y.Z.: Information Theory of Identification, p. 320. Nauka, Moscow (1984) (in Russian)
14. Shteinberg, S.E.: Identification in Control Systems. Energoatomizdat, Moscow (1987) (in Russian)
15. Zadeh, L.A.: The Concept of a Linguistic Variable and its Application to Approximate Reasoning, Part 1-3. Information Sciences 8, 199–251 (1975); 9, 301 - 357, 43 - 80 (1976)
16. Gubinsky, A.I.: Reliability and quality of ergonomic systems functioning. Leningrad, Nauka (1982) (in Russian)
17. Rotshtein, A.: Medical Diagnostics based on Fuzzy Logic, p. 132. Kontinent – PRIM, Vinnitsa (1996) (in Russian)

Chapter 3
Fuzzy Rules Tuning for Direct Inference

The identification of an object consists of the construction of its mathematical model, i.e., an operator of connection between input and output variables from experimental data. Modern identification theory [1 – 3], based on modeling dynamical objects by equations (differential, difference, etc.), is poorly suited for the use of information about an object in the form of expert IF-THEN statements. Such statements are concentrated expertise and play an important role in the process of human solution of various cybernetic problems: control of technological processes, pattern recognition, diagnostics, forecast, etc. The formal apparatus for processing expert information in a natural language is fuzzy set theory [4, 5]. According to this theory, a model of an object is given in the form of a fuzzy knowledge base, which is a set of IF-THEN rules that connect linguistic estimates for input and output object variables. The adequacy of the model is determined by the quality of the membership functions, by means of which linguistic estimates are transformed into a quantitative form. Since membership functions are determined by expert methods [5], the adequacy of the fuzzy model depends on the expert qualification.

A method for identification of nonlinear objects by fuzzy knowledge bases was proposed in [6, 7]. Various theoretical and applied aspects of this method were considered in [8 – 15]. Constructing the object model was realized by two-stage tuning of fuzzy knowledge bases, which can be considered as the stages of structural and parametric identification. The first stage is traditional for fuzzy expert systems [16]. Rough model tuning is realized by constructing knowledge bases using expert information. The passage from a knowledge base to the corresponding logical equations allows one to relate membership functions of output and input variables of the object for identification. For rough tuning of rules weights and membership functions forms in [6] the modified method of Saaty's pairwise comparisons suggested in [17] was used.

The higher the professional level of the expert the higher the adequacy of the fuzzy model simulated at the stage of rough tuning. This model is called a pure expert system so that only expert information is used for its construction. Though, no one can guarantee the coincidence of the results of fuzzy logic evidence (theory) with the experimental data. Therefore, the second stage at which fine tuning of the fuzzy model by way of training according to experi-

A.P. Rotshtein et al.: Fuzzy Evidence in Identif., Forecast. and Diagn., STUDFUZZ 275, pp. 55–117.
springerlink.com

mental data is necessary. The essence of the fine tuning stage consists of the selection of such weights of fuzzy IF-THEN rules and such membership functions parameters that provide the least distance between the desired (experimental) and the model (theoretical) behavior of the object. The stage of fine tuning reduces to the optimization problem, for solution of which a genetic algorithm was used in [8]. The drawback of the method is that it is poorly suited for taking into account new data entering the training sample.

In [10] a method of neural-linguistic identification of nonlinear dependencies was proposed. Linguistic information about the object is represented in the form of a special neural fuzzy network isomorphic to a fuzzy knowledge base. To train the network, recursive relations obtained by the gradient method were used. The principal advantage of the method [10] is the ability to learn fuzzy knowledge bases in real time, i.e., in the on-line mode. The drawback of this method lies in the danger of getting into a local extremum.

In this Chapter, we propose a two-stage tuning of the parameters of the fuzzy model. The first off-line stage uses the training sample at hand and the genetic algorithm for rough hitting into a neighborhood of a global minimum of discrepancy between the model and the experimental results. The second stage uses the neural fuzzy network for on-line tuning of the parameters and their adaptive correction as new experimental data arrive.

This chapter is written using original work materials [6 – 15].

3.1 Problems of Fuzzy Rules Tuning

The certainty of an expert in each IF-THEN rule included in the fuzzy knowledge base (2.1) can be of various nature (may vary from one rule to another). The expert can explicitly single out a rule as some indisputable truth and, relative to another rule, the same expert can experience some doubt. To reflect various degrees of the expert's certainty relative to the various rules we introduce the concept of rules weights [6, 7] into the fuzzy knowledge base. A number on the interval [0, 1] which characterizes the expert's certainty in this rule is called the weight of the rule.

Taking into consideration the weights of the rules, fuzzy knowledge base (2.1) about an unknown dependence $y = f_y(x_1, x_2, ..., x_n)$ takes the following form:

IF $(x_1 = a_1^{11})$ AND $(x_2 = a_2^{11})$ AND ... AND $(x_n = a_n^{11})$ (with weight w_{11})

OR $(x_1 = a_1^{12})$ AND $(x_2 = a_2^{12})$ AND ... AND $(x_n = a_n^{12})$ (with weight w_{12})

OR ...

OR $(x_1 = a_1^{1k_1})$ AND $(x_2 = a_2^{1k_1})$ AND ... AND $(x_n = a_n^{1k_1})$ (with weight w_{1k_1}),

THEN $y = d_1$, OTHERWISE

IF $(x_1 = a_1^{21})$ AND $(x_2 = a_2^{21})$ AND ... AND $(x_n = a_n^{21})$ (with weight w_{21})

OR $(x_1 = a_1^{22})$ AND $(x_2 = a_2^{22})$ AND ... AND $(x_n = a_n^{22})$ (with weight w_{22})

OR ...

OR $(x_1 = a_1^{2k_2})$ AND $(x_2 = a_2^{2k_2})$ AND ... AND $(x_n = a_n^{2k_2})$ (with weight w_{2k_2}),

THEN $y = d_2$, OTHERWISE ...

...

IF $(x_1 = a_1^{m1})$ AND $(x_2 = a_2^{m1})$ AND ... AND $(x_n = a_n^{m1})$ (with weight w_{m1})

OR $(x_1 = a_1^{m2})$ AND $(x_2 = a_2^{m2})$ AND ... AND $(x_n = a_n^{m2})$ (with weight w_{m2})

OR ...

OR $\left(x_1 = a_1^{mk_m}\right)$ AND $\left(x_2 = a_2^{mk_m}\right)$ AND ... AND $\left(x_n = a_n^{mk_m}\right)$ (with weight w_{mk_m}),

THEN $y = d_m$, (3.1)

where d_j, $j = \overline{1,m}$, is either one of the solution types if an object with discrete output is meant or subinterval of output variable y values if an object with continuous output is meant;

a_i^{jp} is the linguistic estimation of input variable x_i in the p-th row of the j-th disjunction chosen from the corresponding term-set A_i, $i = \overline{1,n}$, $j = \overline{1,m}$, $p = \overline{1,k_j}$;

k_j is the number of rules defining the output variable $y = d_j$ value;

w_{jp} is the weight of the rule.

The following system of fuzzy logic equations will correspond to the modified fuzzy knowledge base (3.1):

$$\mu^{d_j}(x_1, x_2, ..., x_n) = \mathop{\vee}_{p=1}^{k_j} \left\{ w_{jp} \left[\mathop{\wedge}_{i=1}^{n} \mu^{a_i^{jp}}(x_i) \right] \right\}, \; j = \overline{1,m} . \quad (3.2)$$

Taking into consideration the fact that in fuzzy sets theory the operations *min* and *max* correspond to operations $\vee$ and $\wedge$, from (3.2) we obtain:

$$\mu^{d_j}(x_1, x_2, ..., x_n) = \max_{p=\overline{1,k_j}} \left\{ w_{jp} \min_{i=\overline{1,n}} \left[\mu^{a_i^{jp}}(x_i) \right] \right\}, \; j = \overline{1,m} . \quad (3.3)$$

3.1.1 Object with Continuous Output

The generalized model of the object with continuous output has the following form:

$$y = F(\mathbf{X}, \mathbf{W}, \mathbf{B}, \mathbf{C}) \ , \tag{3.4}$$

where $\mathbf{X} = (x_1, x_2, ..., x_n)$ is the vector of input variables;

$\mathbf{W} = (w_1, w_2, ..., w_N)$ is the rules weights vector from the fuzzy knowledge base (3.1);

$\mathbf{B} = (b_1, b_2, ..., b_q)$ and $\mathbf{C} = (c_1, c_2, ..., c_q)$ are the vectors of membership functions (2.12) tuning parameters;

N is the total number of rows in fuzzy knowledge base (3.1), $N = k_1 + k_2 + ... + k_m$;

q is the total number of terms in (3.1);

F is the inputs-output connection operator, corresponding to relations (3.3), (2.12) and (2.18).

It is assumed that the training data is given in the form of M pairs of experimental data:

$$(\mathbf{X}^l, y^l), \ l = \overline{1, M} \ , \tag{3.5}$$

where $\mathbf{X}^l = (x_1^l, x_2^l, ..., x_n^l)$ and y^l are the vector of the values of the input variables and the corresponding value of the output variable y for l-th pair "inputs – output", $y^l \in [\underline{y}, \overline{y}]$.

In accordance with the least squares method, the problem of optimal tuning of the fuzzy model can be formulated as follows: it is required to find a vector $(\mathbf{W}, \mathbf{B}, \mathbf{C})$, satisfying the restrictions

$$w_i \in [\underline{w}_i, \overline{w}_i] \ , \ i = \overline{1, N} \ , \ b_j \in [\underline{b}_j, \overline{b}_j] \ , \ c_j \in [\underline{c}_j, \overline{c}_j] \ , \ j = \overline{1, q} \ ,$$

which provides

$$\sum_{l=1}^{M} [F(\mathbf{X}^l, \mathbf{W}, \mathbf{B}, \mathbf{C}) - y^l]^2 = \min_{\mathbf{W}, \mathbf{B}, \mathbf{C}} \ . \tag{3.6}$$

3.1.2 Object with Discrete Output

Relations (3.3) allow us to calculate the vector of inferred membership functions of the output variable y to the different decision classes d_j:

$$\left(\mu^{d_j}(\mathbf{X}, \mathbf{W}, \mathbf{B}, \mathbf{C}), j = \overline{1, m}\right) \tag{3.7}$$

where $\mathbf{X}$, $\mathbf{W}$, $\mathbf{B}$ and $\mathbf{C}$ are vectors which have been defined in Section 3.1.1.

Let us define the desirable vector of membership degrees as:

$$\left.\begin{array}{ll}(1,0,\ldots,0) & \textit{for class-decision } d_1\\ (0,1,\ldots,0) & \textit{for class-decision } d_2\\ \ldots & \\ (0,0,\ldots,1) & \textit{for class-decision } d_m\end{array}\right\} . \tag{3.8}$$

It is assumed that the training data is given in the form of M pairs of experimental data:

$$(\mathbf{X}^l, d^l),\ l=\overline{1,M}, \tag{3.9}$$

where $\mathbf{X}^l=(x_1^l, x_2^l, ..., x_n^l)$ and d^l are the vector of the values of the input variables and the corresponding class-decision of the output variable for l-th pair "inputs – output", $d^l \in \{d_1, d_2, ..., d_m\}$.

To find the unknown parameters vector $(\mathbf{W},\mathbf{B},\mathbf{C})$, which minimizes the difference between theory (3.7) and experiment (3.9), we take advantage of the least squares method. Thus, the problem of optimal fuzzy model tuning can be formulated in the following way: it is required to find a vector $(\mathbf{W},\mathbf{B},\mathbf{C})$, satisfying the restrictions

$$w_i \in [\underline{w}_i, \overline{w}_i]\ ,\ i=\overline{1,N}\ ,\ b_j \in [\underline{b}_j, \overline{b}_j]\ ,\ c_j \in [\underline{c}_j, \overline{c}_j]\ ,\ j=\overline{1,q},$$

which provides the minimal distance between desirable and model membership functions vectors

$$\sum_{l=1}^{M}\left[\sum_{j=1}^{m}\left[\mu^{d_j}(\mathbf{X}^l,\mathbf{W},\mathbf{B},\mathbf{C})-\mu^{d_j}(\mathbf{X}^l)\right]^2\right]=\min_{\mathbf{W},\mathbf{B},\mathbf{C}}, \tag{3.10}$$

where

$$\mu^{d_j}(\mathbf{X}^l)=\begin{cases}1, & \text{if } d_j=d^l\\ 0, & \text{if } d_j\neq d^l\end{cases}.$$

In the following sections we design a hybrid genetic-neuro algorithm to solve the problems (3.6) and (3.10) of optimal fuzzy knowledge base tuning.

3.1.3 "Multiple Inputs – Multiple Outputs" Object

If $x_1, x_2, ..., x_n$ - object inputs and $y_1, y_2, ..., y_m$ - object outputs then "inputs – outputs" interconnection can be assigned using a fuzzy knowledge base of the following form:

IF $(x_1 = A_1^l)$ AND $(x_2 = A_2^l)$ AND . . . AND $(x_n = A_n^l)$,
THEN $(y_1 = B_1^l)$ AND $(y_2 = B_2^l)$ AND . . . AND $(y_m = B_m^l)$,

where l is the rule number, $l = \overline{1, N}$, N is the number of rules, A_i^l and B_j^l are the fuzzy terms for the input variable x_i $(i = \overline{1, n})$ and the output variable y_j $(j = \overline{1, m})$ estimation in l-th rule, respectively.

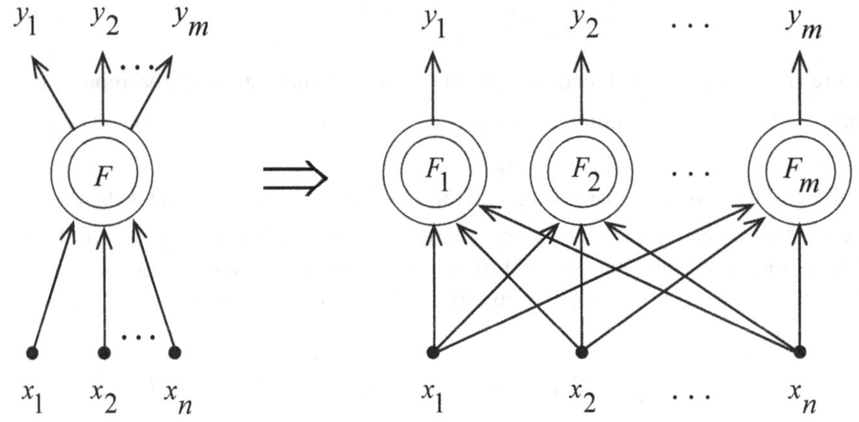

Fig. 3.1. Transformation of knowledge base

After knowledge base transformation (Fig. 3.1) and fuzzy logic inference operation execution we can obtain approximation models of each output variable:

$$y_1 = F_1(\mathbf{X}, \mathbf{W}, \mathbf{B}, \mathbf{C}),$$
$$y_2 = F_2(\mathbf{X}, \mathbf{W}, \mathbf{B}, \mathbf{C}),$$
$$\ldots$$
$$y_m = F_m(\mathbf{X}, \mathbf{W}, \mathbf{B}, \mathbf{C}),$$

where $\mathbf{X} = (x_1.x_2, ..., x_n)$ is the vector of inputs,

$\mathbf{W} = (w_1, w_2, ..., w_N)$ is the vector of rules weights,

$\mathbf{B}$ and $\mathbf{C}$ are the vectors of the fuzzy terms membership functions parameters.

It is assumed that the training data is given in the form of L pairs of experimental data:

$$(\mathbf{X}^p, \hat{y}_1^p, \hat{y}_2^p, ..., \hat{y}_m^p),\ \mathbf{X}^p = (x_1^p, x_2^p, ..., x_n^p),\ p = \overline{1, L}\ .$$

Then optimal model tuning problem for the considered object can be formulated in the following way:

It is required to find such a vector $(\mathbf{B}, \mathbf{C}, \mathbf{W})$, which satisfying change range limitations on parameters provides

$$\sum_{p=1}^{L}[F_1(\mathbf{X}^p, \mathbf{W}, \mathbf{B}, \mathbf{C}) - \hat{y}_1^p]^2 + \sum_{p=1}^{L}[F_2(\mathbf{X}^p, \mathbf{W}, \mathbf{B}, \mathbf{C}) - \hat{y}_2^p]^2 + ...$$

$$... + \sum_{p=1}^{L}[F_m(\mathbf{X}^p, \mathbf{W}, \mathbf{B}, \mathbf{C}) - \hat{y}_m^p]^2 =$$

$$\sum_{j=1}^{m}\sum_{p=1}^{L}[F_j(\mathbf{X}^p, \mathbf{W}, \mathbf{B}, \mathbf{C}) - \hat{y}_j^p]^2 = \min_{\mathbf{W}, \mathbf{B}, \mathbf{C}}\ .$$

The problem of tuning, in case there are discrete outputs, is formulated by analogy.

3.1.4 Criteria of Identification Quality

Object with continuous output. Let $y_F(\mathbf{X}, M)$ be the fuzzy model of the object after tuning by M pairs of training data. To evaluate the quality of the fuzzy inference, the following criterion can be used:

$$R = \frac{1}{|\{\mathbf{X}_i\}|}\sqrt{\sum_{\{\mathbf{X}_i\}}[y_F(\mathbf{X}_i, M) - \hat{y}_i]^2}\ , \tag{3.11}$$

where $y_F(\mathbf{X}_i, M)$ and $\hat{y}_i$ are the inferred and experimental outputs in a point $\mathbf{X}_i = (x_1^i, x_2^i, ..., x_n^i) \in [\underline{x}_1, \overline{x}_1] \times [\underline{x}_2, \overline{x}_2] \times ... \times [\underline{x}_n, \overline{x}_n]$, respectively,

$\{\mathbf{X}_i\}$ is a set of elements of type $\mathbf{X}_i$,

$|\{\mathbf{X}_i\}|$ is the power of set of $\{\mathbf{X}_i\}$.

The proposed criterion (3.11) is similar to a mean-square deviation between the inferred and experimental outputs corresponding to one element of the input space. The dependence of the R (M) criterion (3.11) on the number M of training data pairs can be used to observe the dynamics of fuzzy model learning.

Object with discrete output. Let Q be the total number of situations used for testing the fuzzy model. To evaluate the quality of fuzzy inference in our case of discrete output $y \in \{d_1, d_2, \dots, d_m\}$, it is necessary to make a distribution of Q situations according to the tree shown in Fig. 3.2,

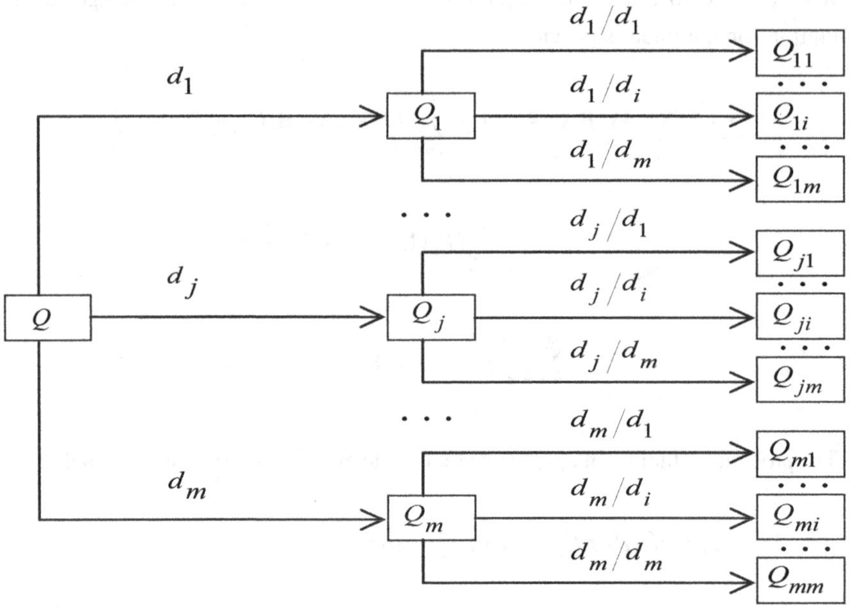

Fig. 3.2. Test sample points distribution tree

where Q_j is the number of situations demanding the decision d_j, that is $Q = Q_1 + Q_2 + \dots + Q_m$,

Q_{ji} is the number of situations demanding the decision d_j, but recognized by fuzzy inference as decision d_i, that is $Q_j = Q_{j1} + Q_{j2} + \dots + Q_{jm}$, $j = \overline{1, m}$.

According to Fig. 3.2 we can evaluate the quality of the fuzzy inference by:

$$\hat{P}_j = \frac{Q_{jj}}{Q_j}, \quad \hat{P}_{ji} = \frac{Q_{ji}}{Q_j}, \quad \hat{P} = \frac{1}{Q}\sum_{j=1}^{m} Q_{jj}, \tag{3.12}$$

where $\hat{P}_j$ is the probability of correct inference of decision d_j;

$\hat{P}_{ji}$ is the probability of incorrect decision d_i when decision d_j was correct;

$\hat{P}$ is the average probability of correct decision inference.

By observing the dependence of the probabilities (3.12) from the number M of training data, we can study the dynamics of fuzzy model learning.

3.2 Genetic Tuning of Fuzzy Rules

3.2.1 Coding

To implement a genetic algorithm one should employ a method of fuzzy models coding [18]. Let us put together all unknown parameters $\mathbf{W}, \mathbf{B}, \mathbf{C}$ into the string (Fig. 3.3):

$$S = (\mathbf{W}, \mathbf{B}, \mathbf{C}) = (w_1, ..., w_N, b_{11}, c_{11}, ..., b_{1l_1}, c_{1l_1}, ..., b_{n1}, c_{n1}, ..., b_{nl_n}, c_{nl_n}) \quad (3.13)$$

where N is the total number of rows in fuzzy knowledge base (3.1);
l_i is the number of input variable x_i term-estimations,

$$l_1 + l_2 + ... + l_n = q \, , \; i = \overline{1, n} \; ;$$

q is the total number of terms in (3.1);

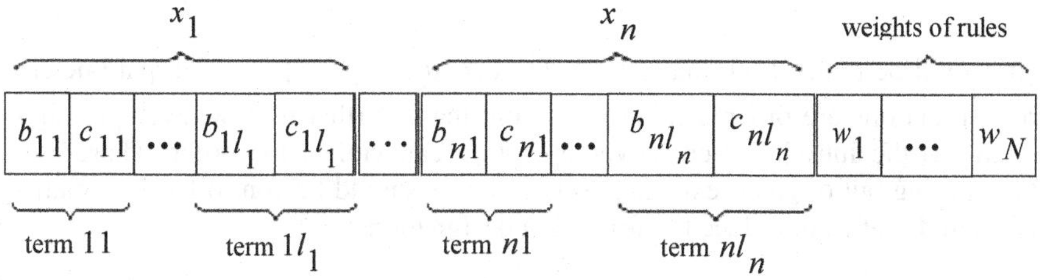

Fig. 3.3. Code of fuzzy model - chromosome

Solely string S defines some fuzzy model $F(\mathbf{X}, \mathbf{W}, \mathbf{B}, \mathbf{C})$ and, vice versa, any model $F(\mathbf{X}, \mathbf{W}, \mathbf{B}, \mathbf{C})$ unambiguously defines some string S. Therefore, string S can be accepted as a code of fuzzy model $F(\mathbf{X}, \mathbf{W}, \mathbf{B}, \mathbf{C})$. Here, model $F(\mathbf{X}, \mathbf{W}, \mathbf{B}, \mathbf{C})$ is understood either as fuzzy model (3.7) of nonlinear object with discrete output or as fuzzy model (3.4) of nonlinear object with continuous output.

3.2.2 Crossover

As crossover is the main operation of the genetic algorithm, its productivity first of all depends upon the productivity of the used crossover operation [18]. Crossover of two chromosomes S_1 and S_2 means obtaining two chromosome-offsprings Ch_1 and Ch_2 by way of genes exchange in parent chromosomes relative to $(n+1)$ crossing points (Fig. 3.4).

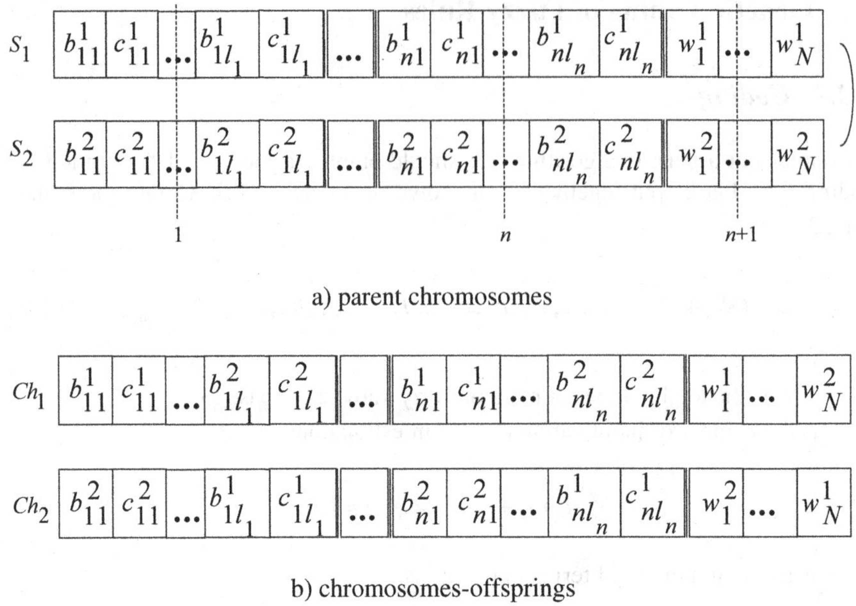

Fig. 3.4. Crossover operation

It should be pointed out that so as sets $A_i = \{a_i^1, a_i^2, ..., a_i^{l_i}\}$ of input parameters term-estimations are ordered in an ascending manner (that is: low, average, high and so on) the introduced crossover operation can violate the order. Therefore, after carrying out of genes exchange sets of terms should be controlled as remaining ordered. Let us introduce the following designations:

$w_j^{S_1}$ – j-th rule weight in parent chromosome S_1,

$w_j^{S_2}$ – j-th rule weight in parent chromosome S_2,

$w_j^{Ch_1}$ – j-th rule weight in chromosome-offspring Ch_1,

$w_j^{Ch_2}$ – j-th rule weight in chromosome-offspring Ch_2, $j = \overline{1, N}$,

$b_{ip}^{S_1}$ – ip-th parameter b in parent chromosome S_1,

$b_{ip}^{S_2}$ – ip-th parameter b in parent chromosome S_2,

$b_{ip}^{Ch_1}$ – ip-th parameter b in chromosome-offspring Ch_1,

$b_{ip}^{Ch_2}$ – ip-th parameter b in chromosome-offspring Ch_2.

A crossover operation algorithm of two parent chromosomes S_1 and S_2, which yields offsprings Ch_1 and Ch_2 has the following form:

1°. Generate $n+1$ random integer numbers z_i, $1 \le z_i < l_i$, where l_i is the number of input variable x_i, $i=\overline{1,n}$, terms-estimations; $1 \le z_{n+1} < N$, where N is the total number of rows in fuzzy knowledge base (3.1).

2°. Exchange genes according to found values of exchange points z_i by using the following rules:

$$b_{ip}^{Ch_1} = \begin{cases} b_{ip}^{S_1}, & p \le z_i \\ b_{ip}^{S_2}, & p > z_i \end{cases}, \quad b_{ip}^{Ch_2} = \begin{cases} b_{ip}^{S_2}, & p \le z_i \\ b_{ip}^{S_1}, & p > z_i \end{cases}, \quad 1 \le p < l_i,\ i=\overline{1,n}, \tag{3.14}$$

$$w_j^{Ch_1} = \begin{cases} w_j^{S_1}, & j \le z_{n+1} \\ w_j^{S_2}, & j > z_{n+1} \end{cases}, \quad w_j^{Ch_2} = \begin{cases} w_j^{S_2}, & j \le z_{n+1} \\ w_j^{S_1}, & j > z_{n+1} \end{cases}, \quad 1 \le j < N. \tag{3.15}$$

3°. Control the order of terms:

$$(b_{i\xi} > b_{i\eta}) \wedge (\xi < \eta) \Rightarrow b_{i\xi} \leftrightarrow b_{i\eta},\ c_{i\xi} \leftrightarrow c_{i\eta},\ 1 \le \xi, \eta \le l_i,\ i=\overline{1,n}, \tag{3.16}$$

where $\leftrightarrow$ is the operation of ordering.

3.2.3 Mutation

Each string S element can undergo a mutation operation with probability p_m. Let us designate the mutation of element s by $Mu(s)$:

$$Mu(w_j) = RANDOM([\underline{w}, \overline{w}]),\ j=\overline{1,N}, \tag{3.17}$$

$$Mu(b_{ip}) = RANDOM([\underline{x}_i, \overline{x}_i]), \tag{3.18}$$

$$Mu(c_{ip}) = RANDOM([\underline{c}_i, \overline{c}_i]), \tag{3.19}$$

where $\underline{w}$ ($\overline{w}$) is the lower (upper) interval bound of possible rule weight value, $[\underline{w}, \overline{w}] \subset [0, 1]$;

$[\underline{c}_i, \overline{c}_i]$ is the interval of possible values of input variable x_i terms-estimations membership function concentration-extension ratio, $[\underline{c}_i, \overline{c}_i] \subset (0, +\infty]$, $i=\overline{1,n}$;

$RANDOM([\underline{\xi}, \overline{\xi}])$ designates the operation of finding uniformly distributed on the interval $[\underline{\xi}, \overline{\xi}]$ random number.

Mutation operation algorithm will have this form:

1°. For each element $s \in S$ in string (3.13) we generate random number $z = RANDOM([0,1])$.

If $z > p_m$ then mutation is not carried out, otherwise we go to 2°.

2°. We carry out mutation operation of element $s \in S$ according to formulae (3.17) – (3.19).

3°. We control ordering of terms according to (3.16).

3.2.4 Fitness Function

Let us designate the chromosome S fitness function using letters $FF(S)$. We use optimization criterion taken with the negative sign as fitness function. The fitness function of chromosome S (3.13) obtained from criterion (3.10) for fuzzy models $F(\mathbf{X}, \mathbf{W}, \mathbf{B}, \mathbf{C})$ of discrete output objects will have the following form:

$$FF(S) = -\sum_{l=1}^{M} \left[\sum_{j=1}^{m} \left[\mu^{d_j}(\mathbf{X}^l, \mathbf{W}, \mathbf{B}, \mathbf{C}) - \mu^{d_j}(\mathbf{X}^l) \right]^2 \right] . \tag{3.20}$$

The fitness function of chromosome S (3.13) obtained from criterion (3.6) for fuzzy models $F(\mathbf{X}, \mathbf{W}, \mathbf{B}, \mathbf{C})$ of continuous output objects will have the following form:

$$FF(S) = -\sum_{l=1}^{M} [F(\mathbf{X}^l, \mathbf{W}, \mathbf{B}, \mathbf{C}) - y^l]^2 . \tag{3.21}$$

To not change fitness function, we use the negative sign; that is, the worse the fuzzy model describes the training sample the lesser the fitness of the model.

3.2.5 Choice of Parents

According to the principles of genetic algorithms the choice of parents for a crossover operation should not be carried out randomly. The greater the fitness function of some chromosome the greater the probability for the given chromosome to yield offsprings [18].

The method of defining parents is based on the fact that each chromosome S_i from a population must be in correspondence with a number p_i such that:

$$p_i \geq 0, \ \sum_{i=1}^{K} p_i = 1, \ FF(S_i) > FF(S_j) \ \Rightarrow \ p_i > p_j,$$

where K is the number of chromosomes in the population.

The set of p_i numbers can be interpreted as the law of discrete random value distribution. A series of numbers p_i is defined using the fitness function in the following way:

$$p_i = \frac{\overline{FF}(S_i)}{\sum_{j=1}^{K} \overline{FF}(S_j)}, \tag{3.22}$$

where $\overline{FF}(S_i) = FF(S_i) - \min_{j=\overline{1,K}} FF(S_j)$.

Using a series of numbers p_i, chromosomes-parents for a crossover operation are found according to the following algorithm:

1°. Let us mark off series p_i on the horizontal axis (Fig. 3.5).

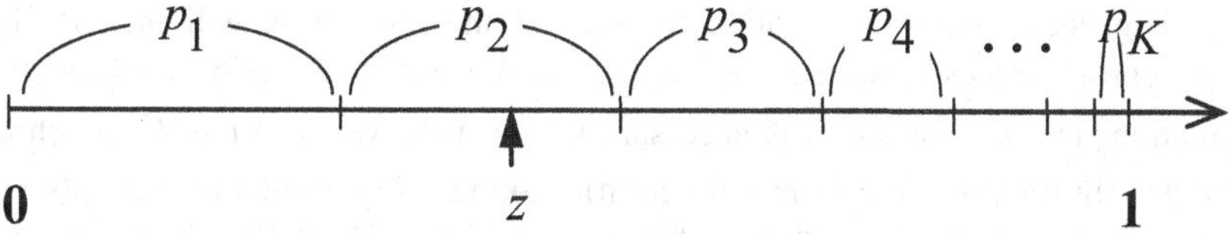

Fig. 3.5. Choice of chromosome-parents

2°. Generate random number z (Fig. 3.5) of uniform distribution law on interval [0, 1].

3°. Choose the chromosome S_i, which corresponds to subinterval p_i into which number z finds itself. In Fig. 3.5 generated number z defines chromosome S_2 as the parent.

4°. Repeat steps 1°-3° to define the second chromosome-parent.

3.2.6 Population Initialization

Chromosomes of initial population can be assigned by various methods. The best method consists of taking available zero variants of fuzzy models as initial chromosomes. Some hybrid decision incorporating the best features of its parents will be derived in the genetic algorithm run.

Some part of chromosomes or all source information as a whole can be assigned randomly according to the formulae:

$$w_i^0 = RANDOM([\underline{w}_i, \overline{w}_i]) \ , \tag{3.23}$$

$$b_i^0 = RANDOM([\underline{x}_i, \overline{x}_i]) \ , \tag{3.24}$$

$$c_i^0 = RANDOM([\underline{c}_i, \overline{c}_i]) \ , \tag{3.25}$$

where $RANDOM([\underline{\xi}, \overline{\xi}])$ designates the operation of finding a uniformly distributed random number on the interval $[\underline{\xi}, \overline{\xi}]$.

After random definition of initial chromosome variants they should be subjected to control (3.16) with the aim of preserving term ordering. It is supposed that the initial population contains K parent chromosomes.

It should be noted that the random method of initial population definition considerably slows down the genetic algorithm convergence process.

3.2.7 Genetic Algorithm

At each genetic algorithm iteration the size of the population will increase by $K \cdot p_c$ chromosomes-offsprings, where p_c is the crossover ratio. To keep the population size K constant it is necessary to reject the worst of the $K \cdot p_c$ chromosomes (in regard to the fitness function). Taking into consideration all abovementioned the genetic algorithm of fuzzy model $F(\mathbf{X}, \mathbf{W}, \mathbf{B}, \mathbf{C})$ optimal tuning will have the following form:

1°. Form initial population by formulae (3.23) – (3.25).

2°. Find fitness function value for each chromosome $FF(S_i)$, $i = \overline{1, K}$, using relations (3.20), (3.21).

3°. Define $\dfrac{K \cdot p_c}{2}$ pairs of chromosome-parents using the algorithm from section 3.2.5.

4°. Perform crossover operation of each chromosome-parents pair according to the algorithm from section 3.2.2.

5°. Carry out mutation of obtained chromosome-offsprings with probability p_m according to the algorithm from section 3.2.3.

6°. Reject $K \cdot p_c$ chromosomes from the obtained population of $K + K \cdot p_c$ chromosomes size as having the worst values of fitness function $FF(S_i)$.

7°. If we obtain chromosome S_i, for which $FF(S_i)=0$ (optimal solution), then it will be algorithm end otherwise go to step 8.

8°. If the preset number of iterations has not yet exhausted then go to step 2° otherwise the chromosome having the highest value of fitness function $FF(S_i)$ represents the found suboptimal solution.

A fuzzy model which is defined by the chromosome obtained by the genetic algorithm can be further optimized using usual optimization methods, the most universal of which is the quickest descent algorithm. The use of the given genetic algorithm is described in consequent sections.

3.3 Neural Tuning of Fuzzy Rules

3.3.1 *Structure of the Network*

In this section, we propose a method of representation of linguistic information about object (3.1) in the form of the specific neuro-fuzzy network which is isomorphic to knowledge base (3.2) [10]. The structure of such a neural fuzzy network is represented in Fig. 3.6, and the elements functions are shown in Table 3.1.

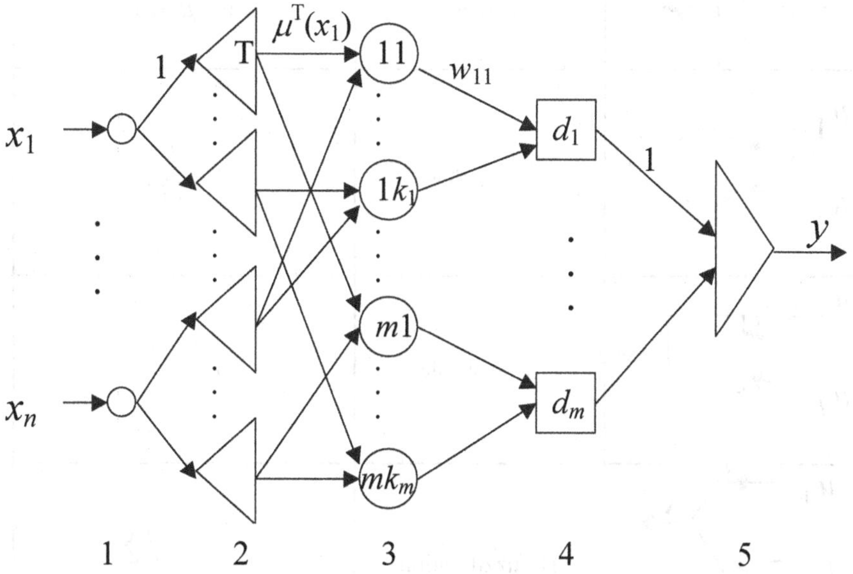

Fig. 3.6. Structure of neuro-fuzzy network

As we can see from Fig. 3.6, the neuro-fuzzy network has five layers:

layer 1 – for object identification inputs;
layer 2 – for fuzzy terms used in knowledge base (3.2);
layer 3 – for strings-conjunctions of the fuzzy knowledge base (3.2);
layer 4 – for fuzzy rules making classes d_j, $j=\overline{1,m}$;
layer 5 – for defuzzification operation (2.18), that is, transformation of the fuzzy logical inference results into some crisp number.

The number of elements of the neuro-fuzzy network is defined in this way:
layer 1 – accordingly to the number of the object identification inputs;
layer 2 – accordingly to the number of fuzzy terms in knowledge base (3.2);
layer 3 – accordingly to the number of strings-conjunctions in knowledge base;
layer 4 – accordingly to the number of classes dividing the output variable range.

Table 3.1. Elements of neuro-fuzzy network

Element	Name	Function
	Input	$v=u$
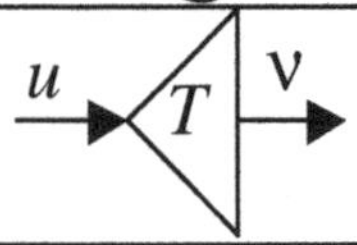	Fuzzy term	$v=\mu^T(u)$
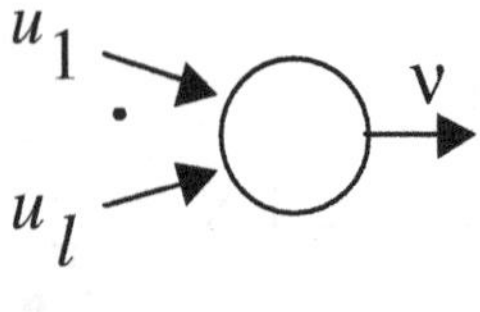	Fuzzy rule	$v=\bigwedge_{i=1}^{l} u_i$
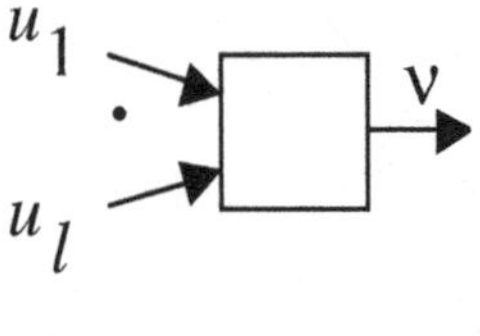	Class of rules	$v=\bigvee_{i=1}^{l} u_i$
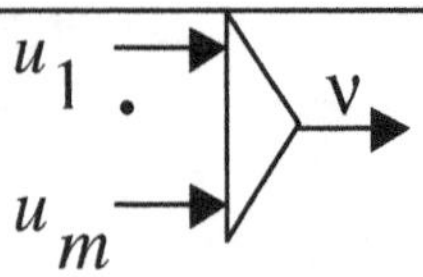	Defuzzification	$v=\sum_{j=1}^{m} u_j \bar{d}_j \Big/ \sum_{j=1}^{m} u_j$

We use the following weights in the proposed neuro-fuzzy network:

- unity for arcs between the first and the second layers;
- input membership functions of fuzzy terms for arcs between the second and the third layers;
- rules weights for arcs between the third and the forth layers;
- unity for arcs between the forth and the fifth layers.

In Table 3.1 we denote:

$\mu^T(u)$ is the membership function of variable u relative to fuzzy term T;

$\bar{d}_j$ is a centre of class $d_j \in [\underline{y}, \overline{y}]$.

For determination of the "fuzzy rule" and the "class of rules" elements shown in Table 3.1 the fuzzy-logical operations *min* and *max* from (3.2) are changed for arithmetical operations of multiplication and addition, respectively. The possibility of such change is given foundation in [5]. It allows us to obtain analytical expressions convenient to differentiation.

3.3.2 Recursive Relations

The essence of the tuning is in such arcs weights selection which minimizes the difference between the results of the neuro-fuzzy approximation and real object behavior. The system of recurrent relations

$$w_{jp}(t+1) = w_{jp}(t) - \eta \frac{\partial E_t}{\partial w_{jp}(t)} , \tag{3.26}$$

$$c_i^{jp}(t+1) = c_i^{jp}(t) - \eta \frac{\partial E_t}{\partial c_i^{jp}(t)} , \tag{3.27}$$

$$b_i^{jp}(t+1) = b_i^{jp}(t) - \eta \frac{\partial E_t}{\partial b_i^{jp}(t)} , \quad j=\overline{1,m}, \ i=\overline{1,n} \ , \ p=k_j , \tag{3.28}$$

minimizing criterion

$$E_t = \frac{1}{2}(\hat{y}_t - y_t)^2 , \tag{3.29}$$

applied in the neural networks theory, is used for training, where:

$\hat{y}_t$ and y_t are the experimental and the theoretical outputs of object (3.1) at the t-th step of training;

$w_{jp}(t)$, $c_i^{jp}(t)$, $b_i^{jp}(t)$ are rules weights (w) and membership functions parameters (b, c) at the t-th step of training;

η is a tuning parameter which is selected accordingly to the recommendations in [2].

Partial derivatives making the part of the relations (3.26) – (3.28), characterize the sensitivity of the error (E_t) relative to the change of the neuro-fuzzy network parameters and are calculated in the following way:

$$\frac{\partial E_t}{\partial w_{jp}} = \varepsilon_1 \varepsilon_2 \varepsilon_3 \frac{\partial \mu^{d_j}(y)}{\partial w_{jp}},$$

$$\frac{\partial E_t}{\partial c_i^{jp}} = \varepsilon_1 \varepsilon_2 \varepsilon_3 \varepsilon_4 \frac{\partial \mu^{jp}(x_i)}{\partial c_i^{jp}}, \quad \frac{\partial E_t}{\partial b_i^{jp}} = \varepsilon_1 \varepsilon_2 \varepsilon_3 \varepsilon_4 \frac{\partial \mu^{jp}(x_i)}{\partial b_i^{jp}},$$

where

$$\varepsilon_1 = \frac{\partial E_t}{\partial y} = y_t - \hat{y}_t, \quad \varepsilon_2 = \frac{\partial y}{\partial \mu^{d_j}(y)} = \frac{\bar{d}_j \sum_{j=1}^{m} \mu^{d_j}(y) - \sum_{j=1}^{m} \bar{d}_j \mu^{d_j}(y)}{\left(\sum_{j=1}^{m} \mu^{d_j}(y) \right)^2},$$

$$\varepsilon_3 = \frac{\partial \mu^{d_j}(y)}{\partial \left(\prod_{i=1}^{n} \mu^{jp}(x_i) \right)} = w_{jp},$$

$$\varepsilon_4 = \frac{\partial \left(\prod_{i=1}^{n} \mu^{jp}(x_i) \right)}{\partial \mu^{jp}(x_i)} = \frac{1}{\mu^{jp}(x_i)} \prod_{i=1}^{n} \mu^{jp}(x_i),$$

$$\frac{\partial \mu^{d_j}(y)}{\partial w_{jp}} = \prod_{i=1}^{n} \mu^{jp}(x_i),$$

$$\frac{\partial \mu^{jp}(x_i)}{\partial c_i^{jp}} = \frac{2 c_i^{jp} (x_i - b_i^{jp})^2}{((c_i^{jp})^2 + (x_i - b_i^{jp})^2)^2}, \quad \frac{\partial \mu^{jp}(x_i)}{\partial b_i^{jp}} = \frac{2 (c_i^{jp})^2 (x_i - b_i^{jp})}{((c_i^{jp})^2 + (x_i - b_i^{jp})^2)^2}.$$

In analogy to the back-propagation rule, the neuro-fuzzy network tuning algorithm is made of two phases. The first phase involves the computation of an output model value of object (y) which corresponds to the given network structure. The second phase involves the computation of error value (E_t) and the interneuron connections weights are recalculated according to (3.26) – (3.28).

3.4 Computer Simulations

3.4.1 Computer Experiment Methods

In this section we describe the results of computer experiments directed at testing the possibility of identification of nonlinear objects with continuous and discrete output using models and algorithms simulated and designed in Chapters 2 and 3. Experiment methods consist of carrying out the following steps.

1°. Nonlinear object was given by some standard model in the form of analytical formulae.

2°. Expert knowledge base was generated from the given standard model.

3°. Identification of nonlinear object was carried out over expert knowledge base using models and algorithms developed in Chapters 2 and 3.

4°. The results of identification carried out using fuzzy knowledge base obtained at step 3° were compared with standard models chosen at step 1°. In this case curves of training dynamics were traced showing identification quality with the increase of training sample volume.

Two pairs of objects with continuous and discrete outputs, respectively, with parametric assignment of membership functions were investigated in sections 3.4.2 and 3.4.3.

3.4.2 Objects with Continuous Output

Experiment 1. Let us consider an object with one input $x \in [0, 1]$ and one output $y \in [0.05,\ 0.417]$. Model-standard has this form:

$$y = f(x) = \frac{(5x-1.1)(4x-2.9)(3x-2.1)(11x-11)(3x-0.05)+10}{40}.$$

This formula was chosen with the aim of studying the possibility of modeling the fifth order object the behavior of which is shown in Fig. 3.7.

Let us divide the interval of change of output variable y into four subintervals:

$$[0.05, 0.417] = [0.05, 0.14) \cup [0.14, 0.23) \cup [0.23, 0.32) \cup [0.32, 0.417].$$

Then the behavior of the investigated object can be described using the following rules:

IF $x = P_1$, THEN $y \in [0.14,\ 0.23]$ (with weight w_1),
IF $x = P_2$, THEN $y \in [0.32,\ 0.42]$ (with weight w_2),
IF $x = P_3$, THEN $y \in [0.05,\ 0.14]$ (with weight w_3),
IF $x = P_4$, THEN $y \in [0.14,\ 0.23]$ (with weight w_4),
IF $x = P_5$, THEN $y \in [0.05,\ 0.14]$ (with weight w_5),
IF $x = P_6$, THEN $y \in [0.23,\ 0.32]$ (with weight w_6),

where $P_1 =$ *about* 0, $P_2 =$ *about* 0.09, $P_3 =$ *about* 0.4, $P_4 =$ *about* 0.71, $P_5 =$ *about* 0.92, $P_6 =$ *about* 1.0 are fuzzy terms of input variable x with membership functions shown in Fig. 3.8b.

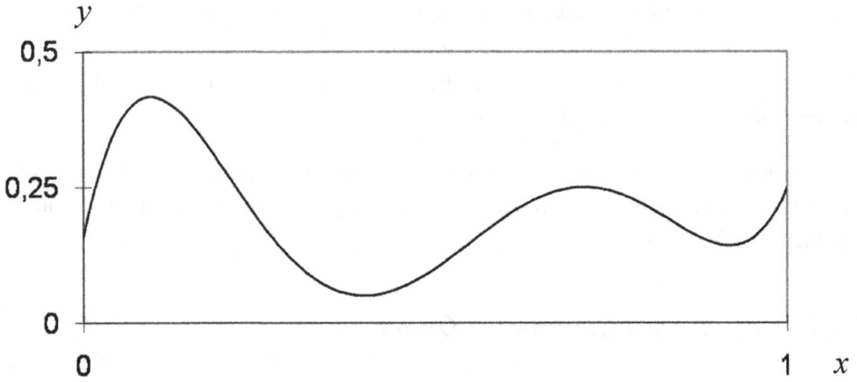

Fig. 3.7. Model-standard for one input – one output object

Before tuning all the rules weights w_j were alike, that is $w_j = 1$, $j = \overline{1,6}$. It provided fuzzy model shown in Fig. 3.8a. After genetic and neuro tuning, which consisted of solving optimization task (3.6), the new fuzzy models shown in Fig. 3.9, 3.10 were obtained. Fuzzy terms membership functions parameters as well as rules weights before and after tuning are presented in Tables 3.2 and 3.3. The dynamics of object $y = f(x)$ fuzzy model learning obtained using formula (3.11) is shown in Fig. 3.11.

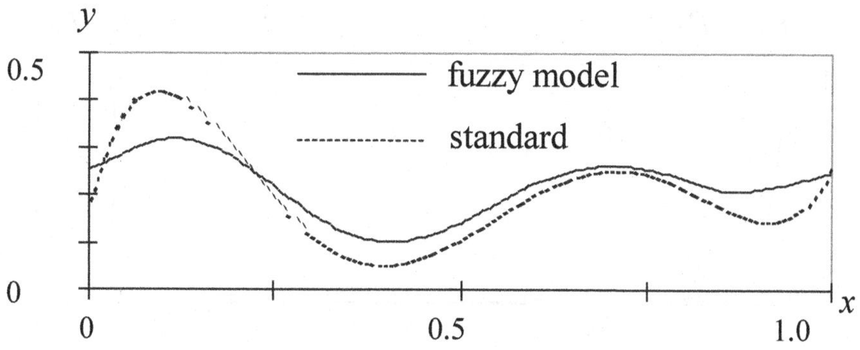

(a) comparison of standard with fuzzy model

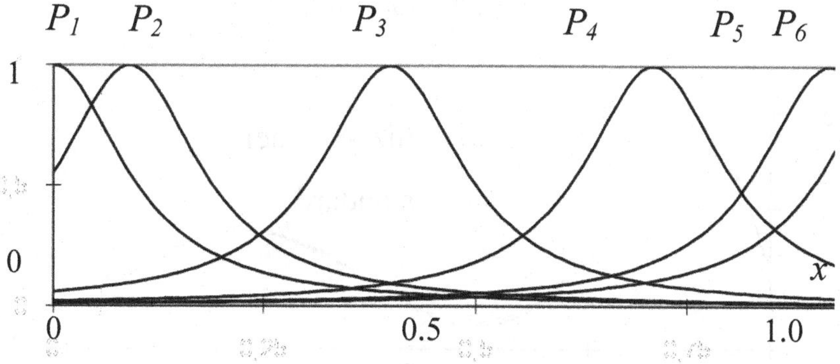

(b) fuzzy terms membership functions

Fig. 3.8. One input – one output object fuzzy model before tuning

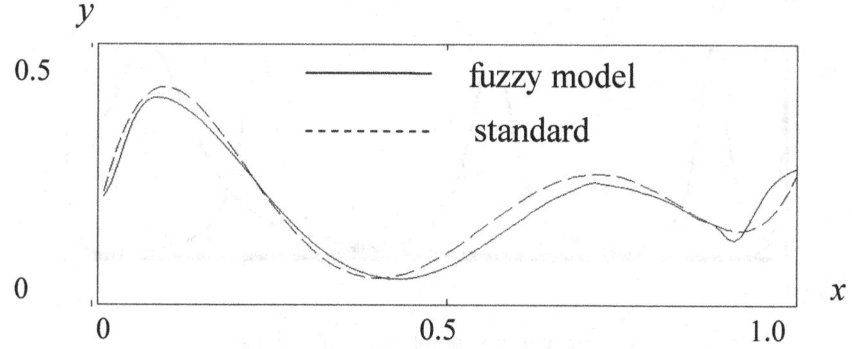

(a) comparison of standard with fuzzy model

Fig. 3.9. One input - one output object fuzzy model after genetic tuning

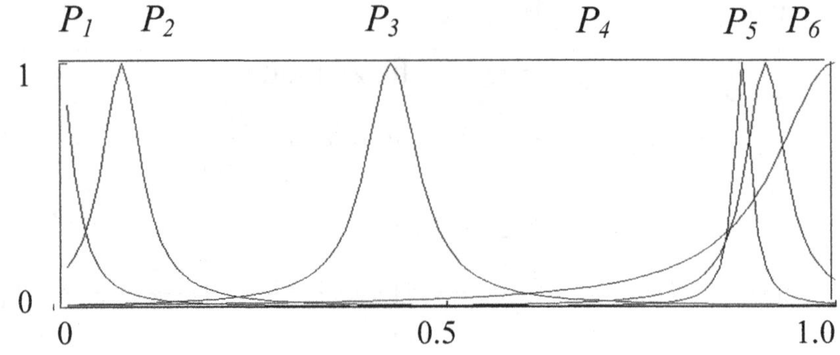

(b) fuzzy terms membership functions

Fig. 3.9. *(continued)*

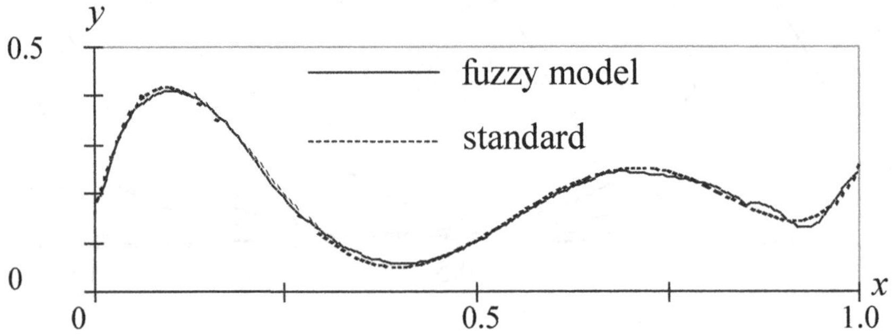

(a) comparison of standard with fuzzy model

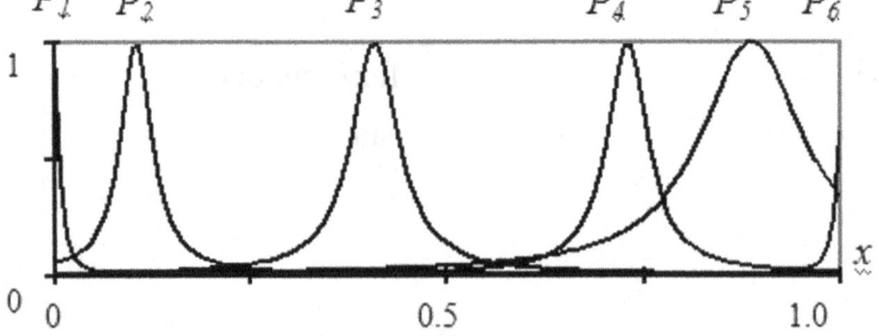

(b) fuzzy terms membership functions

Fig. 3.10. One input - one output object fuzzy model after neural tuning

Table 3.2. Linguistic terms membership functions parameters for object $y = f(x)$ fuzzy model

Term	Parameters					
	before tuning		genetic tuning		neural tuning	
	$в$	c	$в$	c	$в$	c
P_1	0	0.1	0	0.021	0	0.006
P_2	0.09	0.1	0.08	0.030	0.10	0.020
P_3	0.4	0.1	0.43	0.040	0.40	0.030
P_4	0.71	0.1	0.88	0.012	0.85	0.007
P_5	0.92	0.1	0.91	0.030	0.93	0.034
P_6	1.0	0.1	1.0	0.092	1.0	0.090

Table 3.3. Rules weights for object $y = f(x)$ fuzzy model

Rules weights	w_1	w_2	w_3	w_4	w_5	w_6
Before tuning	1	1	1	1	1	1
Genetic tuning	1	0.90	0.92	1	0.89	0.78
Neural tuning	1	0.72	0.71	1	0.68	0.70

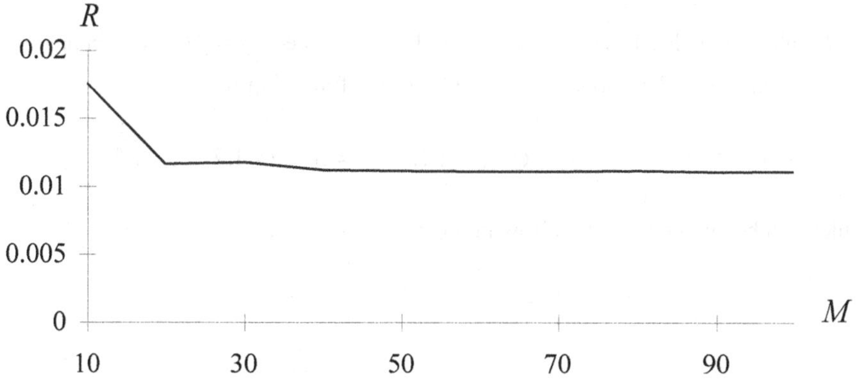

Fig. 3.11. One input - one output object fuzzy model learning dynamics

Experiment 2. We consider an object with two inputs $x_1 \in [0,6]$, $x_2 \in [0,6]$ and one output $y \in [-5.08, 0.855]$, having standard analytical model of this form:

$$y = f(x_1, x_2) = \frac{1}{40}(2z - 0.9)\ (7z - 1)\ (17z - 19)\ (15z - 2),$$

where $z = \dfrac{(x_1 - 3)^2 + (x_2 - 3)^2}{18}$.

The target model is shown in Fig. 3.12

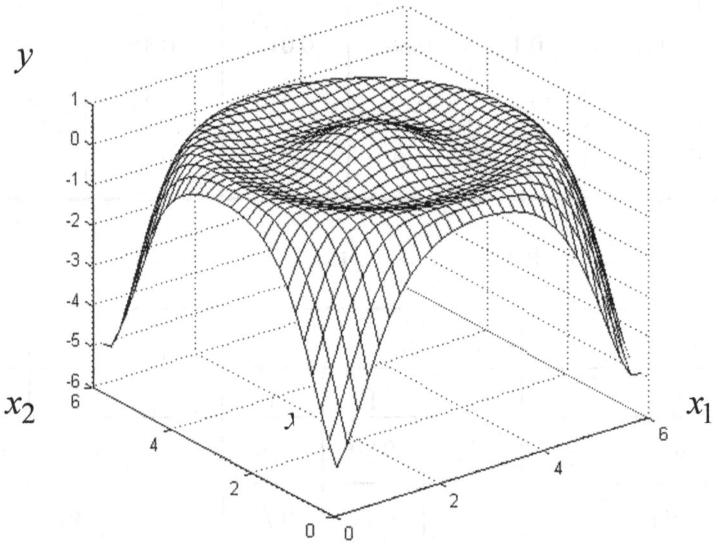

Fig. 3.12. Target model for "two inputs – one output" object

Rough knowledge base formed on the basis of object $y = f(x_1, x_2)$ behavior observation in Fig. 3.12 consists of 49 rules of the following form:

$$\text{IF } x_1 = P_i \text{ AND } x_2 = Q_i \text{ , THEN } y = B_j \text{ , } i = \overline{1,7} \text{ , } j = \overline{1,5} \text{ ,}$$

which can be depicted in the following matrix 7×7 form:

	P_1	P_2	P_3	P_4	P_5	P_6	P_7
Q_1	B_2	B_1	B_3	B_4	B_3	B_1	B_2
Q_2	B_1	B_3	B_4	B_5	B_4	B_3	B_1
Q_3	B_3	B_4	B_4	B_4	B_4	B_4	B_3
Q_4	B_4	B_5	B_4	B_5	B_4	B_5	B_4
Q_5	B_3	B_4	B_4	B_4	B_4	B_4	B_3
Q_6	B_1	B_3	B_4	B_5	B_4	B_3	B_1
Q_7	B_2	B_1	B_3	B_4	B_3	B_1	B_2

where $P_1 = Q_1 = about$ 0, $P_2 = Q_2 = about$ 0.5, $P_3 = Q_3 = about$ 1.5, $P_4 = Q_4 = about$ 3, $P_5 = Q_5 =$ *about* 4.5, $P_6 = Q_6 =$ *about* 5.5, $P_7 = Q_7 =$ *about* 6 are fuzzy terms of input variables x_1 and x_2 with membership functions shown in Fig. 3.13b;

$B_1 = [-5.08, \ -4.5)$, $B_2 = [-4.5, \ -3.0)$, $B_3 = [-3.0, \ -0.5)$, $B_4 = [-0.5, \ 0)$, $B_5 = [0, \ 0.855]$ are the classes of the output variable y.

Weights of all rules before tuning were considered to be identical and were equal to 1. This provided the fuzzy model shown in Fig. 3.13a.

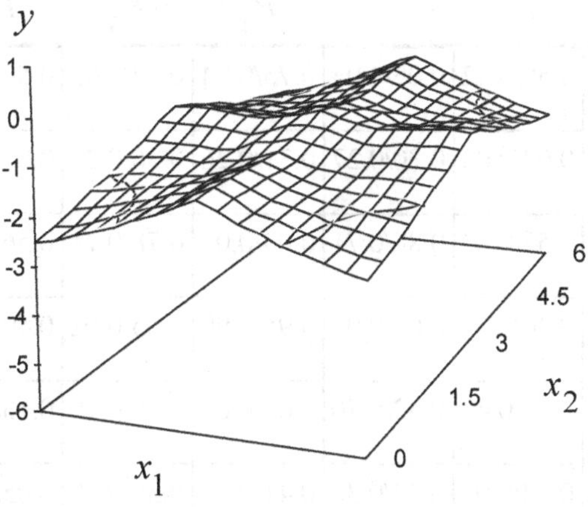

(a) fuzzy model

Fig. 3.13. Fuzzy model of two inputs - one output object before tuning

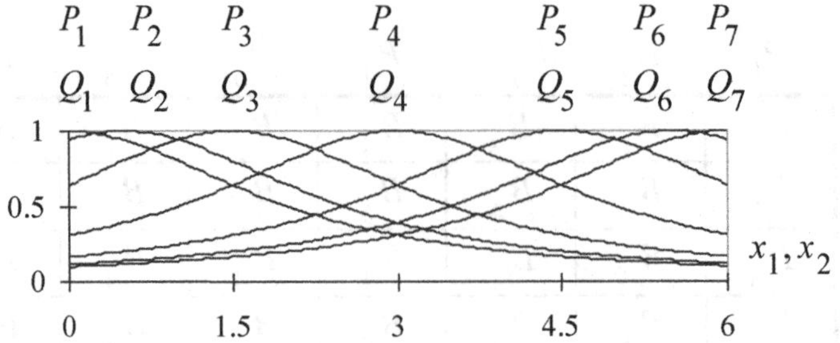

(b) fuzzy terms membership functions

Fig. 3.13. *(continued)*

Table 3.4. Fuzzy model $y = f(x_1, x_2)$ linguistic terms membership functions parameters b and c before tuning

Function	$\mu^{P_1}(x_1)$, $\mu^{Q_1}(x_2)$	$\mu^{P_2}(x_1)$, $\mu^{Q_2}(x_2)$	$\mu^{P_3}(x_1)$, $\mu^{Q_3}(x_2)$	$\mu^{P_4}(x_1)$, $\mu^{Q_4}(x_2)$	$\mu^{P_5}(x_1)$, $\mu^{Q_5}(x_2)$	$\mu^{P_6}(x_1)$, $\mu^{Q_6}(x_2)$	$\mu^{P_7}(x_1)$, $\mu^{Q_7}(x_2)$
b	0	0.5	1.5	3	4.5	5.5	6
c	2	2	2	2	2	2	2

Membership functions parameters before tuning are shown in Table. 3.4.

As the result of genetic and neuro tuning, an improved fuzzy model is shown in Fig. 3.14 and 3.15. The change of rules weights in the process of tuning can be represented in the form of the following matrix:

	P_1	P_2	P_3	P_4	P_5	P_6	P_7
Q_1	0.01/0.01	0.50/0.42	0.05/0.03	0.75/0.64	0.07/0.05	0.50/0.42	0.05/0.03
Q_2	0.73/0.61	0.02/0.03	0.30/0.27	0.31/0.26	0.36/0.27	0.05/0.05	0.70/0.61
Q_3	0.04/0.05	0.52/0.44	0.88/0.71	0.04/0.03	0.78/0.71	0.56/0.44	0.12/0.09
Q_4	0.92/0.87	0.40/0.34	0.02/0.01	0.95/0.90	0.05/0.02	0.40/0.34	0.92/0.87
Q_5	0.02/0.03	0.55/0.43	0.82/0.70	0.05/0.05	0.78/0.70	0.55/0.43	0.03/0.01
Q_6	0.71/0.60	0.02/0.03	0.44/0.30	0.41/0.29	0.38/0.30	0.05/0.05	0.71/0.60
Q_7	0.01/0.03	0.55/0.48	0.06/0.05	0.87/0.83	0.12/0.09	0.61/0.48	0.03/0.01

where numerators are weights after genetic tuning, and denominators are weights after neural tuning. Fuzzy terms membership functions parameters b and c values are shown in Table 3.5.

The dynamics of fuzzy model learning is shown in Fig. 3.16.

Table 3.5. Fuzzy model $y = f(x_1, x_2)$ linguistic terms membership functions parameters b and c after genetic and neural tuning

Function	$\mu^{P_1}(x_1)$	$\mu^{P_2}(x_1)$	$\mu^{P_3}(x_1)$	$\mu^{P_4}(x_1)$	$\mu^{P_5}(x_1)$	$\mu^{P_6}(x_1)$	$\mu^{P_7}(x_1)$
b	0 (0.07)	0.15 (0.00)	2.37 (2.39)	3.0 (3.00)	3.72 (3.63)	5.86 (5.99)	5.99 (5.92)
c	0.35 (0.09)	1.26 (0.81)	0.61 (0.32)	0.88 (0.65)	0.47 (0.29)	1.34 (0.83)	0.28 (0.09)
Function	$\mu^{Q_1}(x_2)$	$\mu^{Q_2}(x_2)$	$\mu^{Q_3}(x_2)$	$\mu^{Q_4}(x_2)$	$\mu^{Q_5}(x_2)$	$\mu^{Q_6}(x_2)$	$\mu^{Q_7}(x_2)$
b	0 (0.00)	0.17 (0.17)	2.32 (2.06)	3.0 (2.99)	3.85 (3.94)	5.89 (5.82)	5.99 (5.99)
c	0.22 (0.09)	1.35 (1.10)	0.77 (0.49)	0.97 (0.72)	0.65 (0.49)	1.25 (1.10)	0.36 (0.09)

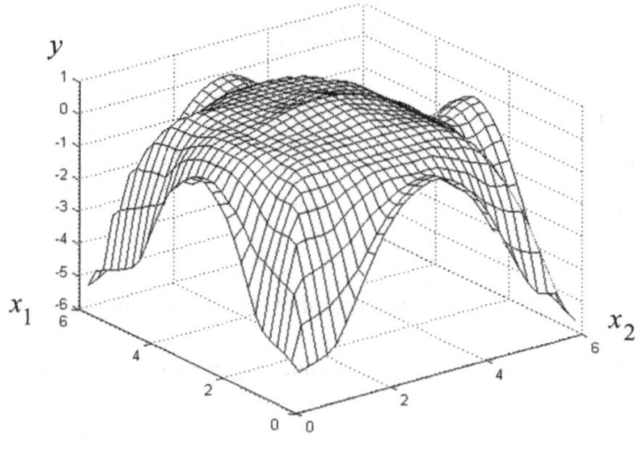

(a) fuzzy model

Fig. 3.14. Fuzzy model of two inputs - one output object after genetic tuning

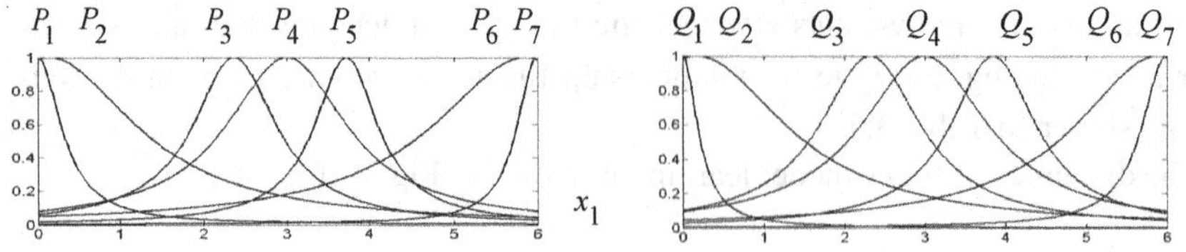

(b) fuzzy terms membership functions

Fig. 3.14. *(continued)*

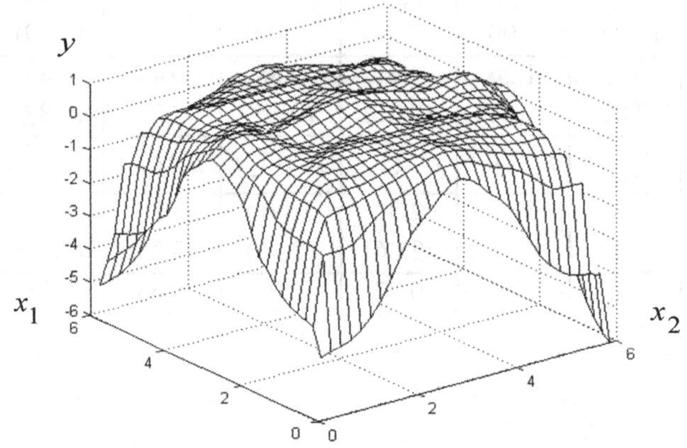

(a) fuzzy model

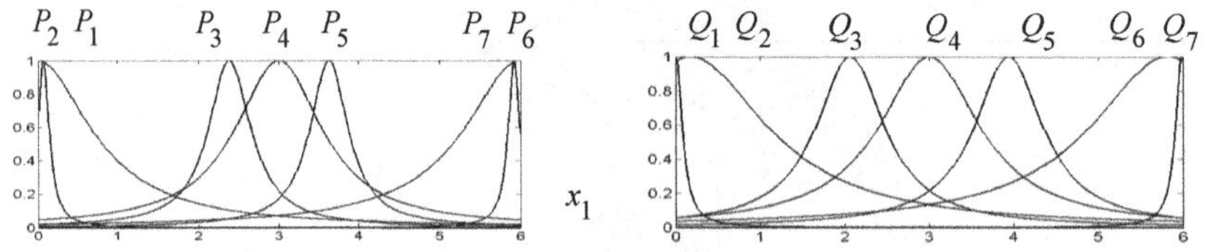

(b) fuzzy terms membership functions

Fig. 3.15. Fuzzy model of two inputs - one output object after neural tuning

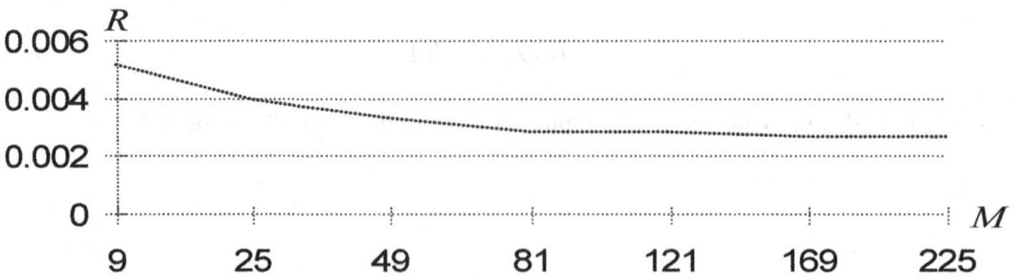

Fig. 3.16. Two inputs - one output object fuzzy model learning dynamics

3.4.3 Objects with Discrete Output

Experiment 1. An object with two inputs $x_1 \in [0,1]$, $x_2 \in [0,1]$ and one output y is considered where there can occur 5 values: $\{d_1, d_2, d_3, d_4, d_5\}$. The investigated object represents itself two-dimensional space divided into 5 regions-classes of decision (Fig. 3.17a). The target bounds of the region-classes have the following form:

①: $x_1 \in [0.5, 0.625]$, $x_2 = 32(x_1 - 0.5)^2 + 0.5$;

②: $x_1 \in [0, 0.25]$, $x_2 = -16(x_1 - 0.25)^4 + 0.8125$;

③: $x_1 \in [0.25, 0.5]$, $x_2 = -80(x_1 - 0.25)^4 + 0.8125$;

④: $x_2 \in [0, 0.5]$ $x_1 = (0.5 - x_2)(2 - x_2)(-0.1 - x_2) + 0.5$;

⑤: $x_2 \in [0.25, 0.5]$, $x_1 = \sqrt{(0.5 - x_2)} + 0.5$;

⑥: $x_1 \in [0.75, 1.0]$, $x_2 = -80(x_1 - 1)^4 + 0.75$

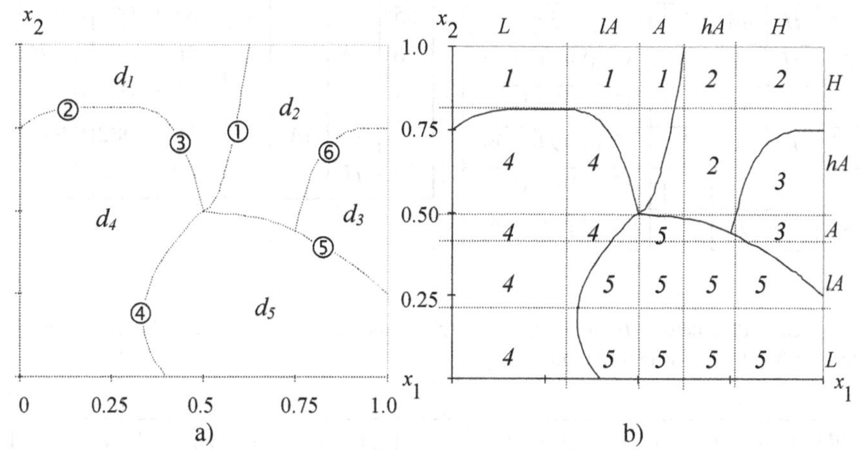

Fig. 3.17. Two inputs - one output object - 1: a) five decision classes, b) fuzzy IF-THEN rules acquisition

In Fig. 3.17b it is shown the way fuzzy IF-THEN rules were obtained and used to form a fuzzy knowledge base in Table 3.6, where L – low, lA – lower than average, A – average, hA – higher than average, H – High.

Fuzzy terms membership functions from knowledge base (Table 3.6) are shown in Fig. 3.18b, and corresponding parameters are presented in Table 3.7. Rules weights before tuning were considered identical and equal to 1 that provided the rough fuzzy model depicted in Fig. 3.18a.

Table 3.6. Knowledge matrix and corresponding rules weights after genetic (w_g) and neural (w_n) tuning

N	x_1	x_2	d	N	w_g	w_n
11	*L*	*H*		11	0.943	0.997
12	*lA*	*H*	d_1	12	0.987	0.995
13	*A*	*H*		13	0.935	0.995
21	*hA*	*H*		21	1.000	0.995
22	*hA*	*hA*	d_2	22	1.000	0.993
23	*H*	*H*		23	1.000	0.997
31	*H*	*hA*	d_3	31	0.909	0.996
32	*H*	*A*		32	0.984	0.996
41	*L*	*hA*		41	1.000	0.995
42	*L*	*A*	d_4	42	0.999	0.995
43	*L*	*lA*		43	0.998	0.996
44	*L*	*L*		44	0.998	0.997
45	*lA*	*hA*	d_4	45	0.984	0.994
46	*lA*	*A*		46	1.000	0.994
51	*lA*	*lA*		51	0.978	0.995
52	*lA*	*L*		52	0.988	0.996
53	*A*	*A*		53	0.965	0.993
54	*A*	*lA*		54	0.986	0.994
55	*A*	*L*	d_5	55	1.000	0.996
56	*hA*	*lA*		56	0.977	0.995
57	*hA*	*L*		57	1.000	0.996
58	*H*	*lA*		58	0.982	0.997
59	*H*	*L*		59	1.000	0.997

Table 3.7. Parameters b and c of the fuzzy model $y = f(x_1, x_2)$ linguistic terms membership functions before tuning

Function	$\mu^L(x_1)$	$\mu^{lA}(x_1)$	$\mu^A(x_1)$	$\mu^{hA}(x_1)$	$\mu^H(x_1)$
b	0.100	0.375	0.550	0.650	0.950
c	0.250	0.250	0.250	0.250	0.250
Function	$\mu^L(x_2)$	$\mu^{lA}(x_2)$	$\mu^A(x_2)$	$\mu^{hA}(x_2)$	$\mu^H(x_2)$
b	0.100	0.375	0.460	0.620	0.900
c	0.250	0.250	0.250	0.250	0.250

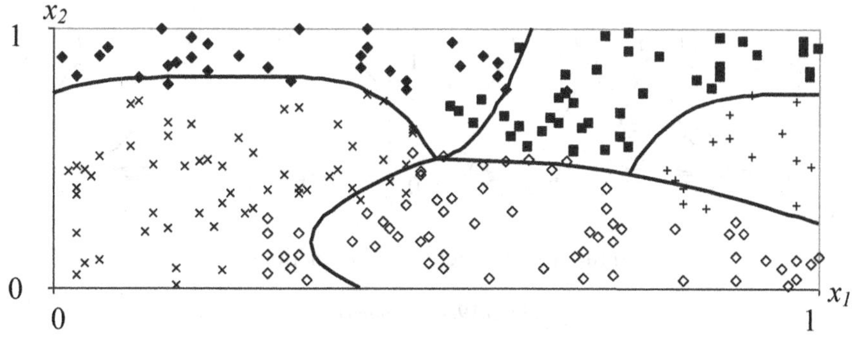

a) results of decision making (♦ - d_1, ■ - d_2, + - d_3, × - d_4, ◊ - d_5)

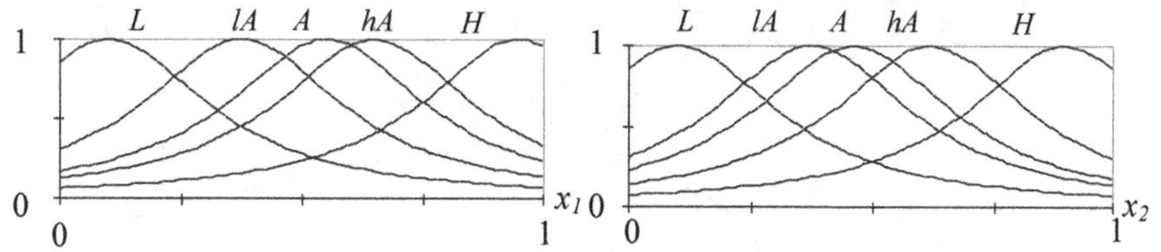

b) fuzzy terms membership functions

Fig. 3.18. Two inputs - one output object-1 fuzzy model before tuning

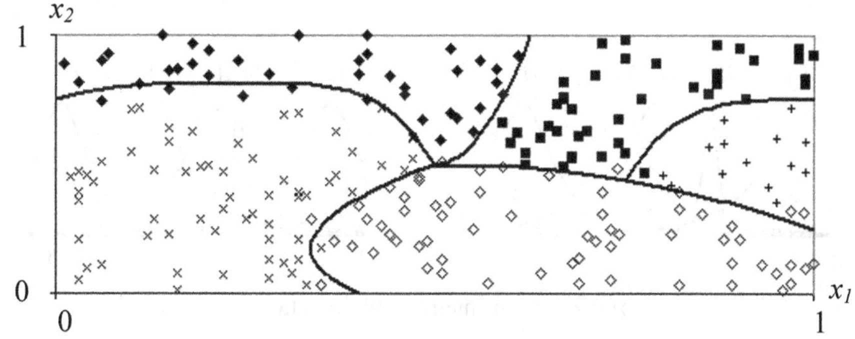

a) results of decision making (♦ - d_1, ■ - d_2, + - d_3, × - d_4, ◊ - d_5)

Fig. 3.19. Two inputs - one output object-1 fuzzy model after genetic tuning

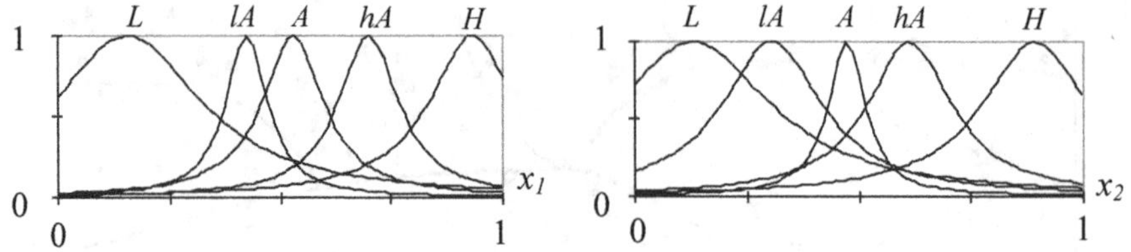

b) fuzzy terms membership functions

Fig. 3.19. *(continued)*

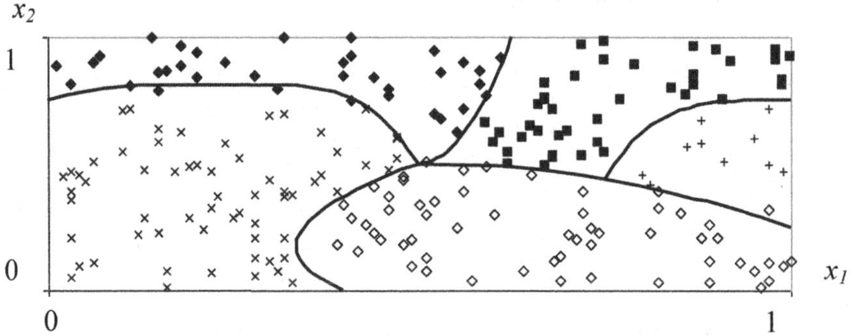

a) results of decision making (♦ - d_1, ■ - d_2, + - d_3, × - d_4, ◊ - d_5)

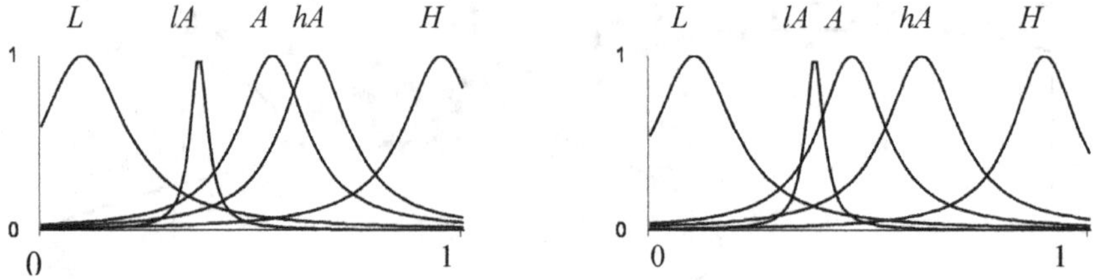

b) fuzzy terms membership functions

Fig. 3.20. Two inputs - one output object-1 fuzzy model after neural tuning

After genetic and neuro tuning, which consisted of solving optimization problem (3.10), the new fuzzy models were obtained which are shown in Fig. 3.19a, 3.20a.

Membership functions, obtained as the result of tuning, are shown in Fig. 3.19b, 3.20b, and their parameters are presented in Table. 3.8.

Table 3.8. Parameters b and c of the fuzzy model $y = f(x_1, x_2)$ linguistic terms membership functions after genetic (neuro) tuning

Function	$\mu^L(x_1)$	$\mu^{lA}(x_1)$	$\mu^A(x_1)$	$\mu^{hA}(x_1)$	$\mu^H(x_1)$
b	0.156 (0.100)	0.423 (0.375)	0.528 (0.550)	0.699 (0.650)	0.931 (0.950)
c	0.201 (0.122)	0.056 (0.025)	0.085 (0.098)	0.081 (0.098)	0.117 (0.107)
Function	$\mu^L(x_2)$	$\mu^{lA}(x_2)$	$\mu^A(x_2)$	$\mu^{hA}(x_2)$	$\mu^H(x_2)$
b	0.130 (0.100)	0.305 (0.375)	0.472 (0.460)	0.612 (0.620)	0.893 (0.901)
c	0.210 (0.109)	0.135 (0.025)	0.050 (0.092)	0.115 (0.100)	0.143 (0.089)

Tuned rules weights are shown in Table. 3.6. Object $y = f(x_1, x_2)$ fuzzy model learning dynamics can be seen in Fig. 3.21, where $\hat{P}_i$ $(i = \overline{1,5})$ and $\hat{P}$ are criteria of identification quality (3.12).

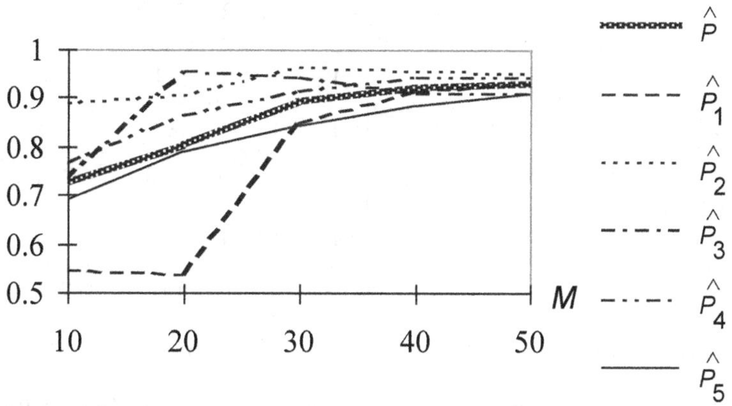

Fig. 3.21. Two inputs - one output object-1 fuzzy model learning dynamics

Experiment 2. Another object with two inputs $x_1 \in [0, 1]$, $x_2 \in [0, 1]$ and one discrete output y is considered here. Let number of classes - decisions be equal to five as before. Let us change standard models of region bounds (Fig. 3.22a):

①: $x_2 \in [0.5, 0.75]$, $x_1 = \frac{1}{2}\sqrt[3]{4(0.75 - x_2)} + 0.25$

②: $x_2 \in [0.25, 0.75]$, $x_1 = (x_2 - 0.5)^2 + 0.1875$

③: $x_1 \in [0.25, 0.75]$, $x_2 = 0.5 - 0.0625(4(0.75 - x_1))^2$

④: $x_1 \in [0.25, 0.5]$, $x_2 = 0.25(4(x_1 - 0.25))^{0.6} + 0.75$

⑤: $x_1 \in [0, 0.1875]$, $x_2 = 0.25(1 - \frac{16}{3}x_1)^{0.34} + 0.5$

⑥: $x_1 \in [0.25, 0.5]$, $x_2 = 0.25 - 0.125\sqrt{x_1 - 0.25}$

⑦: $x_1 \in [0.5, 0.8125]$, $x_2 = 0.1875 + 0.25\left(1 - 0.0625^{\frac{4}{5}(x_1 - 0.5)}\right)$

⑧: $x_1 \in [0.5, 1.0]$, $x_2 = 0.3125 + 0.125(\frac{16}{5}(x_1 - 0.8125))^2$.

Obtaining of fuzzy IF-THEN rules describing the object behavior is depicted in Fig. 3.22b. These rules are brought together in fuzzy knowledge base (Table 3.9).

Fuzzy terms membership functions parameters which are used in knowledge base in Table 3.9 before tuning are shown in Table 3.10.

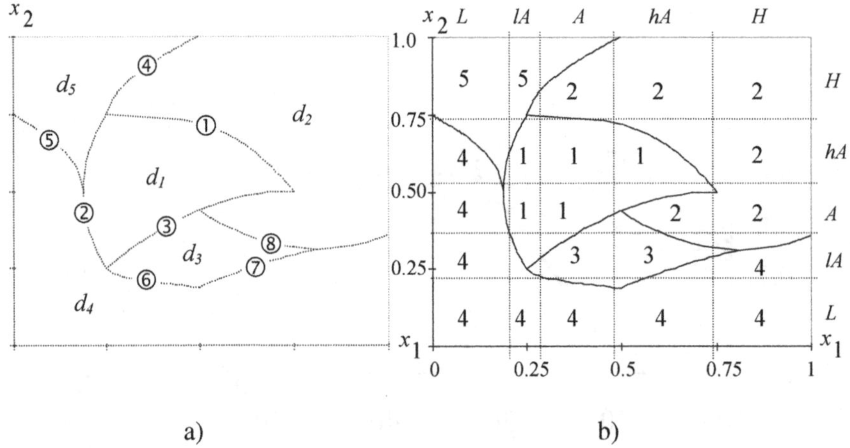

Fig. 3.22. Two inputs - one output object - 2: a) five classes-decisions, b) fuzzy IF-THEN rules acquisition

Table 3.9. Knowledge matrix and corresponding rules weights after genetic (w_g) and neural (w_n) tuning

N	x_1	x_2	d	w_g	w_n
11	*lA*	*hA*		0.998	0.999
12	*A*	*hA*		0.998	0.998
13	*hA*	*hA*	d_1	0.997	0.999
14	*lA*	*A*		1.000	0.997
15	*A*	*A*		0.997	0.995
21	*A*	*H*		1.000	0.996
22	*hA*	*H*		1.000	0.999
23	*H*	*H*	d_2	1.000	0.998
24	*H*	*hA*		1.000	0.999
25	*hA*	*A*		1.000	0.999
26	*H*	*A*		1.000	0.997
31	*A*	*lA*	d_3	0.999	0.997
32	*hA*	*lA*	d_3	0.995	0.999
41	*L*	*hA*		0.998	0.999
42	*L*	*A*		1.000	0.998
43	*L*	*lA*		1.000	0.998
44	*L*	*L*		1.000	0.999
45	*lA*	*L*	d_4	1.000	0.998
46	*A*	*L*		1.000	0.997
47	*hA*	*L*		1.000	0.999
48	*H*	*L*		1.000	0.998
49	*H*	*lA*		0.998	0.999
51	*L*	*H*	d_5	1.000	0.998
52	*lA*	*H*		0.998	0.998

Table 3.10. Parameters b and c of the fuzzy model $y = f(x_1, x_2)$ linguistic terms membership functions before tuning

Function	$\mu^L(x_1)$	$\mu^{lA}(x_1)$	$\mu^A(x_1)$	$\mu^{hA}(x_1)$	$\mu^H(x_1)$
b	0.00	0.25	0.50	0.75	1.00
c	0.10	0.10	0.10	0.10	0.10
Function	$\mu^L(x_2)$	$\mu^{lA}(x_2)$	$\mu^A(x_2)$	$\mu^{hA}(x_2)$	$\mu^H(x_2)$
b	0.00	0.25	0.50	0.75	1.00
c	0.10	0.10	0.10	0.10	0.10

Fuzzy terms membership functions from knowledge base (Table 3.9) are shown in Fig. 3.23b, and corresponding parameters are presented in Table 3.10. Rules weights before tuning were considered identical and equal to 1. The described fuzzy model is shown in Fig. 3.23a.

After genetic and neuro tuning, which consisted of solving optimization problem (3.10), the new fuzzy models were obtained which are shown in Fig. 3.24a, 3.25a. Tuned membership functions are depicted in Fig. 3.24b, 3.25b.

Fuzzy model membership functions parameters, obtained as the result of tuning, are shown in Table 3.11

Table 3.11. Parameters b and c of the fuzzy model $y = f(x_1, x_2)$ linguistic terms membership functions after genetic (neural) tuning

Function	$\mu^L(x_1)$	$\mu^{lA}(x_1)$	$\mu^A(x_1)$	$\mu^{hA}(x_1)$	$\mu^H(x_1)$
b	0.033 (0.025)	0.275 (0.273)	0.503 (0.492)	0.735 (0.743)	0.960 (0.972)
c	0.149 (0.114)	0.073 (0.078)	0.139 (0.118)	0.010 (0.014)	0.313 (0.119)
Function	$\mu^L(x_2)$	$\mu^{lA}(x_2)$	$\mu^A(x_2)$	$\mu^{hA}(x_2)$	$\mu^H(x_2)$
b	0.031 (0.035)	0.275 (0.275)	0.529 (0.515)	0.733 (0.725)	0.967 (0.977)
c	0.188 (0.139)	0.062 (0.069)	0.136 (0.118)	0.010 (0.017)	0.243 (0.202)

Rules weights after tuning are shown in Table 3.9. Object $y = f(x_1, x_2)$ fuzzy model learning dynamics can be seen in Fig. 3.26.

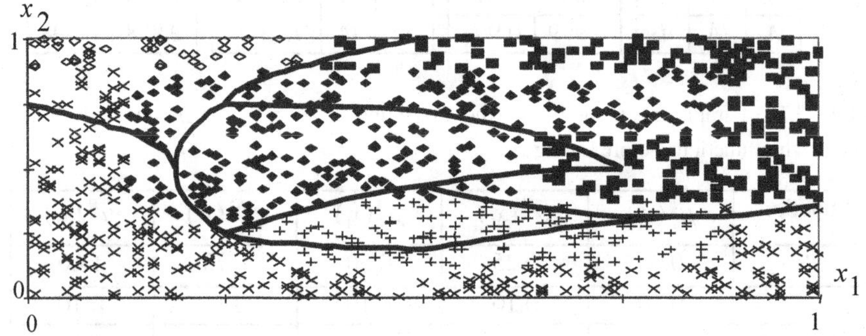

a) results of decision making (♦ - d_1, ■ - d_2, + - d_3, × - d_4, ◊ - d_5)

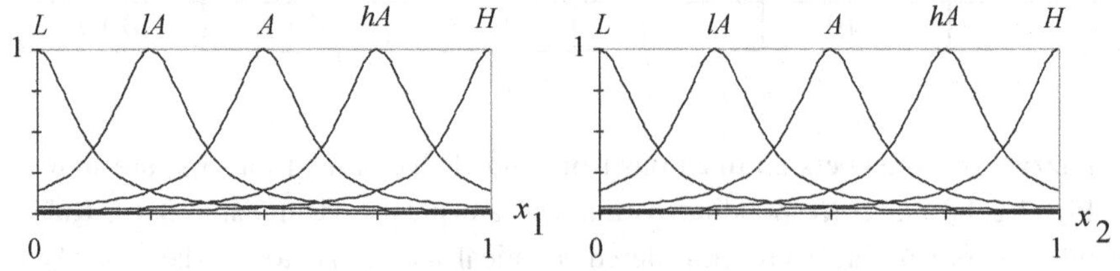

b) fuzzy terms membership functions

Fig. 3.23. Two inputs - one output object-2 fuzzy model before tuning

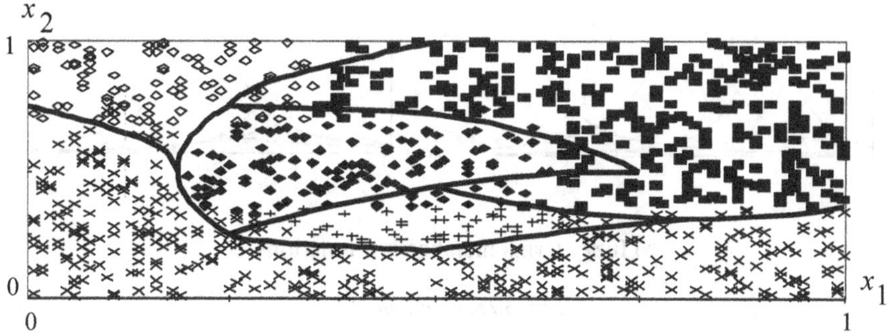

a) results of decision making (♦ - d_1, ■ - d_2, + - d_3, × - d_4, ◊ - d_5)

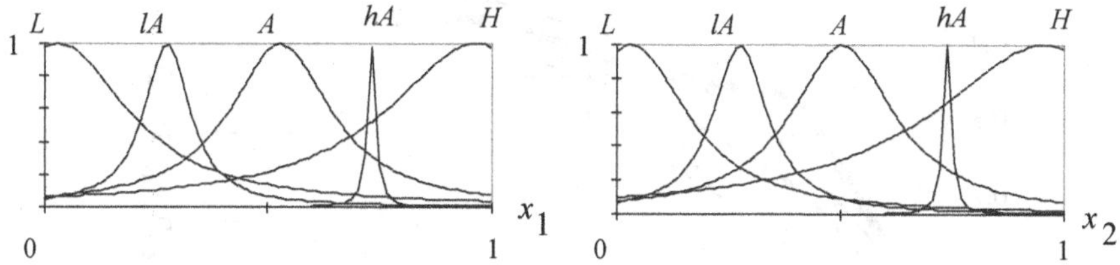

b) fuzzy terms membership functions

Fig. 3.24. Two inputs - one output object-2 fuzzy model after genetic tuning

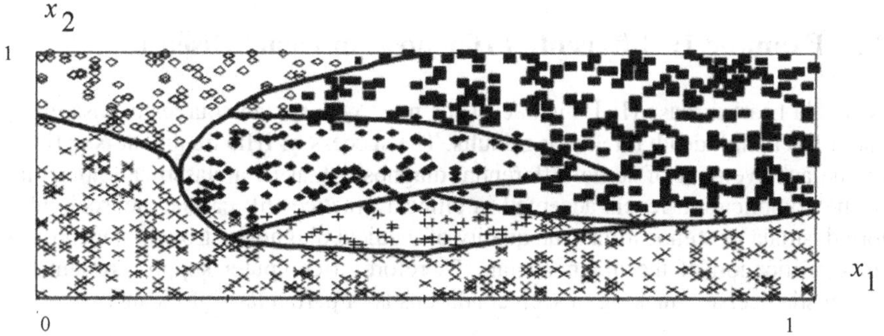

a) results of decision making (♦ - d_1, ■ - d_2, + - d_3, × - d_4, ◊ - d_5)

Fig. 3.25. Two inputs - one output object-2 fuzzy model after neural tuning

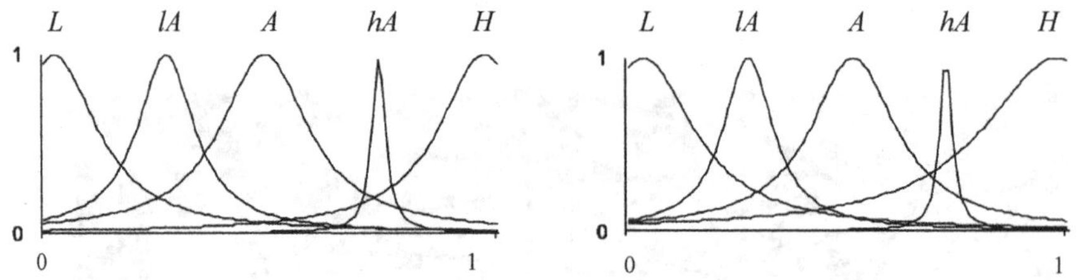

b) fuzzy terms membership functions

Fig. 3.25. *(continued)*

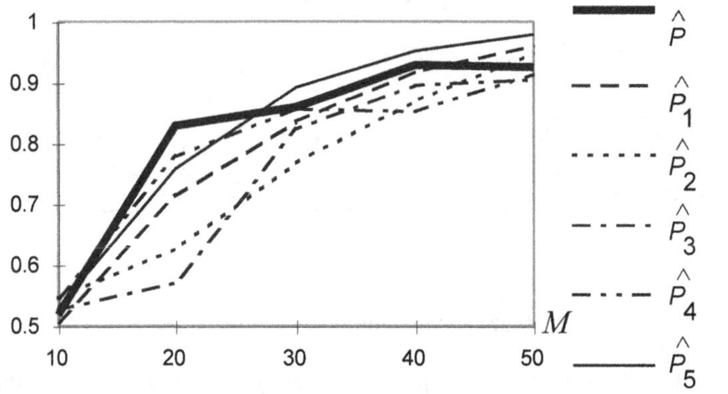

Fig. 3.26. Two inputs – one output object-2 fuzzy model learning dynamics

3.5 Example 1: Differential Diagnosis of Heart Disease

Ischemia heart disease (IHD) is one of the most widespread sources of disability, and it has a high death rate among adults. The success of IHD treatment is defined by the achievement of some differential diagnosis, that is, a classification as one of the complication levels accepted in clinical practice [6]: cardiac neurocirculatory dystonia or stenocardia. The quality of medical diagnosis strongly depends on the qualification of the diagnostician. Therefore, a computer support system for diagnostic decision making in such conditions is of particular significance.

3.5.1 Diagnosis Types and Parameters of Patient's State

According to current clinical practice, the complication of IHD will be defined at the levels as follows (from the lowest to the highest):

d_1 – neurocirculatory dystonia (NCD) of light complication;
d_2 – NCD of average complication;
d_3 – NCD of heavy complication;
d_4 – stenocardia of the first functional disability degree;
d_5 – stenocardia of the second functional disability degree;
d_6 – stenocardia of the third functional disability degree.

The above mentioned levels $d_1 \div d_6$ are considered the types of diagnoses which should be identified. While making the diagnosis of IHD of a specific patient, we should take into consideration the next main parameters defined by laboratory tests (possible variation ranges are indicated in round brackets where c. u. is a conventional unit):

x_1 is the age of the patient (31–57 years),
x_2 is the double product (DP) of pulse and blood pressure (147–405 c.u.),
x_3 is the tolerance to physical loads (90–1200 kgm/min),
x_4 is the increase of DP per kg of the patient body weight (0.6–3.9 c.u.),
x_5 is the increase of DP per kg of load (0.1–0.4 c.u.),
x_6 is the adenosine-triphosphoric acid - ATP (34.5–66.2 mmol/l),
x_7 is the adenosine-diphosphoric acid - ADP (11.9–29.2 mmol/l),
x_8 is the adenosine-monophosphoric acid - AMP (3.6–27.1 mmol/l),
x_9 is the coefficient of phosphorylation (1–5.7 c.u.),
x_{10} is the max. oxygen consumption per kg of patient weight (10.5 – 40.9 mlitre/min$\times$kg),
x_{11} is the increase of DP in response to submaximal load (46–312 c.u.),
x_{12} is the ratio factor of milk and pyruvic acid (3.9–22.8 c.u.).

The aim of the diagnosis is to translate a set of specific parameters $x_1 \div x_{12}$ into a decision d_j $(j = \overline{1,6})$.

3.5.2 *Fuzzy Rules*

The structure of the model for differential diagnosis of IHD is shown in Fig. 3.27, which corresponds to the following hierarchical tree of logic inference:

$$d = f_d(x_1, y, z) \ , \tag{3.30}$$

$$y = f_y(x_2, x_3, x_4, x_5, x_{10}, x_{11}) \ , \tag{3.31}$$

$$z = f_z(x_6, x_7, x_8, x_9, x_{12}) \ , \tag{3.32}$$

where d is the danger of IHD measured by levels $d_1 \div d_6$; y is the instrumental danger depending on parameters $\{x_2, x_3, x_4, x_5, x_{10}, x_{11}\}$; z is the biochemical danger depending on parameters $\{x_6, x_7, x_8, x_9, x_{12}\}$.

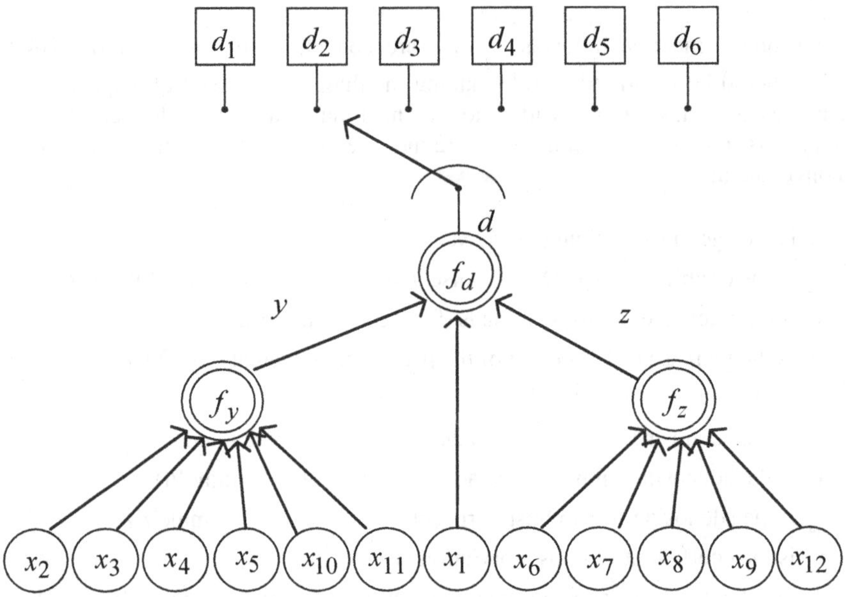

Fig. 3.27. Tree of logic inference

The expert fuzzy if-then rules which correspond to the relations (3.30) – (3.32) are represented in Tables 3.12–3.14 using fuzzy terms as: *L* – low, *lA* – lower than average, *A* – average, *hA* – higer than average, *H* – high. These rules were presented by the therapist of Vinnitsa Cardiology Clinic V. M. Sheverda.

Table 3.12. Knowledge about relation (3.30)

x_1	y	z	d
L	*L*	*L*	
L	*lA*	*lA*	d_1
lA	*lA*	*L*	
lA	*lA*	*lA*	
A	*lA*	*lA*	d_2
lA	*lA*	*A*	

Table 3.12. *(continued)*

A	lA	A	
hA	hA	lA	d_3
hA	A	A	
hA	A	hA	
A	hA	hA	d_4
lA	hA	hA	
A	H	A	
hA	hA	H	d_5
H	hA	hA	
H	H	H	
hA	H	hA	d_6
A	H	hA	

Table 3.13. Knowledge about relation (3.31)

x_2	x_3	x_4	x_5	x_{10}	x_{11}	y
H	H	H	L	H	H	
H	hA	H	lA	H	H	L
hA	H	hA	L	H	H	
hA	hA	H	lA	H	hA	
H	H	hA	A	H	H	lA
hA	hA	H	lA	hA	hA	
A	A	A	A	A	A	
hA	hA	A	lA	hA	A	A
A	hA	hA	A	hA	hA	
lA	A	lA	hA	lA	lA	
lA	lA	A	A	L	lA	hA
A	lA	lA	hA	lA	A	
L	L	L	hA	L	L	
lA	L	lA	H	L	lA	H
L	lA	lA	hA	L	L	

Table 3.14. Knowledge about relation (3.32)

x_6	x_7	x_8	x_9	x_{12}	z
H	H	H	H	H	
hA	H	hA	hA	hA	L
H	hA	H	A	hA	
hA	hA	A	A	hA	
A	hA	A	hA	H	lA
A	H	hA	hA	hA	

Table 3.14. *(continued)*

A	*A*	*A*	*hA*	*hA*	
hA	*hA*	*A*	*A*	*A*	*A*
hA	*A*	*hA*	*hA*	*A*	
lA	*A*	*lA*	*A*	*A*	
hA	*lA*	*A*	*lA*	*lA*	*hA*
L	*A*	*A*	*lA*	*A*	
L	*L*	*L*	*L*	*lA*	
lA	*L*	*lA*	*L*	*L*	*H*
L	*lA*	*lA*	*L*	*lA*	

3.5.3 Fuzzy Logic Equations

Using Tables 3.12–3.14 and operations • (AND - min) and $\vee$ (OR - max), it is easy to write the system of fuzzy logic equations which connect the membership functions of diagnosis and parameters of the patient state as:

$$\mu^{d_1}(d)=[\mu^{L}(x_1)\cdot\mu^{L}(y)\cdot\mu^{L}(z)]\vee[\mu^{L}(x_1)\cdot\mu^{lA}(y)\cdot\mu^{lA}(z)]\vee[\mu^{lA}(x_1)\cdot\mu^{lA}(y)\cdot\mu^{L}(z)],$$
$$\mu^{d_2}(d)=[\mu^{lA}(x_1)\cdot\mu^{lA}(y)\cdot\mu^{lA}(z)]\vee[\mu^{A}(x_1)\cdot\mu^{lA}(y)\cdot\mu^{lA}(z)]\vee[\mu^{lA}(x_1)\cdot\mu^{lA}(y)\cdot\mu^{A}(z)],$$
$$\mu^{d_3}(d)=[\mu^{A}(x_1)\cdot\mu^{lA}(y)\cdot\mu^{A}(z)]\vee[\mu^{hA}(x_1)\cdot\mu^{hA}(y)\cdot\mu^{lA}(z)]\vee[\mu^{hA}(x_1)\cdot\mu^{A}(y)\cdot\mu^{A}(z)],$$
$$\mu^{d_4}(d)=[\mu^{hA}(x_1)\cdot\mu^{A}(y)\cdot\mu^{hA}(z)]\vee[\mu^{A}(x_1)\cdot\mu^{hA}(y)\cdot\mu^{hA}(z)]\vee[\mu^{lA}(x_1)\cdot\mu^{hA}(y)\cdot\mu^{hA}(z)],$$
$$\mu^{d_5}(d)=[\mu^{A}(x_1)\cdot\mu^{H}(y)\cdot\mu^{A}(z)]\vee[\mu^{hA}(x_1)\cdot\mu^{hA}(y)\cdot\mu^{H}(z)]\vee[\mu^{H}(x_1)\cdot\mu^{hA}(y)\cdot\mu^{hA}(z)],$$
$$\mu^{d_6}(d)=[\mu^{H}(x_1)\cdot\mu^{H}(y)\cdot\mu^{H}(z)]\vee[\mu^{hA}(x_1)\cdot\mu^{H}(y)\cdot\mu^{hA}(z)]\vee[\mu^{A}(x_1)\cdot\mu^{H}(y)\cdot\mu^{hA}(z)]. \tag{3.33}$$

$$\begin{aligned}\mu^{L}(y)=\;&\mu^{H}(x_2)\cdot\mu^{H}(x_3)\cdot\mu^{H}(x_4)\cdot\mu^{L}(x_5)\cdot\mu^{H}(x_{10})\cdot\mu^{H}(x_{11})\vee\\ &\vee\mu^{H}(x_2)\cdot\mu^{hA}(x_3)\cdot\mu^{H}(x_4)\cdot\mu^{lA}(x_5)\cdot\mu^{H}(x_{10})\cdot\mu^{H}(x_{11})\vee\\ &\vee\mu^{hA}(x_2)\cdot\mu^{H}(x_3)\cdot\mu^{hA}(x_4)\cdot\mu^{L}(x_5)\cdot\mu^{H}(x_{10})\cdot\mu^{H}(x_{11}),\end{aligned}$$

$$\begin{aligned}\mu^{lA}(y)=\;&\mu^{hA}(x_2)\cdot\mu^{hA}(x_3)\cdot\mu^{H}(x_4)\cdot\mu^{lA}(x_5)\cdot\mu^{H}(x_{10})\cdot\mu^{hA}(x_{11})\vee\\ &\vee\mu^{H}(x_2)\cdot\mu^{H}(x_3)\cdot\mu^{hA}(x_4)\cdot\mu^{A}(x_5)\cdot\mu^{H}(x_{10})\cdot\mu^{H}(x_{11})\vee\\ &\vee\mu^{hA}(x_2)\cdot\mu^{hA}(x_3)\cdot\mu^{H}(x_4)\cdot\mu^{lA}(x_5)\cdot\mu^{hA}(x_{10})\cdot\mu^{hA}(x_{11}),\end{aligned}$$

$$\begin{aligned}\mu^{A}(y)=\;&\mu^{A}(x_2)\cdot\mu^{A}(x_3)\cdot\mu^{A}(x_4)\cdot\mu^{A}(x_5)\cdot\mu^{A}(x_{10})\cdot\mu^{A}(x_{11})\vee\\ &\vee\mu^{hA}(x_2)\cdot\mu^{hA}(x_3)\cdot\mu^{A}(x_4)\cdot\mu^{lA}(x_5)\cdot\mu^{hA}(x_{10})\cdot\mu^{A}(x_{11})\vee\\ &\vee\mu^{A}(x_2)\cdot\mu^{hA}(x_3)\cdot\mu^{hA}(x_4)\cdot\mu^{A}(x_5)\cdot\mu^{hA}(x_{10})\cdot\mu^{hA}(x_{11}),\end{aligned}$$

$$\mu^{hA}(y) = \mu^{lA}(x_2)\cdot\mu^{A}(x_3)\cdot\mu^{lA}(x_4)\cdot\mu^{hA}(x_5)\cdot\mu^{lA}(x_{10})\cdot\mu^{lA}(x_{11}) \vee$$
$$\vee\, \mu^{lA}(x_2)\cdot\mu^{lA}(x_3)\cdot\mu^{A}(x_4)\cdot\mu^{A}(x_5)\cdot\mu^{L}(x_{10})\cdot\mu^{lA}(x_{11}) \vee$$
$$\vee\, \mu^{A}(x_2)\cdot\mu^{lA}(x_3)\cdot\mu^{lA}(x_4)\cdot\mu^{hA}(x_5)\cdot\mu^{lA}(x_{10})\cdot\mu^{A}(x_{11}),$$

$$\mu^{H}(y) = \mu^{L}(x_2)\cdot\mu^{L}(x_3)\cdot\mu^{L}(x_4)\cdot\mu^{hA}(x_5)\cdot\mu^{L}(x_{10})\cdot\mu^{L}(x_{11}) \vee$$
$$\vee\, \mu^{lA}(x_2)\cdot\mu^{L}(x_3)\cdot\mu^{lA}(x_4)\cdot\mu^{H}(x_5)\cdot\mu^{L}(x_{10})\cdot\mu^{lA}(x_{11}) \vee$$
$$\vee\, \mu^{L}(x_2)\cdot\mu^{lA}(x_3)\cdot\mu^{lA}(x_4)\cdot\mu^{hA}(x_5)\cdot\mu^{L}(x_{10})\cdot\mu^{L}(x_{11}) \quad . \tag{3.34}$$

$$\mu^{L}(z) = \mu^{H}(x_6)\cdot\mu^{H}(x_7)\cdot\mu^{H}(x_8)\cdot\mu^{H}(x_9)\cdot\mu^{H}(x_{12})$$
$$\vee\, \mu^{hA}(x_6)\cdot\mu^{H}(x_7)\cdot\mu^{hA}(x_8)\cdot\mu^{hA}(x_9)\cdot\mu^{hA}(x_{12})$$
$$\vee\, \mu^{H}(x_6)\cdot\mu^{hA}(x_7)\cdot\mu^{H}(x_8)\cdot\mu^{A}(x_9)\cdot\mu^{hA}(x_{12}) \ ,$$

$$\mu^{lA}(z) = \mu^{hA}(x_6)\cdot\mu^{hA}(x_7)\cdot\mu^{A}(x_8)\cdot\mu^{A}(x_9)\cdot\mu^{hA}(x_{12})$$
$$\vee\, \mu^{A}(x_6)\cdot\mu^{hA}(x_7)\cdot\mu^{A}(x_8)\cdot\mu^{hA}(x_9)\cdot\mu^{H}(x_{12})$$
$$\vee\, \mu^{A}(x_6)\cdot\mu^{H}(x_7)\cdot\mu^{hA}(x_8)\cdot\mu^{hA}(x_9)\cdot\mu^{hA}(x_{12}) \ ,$$

$$\mu^{A}(z) = \mu^{A}(x_6)\cdot\mu^{A}(x_7)\cdot\mu^{A}(x_8)\cdot\mu^{hA}(x_9)\cdot\mu^{hA}(x_{12})$$
$$\vee\, \mu^{hA}(x_6)\cdot\mu^{hA}(x_7)\cdot\mu^{A}(x_8)\cdot\mu^{A}(x_9)\cdot\mu^{A}(x_{12})$$
$$\vee\, \mu^{hA}(x_6)\cdot\mu^{A}(x_7)\cdot\mu^{hA}(x_8)\cdot\mu^{hA}(x_9)\cdot\mu^{A}(x_{12}) \ ,$$

$$\mu^{hA}(z) = \mu^{lA}(x_6)\cdot\mu^{A}(x_7)\cdot\mu^{lA}(x_8)\cdot\mu^{A}(x_9)\cdot\mu^{A}(x_{12})$$
$$\vee\, \mu^{hA}(x_6)\cdot\mu^{lA}(x_7)\cdot\mu^{A}(x_8)\cdot\mu^{lA}(x_9)\cdot\mu^{lA}(x_{12})$$
$$\vee\, \mu^{L}(x_6)\cdot\mu^{A}(x_7)\cdot\mu^{A}(x_8)\cdot\mu^{lA}(x_9)\cdot\mu^{A}(x_{12}) \ ,$$

$$\mu^{H}(z) = \mu^{L}(x_6)\cdot\mu^{L}(x_7)\cdot\mu^{L}(x_8)\cdot\mu^{L}(x_9)\cdot\mu^{lA}(x_{12})$$
$$\vee\, \mu^{lA}(x_6)\cdot\mu^{L}(x_7)\cdot\mu^{lA}(x_8)\cdot\mu^{L}(x_9)\cdot\mu^{L}(x_{12})$$
$$\vee\, \mu^{L}(x_6)\cdot\mu^{lA}(x_7)\cdot\mu^{lA}(x_8)\cdot\mu^{L}(x_9)\cdot\mu^{lA}(x_{12}) \tag{3.35}$$

The total number of fuzzy logic equations (3.33) – (3.35) is 16. Note that we do not use the weights of rules in (3.33) – (3.35) because all weights before tuning are equal to one.

3.5.4 Rough Membership Functions

Generally, all parameters $x_1 \div x_{12}$ have their own membership functions of the fuzzy terms (*L*, *lA*, *A*, *hA*, *H*), used in equations (3.33) - (3.35). To simplify the modeling, we can use only one shape of the membership functions for all parameters $x_1 \div x_{12}$, as shown in Fig. 3.28:

$$\mu^j(x_i) = \tilde{\mu}^j(u) \ , \quad u = 4\frac{x_i - \underline{x_i}}{\overline{x_i} - \underline{x_i}} \ , \quad j = L, lA, A, hA, H \ ,$$

where $[\underline{x_i}, \overline{x_i}]$ is an interval of parameter x_i changing, $i = \overline{1,12}$.

Analytical expressions of the functions in Fig. 3.28 are of the form:

$$\tilde{\mu}^j(u) = \frac{1}{1+(\frac{u-b}{c})^2} \ , \tag{3.36}$$

where the parameters b and c are given in Table 3.15. Selection of such curves is stipulated by the fact that they are approximations of membership functions gained by the expert method of pairwise comparison [17].

Table 3.15. Parameters of rough membership functions

Term	*L*	*lA*	*A*	*hA*	*H*
b	0	1	2	3	4
c	0.923	0.923	0.923	0.923	0.923

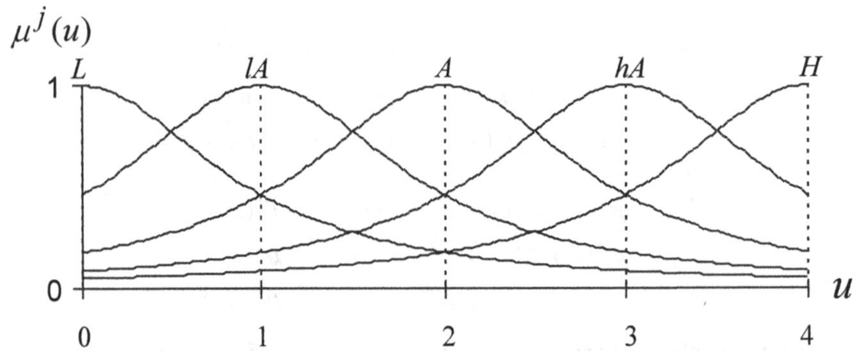

Fig. 3.28. Rough membership functions

3.5.5 Algorithm of Decision Making

Fuzzy logic equations (3.33) – (3.35) with membership functions of fuzzy terms (3.36) allow us to make the decision about the level of IHD according to this algorithm:

1°. Registration of parameters $\mathbf{X}^* = (x_1^*, x_2^*, ..., x_{12}^*)$ values for a specific patient.

2°. Using the model (3.36) and parameters b and c from Table 3.15, we define the values of the membership functions $\mu^j(x_i^*)$ when parameters values x_i^*, $i = \overline{1,12}$, are fixed.

3°. Using fuzzy logic equations (3.33) - (3.35), we calculate membership functions $\mu^{d_j}(x_1^*, x_2^*, ..., x_{12}^*)$ for all diagnoses $d_1, d_2, ..., d_6$. In doing so, according to [4], the logic operations AND ($\wedge$) and OR ($\vee$) are substituted for min and max :

$$\mu(a) \wedge \mu(b) = \min[\mu(a), \mu(b)] \ ,$$

$$\mu(a) \vee \mu(b) = \max[\mu(a), \mu(b)] \ .$$

4°. Let us define the decision d_j^*, for which:

$$\mu^{d_j^*}(x_1^*, x_2^*, ..., x_{12}^*) = \max_{j=\overline{1,6}}[\mu^{d_j}(x_1^*, x_2^*, ..., x_{12}^*)] .$$

Example 1. Let us represent the next values of the parameters of a patient corresponding to her/his state:

$x_1^* = 53$ years, $x_2^* = 175$ c.u., $x_3^* = 507$ kg/min,

$x_4^* = 2.4$ c.u., $x_5^* = 0.25$ c.u., $x_6^* = 60.7$ mmol/l,

$x_7^* = 26.14$ mmol/l, $x_8^* = 10.4$ mmol/l, $x_9^* = 3.9$ c.u.,

$x_{10}^* = 22.4$ mlitre/min$\times$kg, $x_{11}^* = 172$ c.u., $x_{12}^* = 26.1$ c.u..

Using model (3.36) and parameters values b and c from Table 3.15, we find the membership functions values at points x_i^*, $i = \overline{1,12}$ for all fuzzy terms and represent them in Table 3.16.

Substituting the membership functions obtained from equation (3.34), we find:

$$\begin{aligned}\mu^L(y) &= 0.072 \cdot 0.120 \cdot 0.205 \cdot 0.315 \cdot 0.149 \cdot 0.133 \\ &\vee 0.072 \cdot 0.275 \cdot 0.205 \cdot 0.867 \cdot 0.149 \cdot 0.133 \\ &\vee 0.137 \cdot 0.120 \cdot 0.560 \cdot 0.315 \cdot 0.149 \cdot 0.133 = 0.120 .\end{aligned}$$

Similarly, we find: $\mu^{lA}(y)=0.137$, $\mu^{A}(y)=0.328$, $\mu^{hA}(y)=0.241$, $\mu^{H}(y)=0.210$.

According to equation (3.35), we find:

$$\mu^{L}(z)=0.458\cdot 0.635\cdot 0.095\cdot 0.266\cdot 0.687$$
$$\vee 1.000\cdot 0.635\cdot 0.201\cdot 0.751\cdot 0.857$$
$$\vee 0.458\cdot 0.911\cdot 0.095\cdot 0.795\cdot 0.857=0.201 .$$

Similarly: $\mu^{lA}(z)=0.545$, $\mu^{A}(z)=0.343$, $\mu^{hA}(z)=0.176$, $\mu^{H}(z)=0.087$.

Table 3.16 Membership functions values $\mu^{j}(x_i^*)$

№	x_i^*	u^*	$\mu^{L}(x_i^*)$	$\mu^{lA}(x_i^*)$	$\mu^{A}(x_i^*)$	$\mu^{hA}(x_i^*)$	$\mu^{H}(x_i^*)$
1	53	3.259	0.074	0.143	0.349	0.927	0.608
2	175	0.679	0.649	0.892	0.328	0.136	0.072
3	507	1.503	0.274	0.771	0.775	0.275	0.120
4	2.4	2.182	0.152	0.379	0.963	0.560	0.205
5	0.25	1.362	0.315	0.867	0.676	0.241	0.109
6	60.7	2.996	0.087	0.176	0.462	1.000	0.458
7	26.14	3.255	0.074	0.142	0.343	0.911	0.635
8	10.4	1.157	0.389	0.972	0.545	0.201	0.095
9	3.9	2.468	0.123	0.283	0.795	0.751	0.266
10	22.4	1.791	0.210	0.576	0.951	0.368	0.149
11	172	1.647	0.239	0.670	0.872	0.318	0.133
12	26.1	3.376	0.070	0.131	0.310	0.857	0.687

According to equation (3.33) we ultimately find that:

$$\mu^{d_1}(d)=0.074\cdot 0.120\cdot 0.201 \vee 0.074\cdot 0.137\cdot 0.545$$
$$\vee 0.143\cdot 0.137\cdot 0.201=0.137 .$$

Finally, we find:

$\mu^{d_2}(d)=0.137$, $\mu^{d_3}(d)=0.328$, $\mu^{d_4}(d)=0.176$, $\mu^{d_5}(d)=0.210$, $\mu^{d_6}(d)=0.176$

Because the largest membership value corresponds to decision d_3, we select NCD with heavy complication as the patient's diagnosis.

3.5.6 Fine Tuning of the Fuzzy Rules

We used real data related to diseases with verified diagnoses as the training data for the fine tuning of fuzzy rules for differential diagnosis of IHD. The optimization problem was solved by the combination of a genetic algorithm and gradient descent. The parameters b and c of the linguistic terms used in the fuzzy rules and the weights of rules w after tuning are presented in Tables 3.17–3.20. Membership functions after tuning are shown in Fig. 3.29.

The comparison of the inferred and correct diagnosis for 65 patients is shown in Table 3.21. There is only one case (**) in which the inferred decision (d_4) is too far from the real decision (d_2). In 8 cases (*) we have the real and inferred decisions on a boundary between classes of diagnoses. For the rest of the patients, there is full matching of real and inferred decisions. These results are quite satisfactory from a practical point of view, and thus the expert system can be used as a decision support system for the differential diagnosis of IHD.

Table 3.17. Parameters b and c of the membership functions after tuning

	L		lA		A		hA		H	
	b	c	b	c	b	c	b	c	b	c
x_1	32.58	23.33	38.21	9.80	43.39	11.92	51.07	16.01	56.74	22.62
x_2	128.00	57.31	186.39	87.84	235.24	80.39	332.76	109.61	389.33	162.75
x_3	182.46	807.90	509.02	242.91	648.08	575.61	922.18	261.01	1105.75	568.27
x_4	0.600	0.761	1.288	0.985	1.847	0.386	2.852	1.421	3.900	0.064
x_5	0.117	0.055	0.217	0.026	0.322	0.179	0.422	0.079	0.530	0.066
x_6	34.48	7.88	47.34	40.20	51.94	8.90	59.48	21.90	69.49	8.08
x_7	11.90	4.04	16.28	4.04	21.56	11.31	25.03	4.04	27.99	13.19
x_8	3.60	5.42	8.61	10.79	15.93	0.24	19.03	25.09	27.10	5.42
x_9	1.00	1.08	2.18	1.08	3.18	3.37	4.02	1.20	5.70	1.08
x_{10}	9.01	18.62	16.69	8.50	21.63	17.92	32.44	8.88	36.24	10.14
x_{11}	46.00	30.15	144.32	157.80	200.10	147.95	270.43	5.96	335.18	193.27
x_{12}	3.90	6.07	10.48	6.07	19.47	24.04	22.81	10.87	30.20	6.07

Table 3.18 Weights of the rules before (w_b) and after (w_a) tuning in Table 3.13

y	w_b	w_a
	1.000	0.500
L	1.000	0.500
	1.000	0.734
	1.000	0.500
lA	1.000	0.632
	1.000	0.500
	1.000	0.757
A	1.000	0.470
	1.000	0.473
	1.000	0.527
hA	1.000	0.480
	1.000	0.664
	1.000	0.499
H	1.000	0.806
	1.000	0.499

Table 3.19 Weights of the rules before (w_b) and after (w_a) tuning in Table 3.14

z	w_b	w_a
	1.000	0.500
L	1.000	0.744
	1.000	0.500
	1.000	0.500
lA	1.000	0.500
	1.000	0.400
	1.000	0.500
A	1.000	0.500
	1.000	0.565
	1.000	0.771
hA	1.000	0.500
	1.000	0.500
	1.000	0.500
H	1.000	0.500
	1.000	0.500

Table 3.20. Weights of the rules before (w_b) and after (w_a) tuning in Table 3.12

d	w_b	w_a	d	w_b	w_a
	1.000	0.934		1.000	0.663
d_1	1.000	0.500	d_4	1.000	0.449
	1.000	0.419		1.000	0.449
	1.000	0.500		1.000	0.499
d_2	1.000	0.500	d_5	1.000	0.500
	1.000	0.764		1.000	0.770
	1.000	0.428		1.000	0.500
d_3	1.000	0.500	d_6	1.000	0.524
	1.000	0.724		1.000	0.915

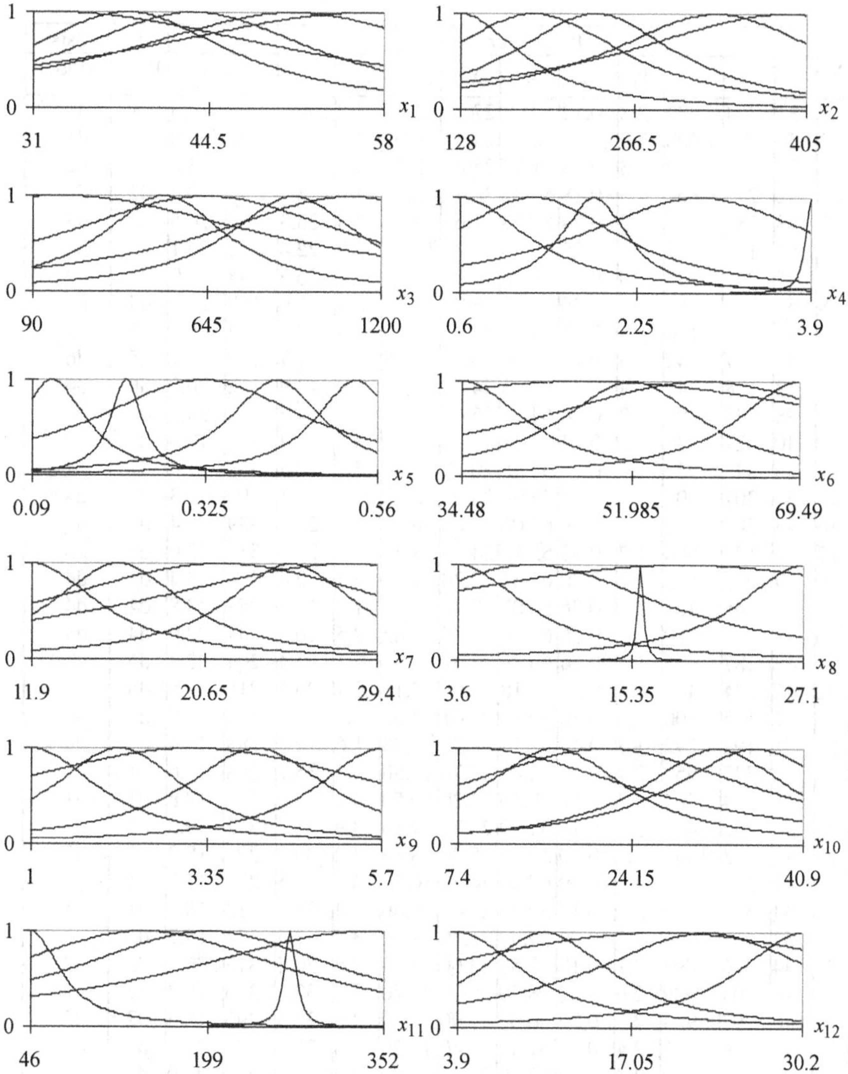

Fig. 3.29. Membership functions of the patient's state parameters after tuning

Table 3.21. Comparison of real and inferred decisions for 65 patients

№	Parameters of state												Diagnosis	
	x_1	x_2	x_3	x_4	x_5	x_6	x_7	x_8	x_9	x_{10}	x_{11}	x_{12}	Real	Model
1	31	324	980	2.8	0.12	50.07	22.76	8.05	3.7	34.2	266	19.3	d1	d1
2	36	330	900	2.9	0.14	56.52	24.33	9.02	4.1	29.7	242	21.0	d1	d1
3	39	260	800	2.3	0.18	51.73	25.62	8.53	4.2	28.5	194	23.8	d2	d2
4	42	272	867	2.5	0.28	59.31	28.44	8.53	4.0	28.7	198	19.4	d2	d2
5	48	287	491	2.2	0.24	52.77	21.61	8.53	3.5	25.3	156	20.5	d3	d3
6	53	175	507	2.4	0.25	60.70	26.14	10.40	3.9	22.4	172	26.1	d3	d3
7	45	247	728	2.0	0.34	62.06	26.14	5.55	2.3	26.5	144	22.9	d4	d4
8	52	231	768	1.5	0.36	62.77	23.01	6.83	2.5	20.0	158	23.8	d4	d4
9	32	151	610	1.3	0.42	54.49	23.91	5.55	2.4	19.8	104	25.7	d5	d5
10	45	177	542	1.6	0.48	62.06	26.14	5.55	2.3	21.7	120	28.1	d5	d6 *
11	38	128	349	1.4	0.48	67.03	24.46	5.20	1.9	13.9	92	30.2	d6	d6
12	38	145	304	1.2	0.56	64.15	25.62	7.11	2.6	14.4	74	25.5	d6	d6
13	40	327	930	2.2	0.24	59.31	25.62	7.56	3.3	35.4	347	18.9	d1	d2 *
14	38	348	952	1.8	0.20	34.48	20.79	9.56	5.7	34.2	352	21.6	d1	d1
15	34	307	800	1.9	0.21	57.90	25.08	6.83	2.9	30.1	304	19.3	d2	d4 **
16	48	284	738	2.0	0.26	62.06	25.08	8.53	3.4	29.7	339	20.4	d2	d2
17	35	174	600	1.7	0.32	55.18	24.46	8.56	3.8	27.2	312	22.0	d3	d3
18	49	229	515	2.1	0.30	61.34	22.20	6.83	2.4	22.4	300	23.4	d3	d4 *
19	58	265	421	2.0	0.26	60.07	22.76	4.08	1.8	17.7	258	23.8	d4	d4
20	49	330	650	1.5	0.25	69.49	25.08	6.83	2.5	20.3	244	22.0	d4	d4
21	48	187	475	1.4	0.34	60.39	23.31	5.55	2.1	21.4	204	22.7	d5	d5
22	42	224	400	1.5	0.39	55.18	21.05	7.11	2.7	20.4	215	22.5	d5	d5
23	32	195	100	1.2	0.48	60.70	21.61	7.52	2.7	22.6	191	25.9	d6	d6
24	51	192	292	1.3	0.45	62.77	23.70	5.55	1.6	19.2	188	24.4	d6	d6
25	36	347	952	2.9	0.10	62.40	23.70	12.50	4.3	35.7	298	19.6	d1	d1
26	48	314	902	3.2	0.14	59.40	24.20	10.50	4.2	33.5	287	18.8	d1	d1
27	42	352	875	3.2	0.16	52.30	22.70	9.50	3.9	38.2	322	19.0	d1	d1
28	40	323	1040	2.7	0.20	59.60	25.20	8.80	3.2	30.4	290	18.2	d1	d2 *
29	41	377	988	2.9	0.09	60.40	24.30	10.20	3.4	32.5	275	17.7	d1	d1
30	34	309	932	3.2	0.15	60.80	25.40	9.40	4.4	31.5	312	18.5	d1	d1
31	52	279	1056	2.7	0.09	59.90	21.30	8.80	3.7	33.4	334	18.7	d1	d1
32	44	376	895	2.7	0.18	61.50	23.60	9.50	3.6	30.4	312	20.1	d2	d2
33	46	304	929	2.6	0.22	58.20	25.10	10.70	3.8	32.5	346	19.2	d2	d2
34	46	292	904	2.2	0.24	56.00	27.90	10.10	4.0	29.3	290	18.5	d2	d2
35	42	276	885	2.4	0.25	61.40	29.40	11.20	3.6	27.8	226	20.8	d2	d2
36	31	311	930	2.7	0.19	62.50	23.80	9.80	2.9	25.6	249	21.0	d2	d1 *
37	44	335	992	2.4	0.22	61.60	24.70	9.90	3.3	24.6	255	20.3	d2	d2
38	47	346	873	2.3	0.18	57.70	22.50	10.60	3.7	28.7	267	18.8	d2	d2
39	48	288	804	2.4	0.27	60.00	22.20	11.50	3.5	20.9	275	19.5	d3	d3
40	50	316	875	2.1	0.31	61.40	24.00	9.30	2.8	22.5	302	21.2	d3	d4 *
41	51	292	774	2.0	0.28	62.50	25.90	8.80	3.0	26.7	277	22.5	d3	d4 *
42	54	315	766	2.2	0.22	53.70	26.20	8.70	2.7	21.4	265	20.5	d3	d4 *

Table 3.21. *(continued)*

№	Parameters of state												Diagnosis	
	x_1	x_2	x_3	x_4	x_5	x_6	x_7	x_8	x_9	x_{10}	x_{11}	x_{12}	Real	Model
43	40	300	865	2.1	0.25	59.40	25.80	9.30	3.5	21.9	303	21.4	d3	d3
44	36	270	777	2.1	0.28	61.00	26.10	9.70	4.1	22.3	316	21.3	d3	d3
45	34	275	859	2.3	0.30	62.50	27.00	9.60	4.2	24.0	295	22.5	d3	d3
46	52	261	776	1.7	0.36	65.00	22.50	8.40	2.7	20.4	204	23.8	d4	d4
47	41	258	785	1.5	0.36	62.70	23.80	7.60	2.5	19.8	225	24.0	d4	d4
48	53	290	845	1.8	0.39	57.10	24.00	7.20	2.5	18.7	268	22.5	d4	d4
49	39	203	723	2.0	0.40	58.50	23.70	6.20	2.8	17.1	209	24.7	d4	d4
50	45	244	802	1.7	0.35	62.00	25.30	6.30	3.0	18.5	212	24.9	d4	d4
51	46	233	795	1.9	0.39	57.90	24.90	5.20	2.4	17.4	251	23.5	d4	d4
52	54	262	805	1.8	0.38	57.90	24.50	7.70	2.2	19.2	244	22.1	d4	d4
53	51	245	595	1.3	0.44	64.20	26.40	5.60	2.1	16.5	204	24.7	d5	d5
54	40	209	772	1.5	0.45	60.20	27.80	5.90	2.4	14.7	195	25.0	d5	d5
55	42	198	621	1.4	0.42	58.80	25.20	6.10	2.6	12.2	225	24.5	d5	d5
56	44	245	523	1.5	0.39	57.50	23.30	6.50	2.2	14.1	207	26.9	d5	d5
57	50	237	652	1.6	0.45	63.70	24.70	6.40	2.1	11.9	262	24.2	d5	d5
58	56	202	744	1.3	0.45	61.80	25.70	5.70	2.4	12.3	226	22.6	d5	d5
59	51	247	723	1.2	0.38	62.50	26.90	5.60	2.3	10.4	230	25.8	d5	d5
60	48	192	516	1.1	0.52	60.10	22.70	5.50	2.0	9.9	200	22.9	d6	d6
61	39	188	446	1.2	0.48	59.00	23.50	5.20	2.4	9.5	212	26.7	d6	d6
62	49	212	406	0.9	0.56	61.70	26.00	5.30	1.9	8.2	225	29.4	d6	d6
63	45	247	527	0.7	0.51	62.60	27.40	5.10	2.0	7.4	197	28.5	d6	d6
64	44	206	448	0.8	0.55	57.40	22.10	6.30	2.1	7.4	188	30.1	d6	d6
65	42	228	512	1.0	0.52	53.90	25.60	5.40	2.3	7.8	204	29.5	d6	d6

3.6 Example 2: Prediction of Disease Rate Evolution

The prediction of the number of diseases of some type or other at the level of a region is a necessary element of organization of medical-preventive measures. From a formal viewpoint, this problem is related to a wide class of problems of predicting discrete sequences (collections of values at some fixed time points [19, 20]) originating not only in medicine but also in engineering, economics, etc. The nontrivial nature of the prediction of discrete sequences is due to the fact that, in contrast to well-algorithmisized interpolation procedures, the prediction requires the extrapolation of data on the past to data on the future. In this case, it is necessary to take into account an unknown law governing a process generating discrete sequences.

A great number of papers deal with the development of mathematical models of prediction [19]. The methods based on probabilistic-statistical means are most widely used; however, their use requires a considerable amount of experimental data, which are not always available under the conditions of even recent events (for example, the Chernobyl accident).

Interest has recently been revived [21] on the use of artificial neural networks for the solution of prediction problems. The networks are considered as universal models akin to the human brain, which are trained to recognize unknown regularities.

However, a large sample of experimental data is required in the case of training neural networks, as well as in the case of using probabilistic-statistical methods. Moreover, a trained neural network does not permit one to explicitly interpret the weights of arcs.

In this section, we propose an approach to the prediction of disease evolution, which combines experimental data of disease numbers with expert-linguistic information on regularities that can be revealed in available experimental data. The use of expert information in the form of the natural language IF-THEN rules formalized by means of fuzzy logic allows us to construct models of prediction in the case of relatively small (in comparison with statistical methods) samples of experimental data. The proposed approach is sufficiently close to the neuro-fuzzy approach [22], which combines learning ability of neural networks with transparency and invariability of fuzzy IF-THEN rules. However, we do not use the neural network for training the fuzzy predictive model. In comparison with [22], we directly train fuzzy IF-THEN rules with the help of the available experimental data [9, 13].

3.6.1 *Linguistic Model of Prediction*

We consider information on the incidence of appendicular peritonitis disease according to the data of the Vinnitsa Clinic of Children's Surgery in 1982-2001 that are presented in Table. 3.22.

Table 3.22. Distribution of the diseases number

Year	1982	1983	1984	1985	1986	1987	1988	1989
Number of diseases	109	143	161	136	161	163	213	220
Year	1990	1991	1992	1993	1994	1995	1996	1997
Number of diseases	162	194	164	196	245	252	240	225
Year	1998	1999	2000	2001				
Number of diseases	160	185	174	207				

Analyzing the disease dynamics in Fig. 3.30, it is easy to observe the presence of four-year cycles the third position of which is occupied by the leap year. These cycles will be denoted as follows:

$$\ldots\ x_4^{i-1}\ \}\ \{\ x_1^i\quad x_2^i\quad \textcircled{x_3^i}\quad x_4^i\ \}\ \{\ x_1^{i+1}\ \ldots$$

Leap year

where i is the number of a four-year cycle,

x_1^i is the number of diseases during two years prior to a leap year,

x_2^i is the number of diseases during one year prior to a leap year,

x_3^i is the number of diseases during a leap year,

x_4^i is the diseases number during the year following the leap year.

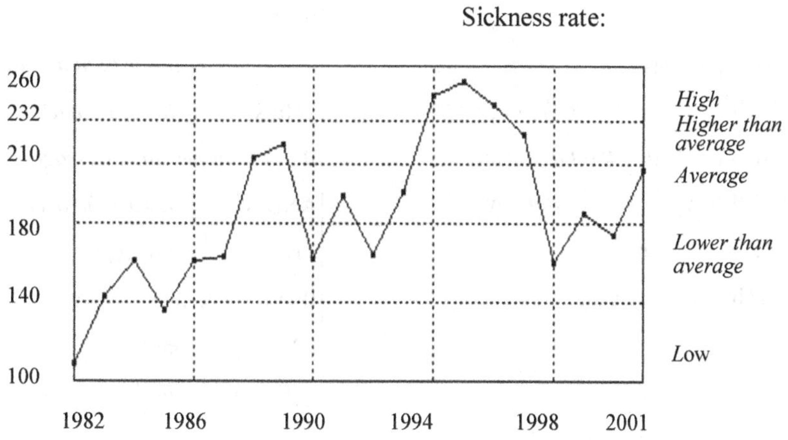

Fig. 3.30. Disease dynamics

The regularities that can be seen in Fig. 3.30 are easily written in the form of four expert opinions in a natural language. These opinions are IF-THEN rules that relate the sickness rates in the i-th and (i+1)-th cycles [9]:

F_1 :

IF x_1^i = *low*
AND x_2^i = *lower than average,*
THEN x_3^i = *lower than average*
IF x_1^i = *lower than average*
AND x_2^i = *lower than average,*
THEN x_3^i = *higher than average*
IF x_1^i = *lower than average*
AND x_2^i = *average,*
THEN x_3^i = *lower than average*
IF x_1^i = *high*
AND x_2^i = *high,*
THEN x_3^i = *high*

F_2 :

IF x_1^i = *low*
AND x_2^i = *lower than average,*
THEN x_4^i = *low*
IF x_1^i = *lower than average*
AND x_2^i = *lower than average,*
THEN x_4^i = *higher than average*
IF x_1^i = *lower than average*
AND x_2^i = *average,*
THEN x_4^i = *average*
IF x_1^i = *high*
AND x_2^i = *high,*
THEN x_4^i = *higher than average*

F_3 :
IF x_4^i = *low*,
THEN x_1^{i+1} = *lower than average*
IF x_4^i = *higher than average*,
THEN x_1^{i+1} = *lower than average*
IF x_4^i = *average*,
THEN x_1^{i+1} = *high*

F_4 :
IF x_4^i = *low*
AND x_1^{i+1} = *lower than average*,
THEN x_2^{i+1} = *lower than average*
IF x_4^i = *higher than average*
AND x_1^{i+1} = *lower than average*,
THEN x_2^{i+1} = *average*
IF x_4^i = *average*
AND x_1^{i+1} = *high*,
THEN x_2^{i+1} = *high*

The network of relations in Fig. 3.31 shows that it is possible to predict the situation for the next four years: for the last two years of the i-th cycle and for the first two years of the succeeding (i+1)-th cycle using the data of the first two years of the i-th cycle.

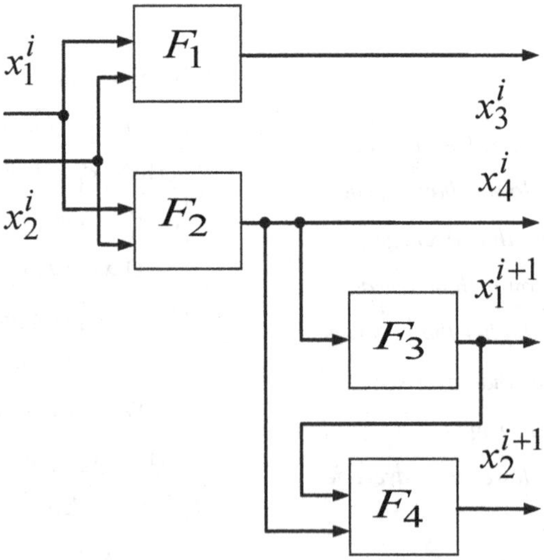

Fig. 3.31. A network of relations for prediction

To use the expert-linguistic opinions $F_1 \div F_4$, we apply the methods of fuzzy sets theory. According to this theory the linguistic estimates "low", "lower than average", "average", and others are formalized with the help of membership functions. The parameters b and c chosen by an expert for various linguistic estimates used in rules $F_1 \div F_4$, are presented in Table. 3.23. The membership functions obtained in this case are shown in Fig. 3.32.

Table 3.23. Membership functions parameters before training

Linguistic estimates of variables $x_1^i \div x_4^i$	Parameter	
	b	c
low (*L*)	100	50
lower than average (*lA*)	160	30
average (*A*)	195	25
higher than average (*hA*)	222	20
high (*H*)	260	30

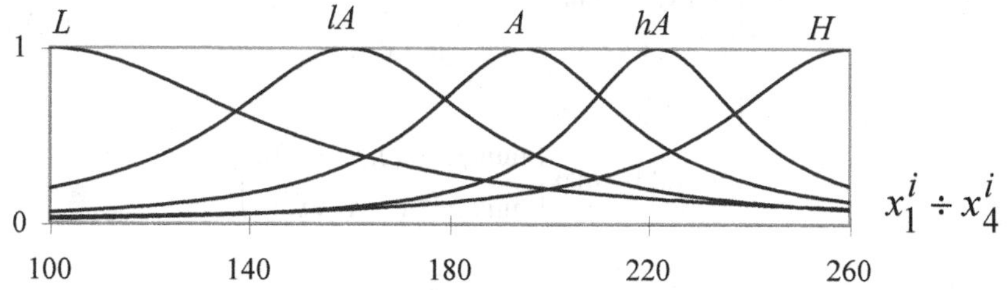

Fig. 3.32. Membership functions of linguistic estimates before training

In addition to the two-parameter membership functions chosen above, other functions can also be used, for example, triangular or trapezoidal ones [5], containing three and four adjustable parameters, respectively.

We denote by $[\underline{x}, \overline{x}]$ the range of all possible values of the number of diseases. Let us subdivide this range into the following five parts:

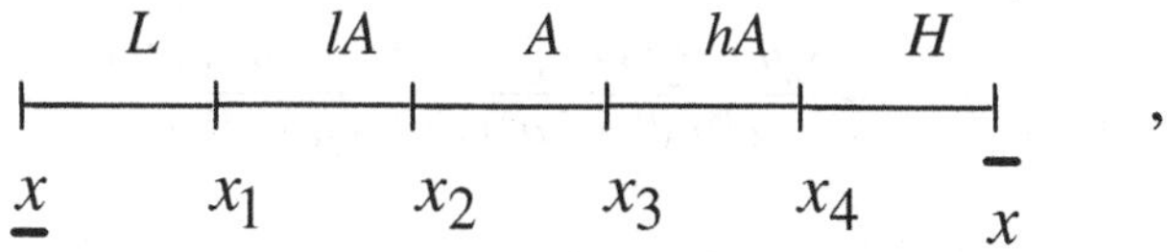

associated with the following linguistic estimations: *low* (*L*), *lower than average* (*lA*), *average* (*A*), *higher than average* (*hA*), *high* (*H*). Then, using the fuzzy-logic operations *min (AND)* and *max (OR)*, and the principle of weighted sum for transformation of a membership function into a precise number, we can write a model of prediction in the following explicit form:

$$F_1:\begin{cases} x_3^i = \dfrac{x_1\mu^{lA}(x_3^i)+x_3\mu^{hA}(x_3^i)+x_4\mu^H(x_3^i)}{\mu^{lA}(x_3^i)+\mu^{hA}(x_3^i)+\mu^H(x_3^i)}, \\ \mu^{lA}(x_3^i)=\max\begin{pmatrix}\min(\mu^L(x_1^i),\mu^{lA}(x_2^i)),\\ \min(\mu^{lA}(x_1^i),\mu^A(x_2^i))\end{pmatrix} \\ \mu^{hA}(x_3^i)=\min(\mu^{lA}(x_1^i),\mu^{lA}(x_2^i)) \\ \mu^H(x_3^i)=\min(\mu^H(x_1^i),\mu^H(x_2^i)) \end{cases} \tag{3.37}$$

$$F_2:\begin{cases} x_4^i = \dfrac{\underline{x}\,\mu^L(x_4^i)+x_2\mu^A(x_4^i)+x_3\mu^{hA}(x_4^i)}{\mu^L(x_4^i)+\mu^A(x_4^i)+\mu^{hA}(x_4^i)} \\ \mu^L(x_4^i)=\min(\mu^L(x_1^i),\mu^{lA}(x_2^i)) \\ \mu^A(x_4^i)=\min(\mu^{lA}(x_1^i),\mu^A(x_2^i)) \\ \mu^{hA}(x_4^i)=\max\begin{pmatrix}\min(\mu^{lA}(x_1^i),\mu^{lA}(x_2^i)),\\ \min(\mu^H(x_1^i),\mu^H(x_2^i))\end{pmatrix} \end{cases} \tag{3.38}$$

$$F_3:\begin{cases} x_1^{i+1} = \dfrac{x_1\mu^{lA}(x_1^{i+1})+x_4\mu^H(x_1^{i+1})}{\mu^{lA}(x_1^{i+1})+\mu^H(x_1^{i+1})} \\ \mu^{lA}(x_1^{i+1})=\max(\mu^L(x_4^i),\mu^{hA}(x_4^i)) \\ \mu^H(x_1^{i+1})=\mu^A(x_4^i) \end{cases} \tag{3.39}$$

$$F_4:\begin{cases} x_2^{i+1} = \dfrac{x_1\mu^{lA}(x_2^{i+1})+x_2\mu^A(x_2^{i+1})+x_4\mu^H(x_2^{i+1})}{\mu^{lA}(x_2^{i+1})+\mu^A(x_2^{i+1})+\mu^H(x_2^{i+1})} \\ \mu^{lA}(x_2^{i+1})=\min(\mu^L(x_4^i),\mu^{lA}(x_1^{i+1})) \\ \mu^A(x_2^{i+1})=\min(\mu^{hA}(x_4^i),\mu^{lA}(x_1^{i+1})) \\ \mu^H(x_2^{i+1})=\min(\mu^A(x_4^i),\mu^H(x_1^{i+1})) \end{cases} \tag{3.40}$$

Using the obtained model we can receive some rough diseases number prediction as shown in Fig. 3.33.

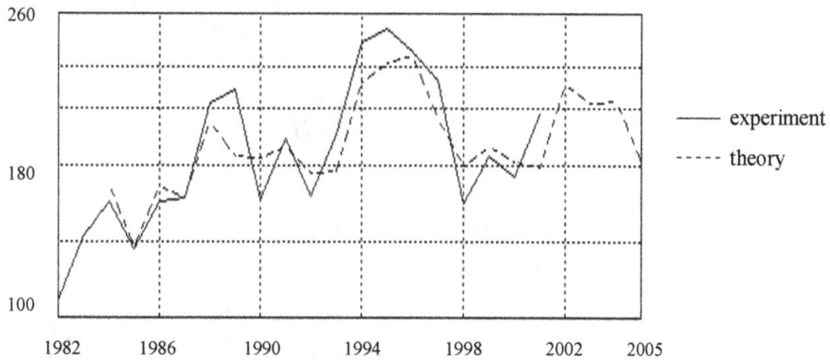

Fig. 3.33. Comparison of the experimental data and the prediction model before training

To increase the precision of prediction, it is necessary to train the model by available experimental data.

3.6.2 *Neuro-fuzzy Model of Prediction*

A neuro-fuzzy model of prediction based on the elements from Table 3.1 is presented in Fig. 3.34.

As is seen from Fig. 3.34 the neuro-fuzzy network has the following five layers:

(1) – the inputs of the model of prediction;

(2) – the fuzzy terms used in knowledge bases $F_1 \div F_4$;

(3) – the conjunctions rows of knowledge bases $F_1 \div F_4$;

(4) – the rules united in the classes $[\underline{x}, x_1, x_2, .., \overline{x}]$;

(5) – the operation of defuzzification, i. e. transformation of the results of fuzzy logic inference into the crisp number.

The following weights are assigned to the edges of the graph: unity (the edges between the first and second layers, fourth and fifth layers); the membership function specifying the grade of membership of an input in a fuzzy term (the edges between the second and third layers); the weights of rules (the edges between the third and fourth layers).

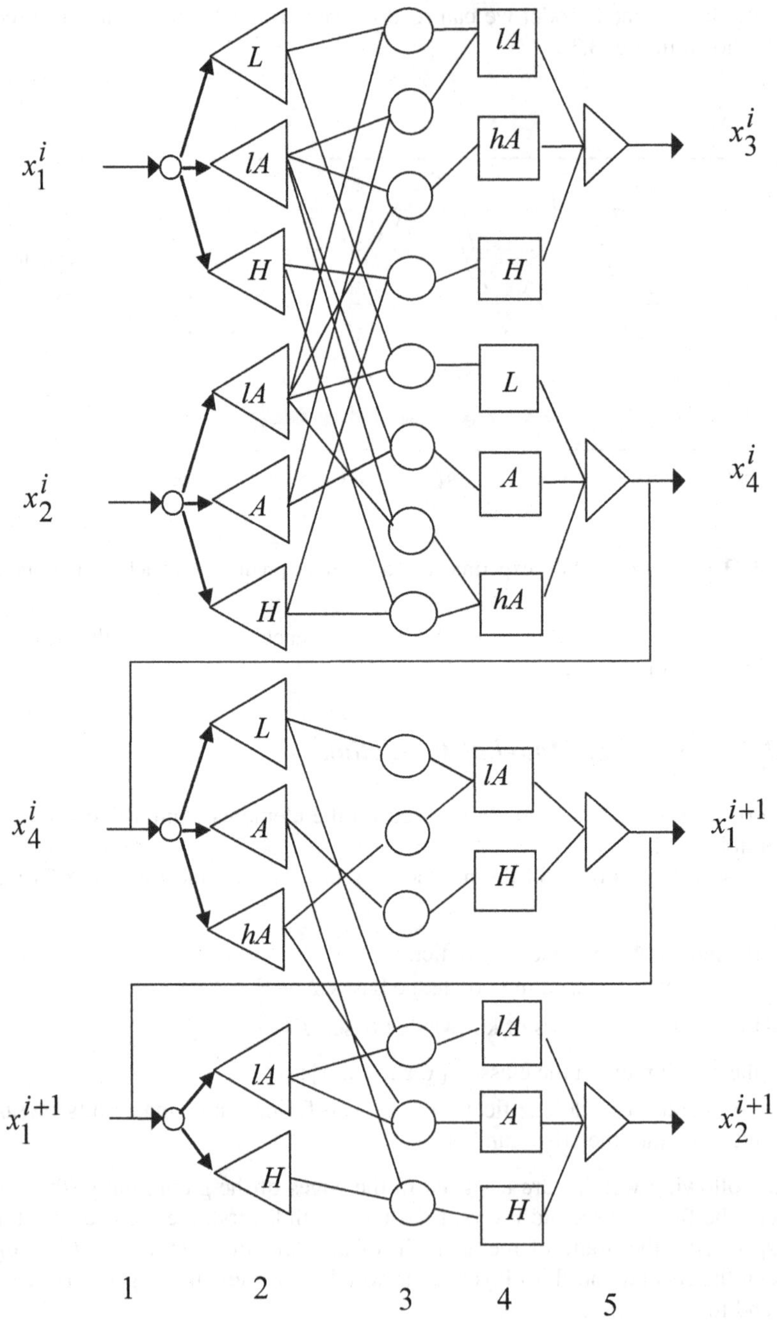

Fig. 3.34. Neuro-fuzzy model of prediction

3.6.3 On-Line Training of the Neuro-fuzzy Model of Prediction

The essence of training consists of the selection of such rules weights (w-) and such parameters of the membership functions (b-, c-) for the linguistic estimates, which provide the least distance between theoretical and experimental numbers of diseases:

$$\sum_{i=1}^{N}(x_3^i-\hat{x}_3^i)^2+\sum_{i=1}^{N}(x_4^i-\hat{x}_4^i)^2+\sum_{i=1}^{N-1}(x_1^{i+1}-\hat{x}_1^{i+1})^2+\sum_{i=1}^{N-1}(x_2^{i+1}-\hat{x}_2^{i+1})^2=\min_{w,b,c} ,$$

where $x_3^i, x_4^i, x_1^{i+1}, x_2^{i+1}$ are predicted numbers of diseases depending on the parameters b and c of the membership functions and rules weights;

$\hat{x}_3^i, \hat{x}_4^i, \hat{x}_1^{i+1}, \hat{x}_2^{i+1}$ are experimental numbers of diseases;

N is the number of cycles used to train the model.

To train the parameters of the neuro-fuzzy network, the following system of recursive relations is used:

$$w_{jk}(t+1)=w_{jk}(t)-\eta\frac{\partial E_t}{\partial w_{jk}(t)} , \tag{3.41}$$

$$c_{1-4}^{ip}(t+1)=c_{1-4}^{ip}(t)-\eta\frac{\partial E_t}{\partial c_{1-4}^{ip}(t)} , \tag{3.42}$$

$$b_{1-4}^{ip}(t+1)=b_{1-4}^{ip}(t)-\eta\frac{\partial E_t}{\partial b_{1-4}^{ip}(t)} , \tag{3.43}$$

that minimize the criterion

$$E_t=\frac{1}{2}(\hat{x}_t-x_t)^2 ,$$

applied in the theory of neural networks, where

$\hat{x}_t$ and x_t are the experimental and the theoretical number of diseases at the t-th step of training;

$w_{jk}(t)$ is the weight of k-th rule, combined data about diseases numbers in relation F_j, $j=\overline{1,4}$;

$c_{1-4}^{ip}(t)$, $b_{1-4}^{ip}(t)$ are parameters of the membership function of variable x_{1-4}^i to p-th fuzzy term at the t-th step of training;

η is the parameter of training which can be chosen in accordance with the recommendations of [2].

3.6.4 Results of Prediction

After training of the prediction model for $N = 4$, which means using the data obtained over the years 1982-1997, the weights of expert-linguistic regularities $F_1 \div F_4$ presented in Table 3.24 were evaluated. It was supposed before training that rules weights were equal to 1. Parameters of membership functions after training are presented in Table. 3.25.

Table 3.24. Weights of the expert-linguistic regularities after training

Rules weights in F_1				Rules weights in F_3		
w_{11}	w_{12}	w_{13}	w_{14}	w_{31}	w_{32}	w_{33}
1.000	0.999	0.564	0.885	1.000	1.000	0.668
Rules weights in F_2				Rules weights in F_4		
w_{21}	w_{22}	w_{23}	w_{24}	w_{41}	w_{42}	w_{43}
1.000	1.000	1.000	1.000	1.000	0.992	0.965

Table 3.25. Parameters of membership functions after training

Linguistic estimates of variables $x_1^i \div x_4^i$	Parameters	
	b	c
low (*L*)	99.944	8.194
lower than average (*lA*)	145.813	19.504
average (*A*)	194.949	6.999
higher than average (*hA*)	234.001	10.636
high (*H*)	249.134	42.742

As is seen from Tables 3.23 and 3.25, after training the neuro-fuzzy network we have the greatest changes in the parameters c of membership functions. This can be explained by the fact that in forming the fuzzy knowledge base the expert has specified sufficiently exact positions of the maxima of membership functions (the parameters b) and weights of the rules (parameters w). The choice of large values of the parameters c by the expert testifies to a considerable uncertainty in estimating fuzzy terms. A decrease in the values of the parameters c in the course of training has resulted in a "concentration" (compression) of membership functions which testifies to the removal of the uncertainty in estimating fuzzy terms. Membership functions after training are presented in Fig. 3.35. The following values were taken into consideration: $\underline{x} = 95$, $x_1 = 150$, $x_2 = 190$, $x_3 = 220$, $x_4 = 242$, $\overline{x} = 260$.

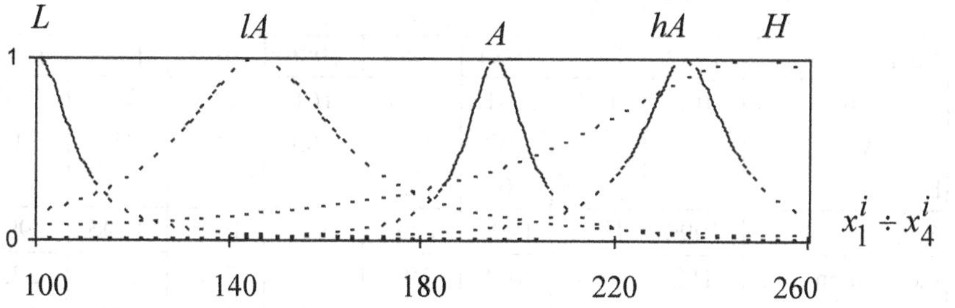

Fig. 3.35. Membership functions of linguistic estimates after training

The training was performed until the prognosis produced by the neuro-fuzzy network was sufficiently close to experimental data. The application of tuned membership functions allows one to obtain a prediction model that is sufficiently close to the experimental data (See Fig. 3.36).

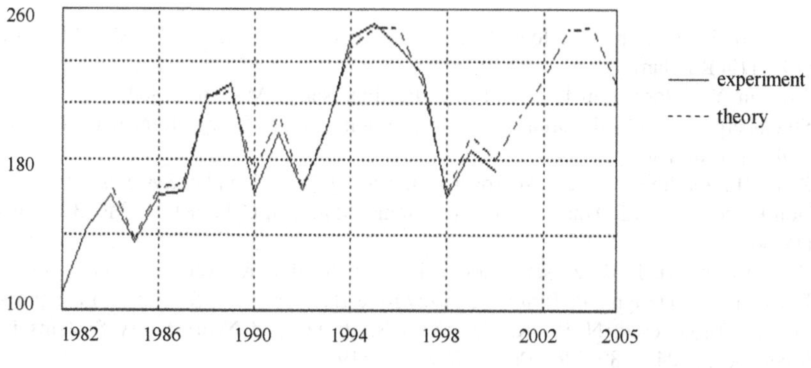

Fig. 3.36. Comparison of the experimental data and the prediction model after training

Since experimental values of the numbers of appendicular peritonitis diseases in 1998-2001 have not been used for fuzzy rules extraction, the proximity of the theoretical and experimental results for these years demonstrates the sufficient quality of the constructed prediction model from the practical viewpoint. A comparison of the results of simulation with the experimental data and also a prediction of the number of appendicular peritonitis diseases until 2005 is presented in Table 3.26.

Table 3.26. Experimental and model number of diseases

Year	1982	1983	1984	1985	1986	1987	1988	1989
Experiment	109	143	161	136	161	163	213	220
Theory			167	138	165	167	214	216
Error			6	2	4	4	1	4
Year	1990	1991	1992	1993	1994	1995	1996	1997
Experiment	162	194	164	196	245	252	240	225
Theory	173	204	165	197	240	250	250	220
Error	11	10	1	1	5	2	10	5
Year	1998	1999	2000	2001	2002	2003	2004	2005
Experiment	160	185	174	207				
Theory	162	193	180	203	223	249	250	220
Error	2	8	6	4				

References

1. Eickhoff, P.: Principles of Identification of Control Systems, p. 321. Mir, Moscow (1975) (in Russian)
2. Tsypkin, Y.: Information Theory of Identification. Nauka, Moscow (1984)
3. Shteinberg, S.E.: Identification in Control Systems, p. 81. Energoatomizdat, Moscow (1987) (in Russian)
4. Zadeh, L.: Outline of a New Approach to the Analysis of Complex Systems and Decision Processes. IEEE Transactions on Systems, Man, and Cybernetics SMC-3, 28–44 (1973)
5. Zimmermann, H.-J.: Fuzzy Set Theory and Its Application. Kluwer, Dordrecht (1991)
6. Rotshtein, A.: Design and Tuning of Fuzzy Rule-Based Systems for Medical Diagnostics. In: Teodorescu, N.-H., Kandel, A. (eds.) Fuzzy and Neuro-Fuzzy Systems in Medicine, pp. 243–289. CRC Press, New York (1998)
7. Rotshtein, A.: Intellectual Technologies of Identification: Fuzzy Sets, Genetic Algorithms, Neural Nets, p. 320. UNIVERSUM, Vinnitsa (1999) (in Russian), `http://matlab.exponenta.ru/fuzzylogic/book5/index.php`
8. Rotshtein, A., Katel'nikov, D.: Identification of Non-linear Objects by Fuzzy Knowledge Bases. Cybernetics and Systems Analysis 34(5), 676–683 (1998)
9. Rotshtein, A., Loiko, E., Katel'nikov, D.: Prediction of the Number of Diseases on the basis of Expert-linguistic Information. Cybernetics and Systems Analysis 35(2), 335–342 (1999)
10. Rotshtein, A., Mityushkin, Y.: Neurolinguistic Identification of Nonlinear Dependencies. Cybernetics and Systems Analysis 36(2), 179–187 (2000)
11. Rotshtein, A., Mityushkin, Y.: Extraction of Fuzzy Knowledge Bases from Experimental Data by Genetic Algorithms. Cybernetics and Systems Analysis 37(4), 501–508 (2001)

12. Rotshtein, A.P., Shtovba, S.D.: Managing a Dynamic System by means of a Fuzzy Knowledge Base. Automatic Control and Computer Sciences 35(2), 16–22 (2001)
13. Rotshtein, A.P., Posner, M., Rakytyanska, H.: Adaptive System for On-line Prediction of Diseases Evolution. In: Mastorakis, D. (ed.) Advances in Neural World. Lectures and Books Series, pp. 178–183. WSEAS, Switzerland (2002)
14. Rotshtein, A.P., Posner, M., Rakytyanska, H.B.: Football Predictions based on a Fuzzy Model with Genetic and Neural Tuning. Cybernetics and Systems Analysis 41(4), 619–630 (2005)
15. Rotshtein, A.P., Rakytyanska, H.B.: Fuzzy Forecast Model with Genetic-Neural Tuning. Journal of Computer and Systems Sciences International 44(1), 102–111 (2005)
16. Terano, T., Asai, K., Sugeno, M. (eds.): Applied Fuzzy Systems. Omsya, Tokyo (1989); Moscow, Mir (1993) (in Russian)
17. Rotshtein, A.: Modification of Saaty Method for the Construction of Fuzzy Set Membership Functions. In: FUZZY 1997 – Int. Conf. Fuzzy Logic and Its Applications in Zichron, Yaakov, Israel, pp. 125–130 (1997)
18. Gen, M., Cheng, R.: Genetic Algorithms and Engineering Design, p. 352 John Wiley & Sons, New York (1997)
19. Ivachnenko, A.G., Lapa, V.G.: Forecasting of Random Processes, p. 416. Naukova dumka, Kiev (1971) (in Russian)
20. Markidakis, S., Wheelwright, S.C., Hindman, R.J.: Forecasting: Methods and Applications, 3rd edn., p. 386. John Wiley & Sons, USA (1998)
21. Mayoraz, F., Cornu, T., Vulliet, L.: Prediction of slope movements using neural networks. Fuzzy Systems Artif. Intel. 4(1), 9–17 (1995)
22. Nie, J.: Nonlinear Time Series Forecasting: A Fuzzy Neural Approach. Neuro Computing 16(1), 63–76 (1997)

Chapter 4
Fuzzy Rules Extraction from Experimental Data

The necessary condition for nonlinear object identification on the basis of fuzzy logic is the availability of IF-THEN rules interconnecting linguistic estimations of input and output variables. Earlier we assumed that IF-THEN rules are generated by an expert who knows the object very well. What is to be done when there is no expert? In this case the generation of IF-THEN rules becomes of interest because it means the generation of fuzzy knowledge base from accessible experimental data [1].

Transformation of experimental information into fuzzy knowledge bases may turn out to be a useful method of data processing in medicine, banking, management and in other fields where persons making decisions instead of strict quantitative relations give preference to the use of transparent easily interpreted verbal rules [2, 3]. In this case proximity of linguistic approximation results and corresponding experimental data is the criterion for the quality of extracted regularities.

Fuzzy-neural networks and genetic algorithms are traditionally used for knowledge extraction from experimental data [4]. Fuzzy-neural network is an excellent approach for automatic rules formation and adjustment due to the mechanisms of pruning redundant membership functions and rules [5 – 7]. However, convergence of the training depends on the initial structure of the fuzzy model. On the other hand, genetic algorithms grow the appropriate structure of fuzzy inference automatically [8, 9]. In this case, the restriction of the total number of fuzzy terms and fuzzy rules prevents the construction of more compact structure of the fuzzy model. Combinations of both paradigms stipulated for the development of a new hybrid approach, which consists of automatic generation of fuzzy-neural network based on the genetic algorithm [10 – 13].

The extraction of fuzzy IF-THEN rules has two phases. In the first phase we define the fuzzy model structure by using the generalized fuzzy approximator proposed in [3, 14]. The second phase consists of finding optimal parameters of rules which provide the least distance between the model and experimental outputs of the object. For solving the optimization problem we use a combination of genetic algorithm and neural network. The genetic algorithm provides a rough finding of the appropriate structure of the fuzzy inference [15, 16]. We use the neural network for fine adjustment and adaptive correction of approximating rules by pruning the redundant membership functions and rules [17].

This chapter is written using original work materials [15 – 17].

A.P. Rotshtein et al.: Fuzzy Evidence in Identif., Forecast. and Diagn., STUDFUZZ 275, pp. 119–148.
springerlink.com

4.1 Fuzzy Rules for "Multiple Inputs – Single Output" Object

Let us consider an object of this form

$$y = f(x_1, x_2, ..., x_n) \tag{4.1}$$

with n inputs and one output for which the following is known:

- intervals of inputs and output change:

$$x_i \in [\underline{x_i}, \overline{x_i}] \ , \ i = \overline{1,n} \, , \ y \in [\underline{y}, \overline{y}] \ ,$$

- classes of decisions d_j $(j = \overline{1,m})$ in case of discrete output:

$$[\underline{y}, \overline{y}] = [\underbrace{\underline{y}, y_1}_{d_1})\cup\ldots\cup[\underbrace{y_{j-1}, y_j}_{d_j})\cup\ldots\cup[\underbrace{y_{m-1}, \overline{y}}_{d_m}] \, .$$

- training sample in the form of M pairs of experimental data "inputs-output":

  $\{\mathbf{X}_p, y_p\}$ - for objects with continuous output,

  $\{\mathbf{X}_p, d_p\}$ - for objects with discrete output,

where $\mathbf{X}_p = \{x_1^p, x_2^p, ..., x_n^p\}$ - input vector in p-th pair, $p = \overline{1,M}$.

It is required: to synthesize knowledge about object (4.1) in the form of fuzzy logical expressions system:

$$\begin{array}{ll}
\text{IF} \left[(x_1 = a_1^{j1}) \text{ AND } (x_2 = a_2^{j1}) \text{ AND} \ldots \ (x_n = a_n^{j1}) \right] & \text{(with weight } w_{j1}\text{)} \\
\text{OR} \left[(x_1 = a_1^{j2}) \text{ AND } (x_2 = a_2^{j2}) \text{ AND} \ldots \ (x_n = a_n^{j2}) \right] & \text{(with weight } w_{j2}\text{)} \ldots \\
\ldots\text{OR} \left[\left(x_1 = a_1^{jk_j}\right) \text{ AND } \left(x_2 = a_2^{jk_j}\right) \text{ AND} \ldots \left(x_n = a_n^{jk_j}\right) \right] & \text{(with weight } w_{jk_j}\text{)}, \\
\text{THEN} \quad y \in d_j = [y_{j-1}, y_j] \text{ , for all } j = \overline{1,m}, &
\end{array} \tag{4.2}$$

where a_i^{jp} is the linguistic term for variable x_i evaluation in the row with number $p = \overline{1,k_j}$,

k_j is the number of conjunction rows corresponding to the class d_j, $j = \overline{1,m}$,

w_{jp} is a number in the range $[0, 1]$, which characterizes the weight of the expression with number jp .

4.2 Rules Extraction as Optimization Problem

It was shown that object (4.1) model in the form of the following calculation relations corresponds to knowledge base (4.2):

$$y = \frac{y\mu^{d_1}(y) + y_1\mu^{d_2}(y) + \ldots + y_{m-1}\mu^{d_m}(y)}{\mu^{d_1}(y) + \mu^{d_2}(y) + \ldots + \mu^{d_m}(y)} \,, \tag{4.3}$$

$$\mu^{d_j}(y) = \max_{p=\overline{1,k_j}} \left\{ w_{jp} \min_{i=\overline{1,n}} [\mu^{jp}(x_i)] \right\} \,, \tag{4.4}$$

$$\mu^{jp}(x_i) = \frac{1}{1 + \left(\frac{x_i - b_i^{jp}}{c_i^{jp}} \right)} \,, \quad i = \overline{1,n} \,, \; p = \overline{1,k_j} \,, \; j = \overline{1,m} \,, \tag{4.5}$$

where $\mu^{d_j}(y)$ is the membership function of the output y to the class d_j,

$\mu^{jp}(x_i)$ is the membership function of the input x_i to the term a_i^{jp},

b_i^{jp} and c_i^{jp} are the tuning parameters for the input variables x_i membership functions.

Relations (4.3) - (4.5) define the model of the object (4.1) which is written down in this form:

$y = F(\mathbf{X}, \mathbf{W}, \mathbf{B}, \mathbf{C})$ - for continuous output,

$\mu^{d_j}(y) = \mu^{d_j}(\mathbf{X}, \mathbf{W}, \mathbf{B}, \mathbf{C})$ - for discrete output,

where $\mathbf{X} = (x_1, x_2, ..., x_n)$ is the input vector, $\mathbf{W} = (w_1, w_2, ..., w_N)$ is the vector of rules-rows in the fuzzy knowledge base (4.2), $\mathbf{B} = (b_1, b_2, ..., b_q)$ and $\mathbf{C} = (c_1, c_2, ..., c_q)$ are the vectors of fuzzy terms membership functions tuning parameters in (4.5), N is the total number of rules-rows, q is the total number of terms, F is the operator of inputs-output connection corresponding to relations (4.3) - (4.5).

Let us impose limitations on the knowledge base (4.2) volume in one of the following forms:

a) $N = k_1 + k_2 + ... + k_m \le \overline{N}$,

b) $k_1 \le \overline{k_1}$, $k_2 \le \overline{k_2}$, ..., $k_m \le \overline{k_m}$,

where $\overline{N}$ is the maximum permissible total number of conjunction rows in (4.2),

$\overline{k_j}$ is the maximum permissible number of conjunction rows in rules of j-th decision class, $j = \overline{1,m}$.

So as the content and number of linguistic terms a_i^{jp} ($i = \overline{1,n}$, $p = \overline{1,k_j}$, $j = \overline{1,m}$), used in the knowledge base (4.2), are not known beforehand then it is suggested to interpret them on the basis of membership functions (4.5) parameter values (b_i^{jp}, c_i^{jp}). Therefore, knowledge base (4.2) synthesis is reduced to obtaining the parameter matrix shown in Table 4.1.

Table 4.1. Knowledge base parameters matrix

Rule	IF			Weight	THEN
	x_1	... x_i ...	x_n		y
11	(b_1^{11}, c_1^{11})	(b_i^{11}, c_i^{11})	(b_n^{11}, c_n^{11})	w_{11}	
12	(b_1^{12}, c_1^{12})	(b_i^{12}, c_i^{12})	(b_n^{12}, c_n^{12})	w_{12}	d_1
...	...	...	...	...	
$1\,k_1$	$(b_1^{1k_1}, c_1^{1k_1})$	$(b_i^{1k_1}, c_i^{1k_1})$	$(b_n^{1k_1}, c_n^{1k_1})$	w_{1k_1}	
...	...	...	...	...	
$j\,1$	(b_1^{j1}, c_1^{j1})	(b_i^{j1}, c_i^{j1})	(b_n^{j1}, c_n^{j1})	w_{j1}	
$j\,2$	(b_1^{j2}, c_1^{12})	(b_i^{j2}, c_i^{j2})	(b_n^{j2}, c_n^{j2})	w_{j2}	
...	...	...	...	...	d_j
$j\,k_j$	$\left(b_1^{jk_j}, c_1^{jk_j}\right)$	$\left(b_i^{jk_j}, c_i^{jk_j}\right)$	$\left(b_n^{jk_j}, c_n^{jk_j}\right)$	w_{jk_j}	
...	...	...	...	...	
$m\,1$	(b_1^{m1}, c_1^{m1})	(b_i^{m1}, c_i^{m1})	(b_n^{m1}, c_n^{m1})	w_{m1}	
$m\,2$	(b_1^{m2}, c_1^{m2})	(b_i^{m2}, c_i^{m2})	(b_n^{m2}, c_n^{m2})	w_{m2}	
...	...	...	...	...	d_m
$m\,k_m$	$\left(b_1^{mk_m}, c_1^{mk_m}\right)$	$\left(b_i^{mk_m}, c_i^{mk_m}\right)$	$\left(b_n^{mk_m}, c_n^{mk_m}\right)$	w_{mk_m}	

In terms of mathematical programming this problem can be formulated in the following way. It is required to find such matrix (Table 4.1) which satisfying limitations imposed on parameters $(\mathbf{W}, \mathbf{B}, \mathbf{C})$ change ranges and number of rows provides for:

$$\sum_{p=1}^{M} [F(\mathbf{X}_p, \mathbf{W}, \mathbf{B}, \mathbf{C}) - y_p]^2 = \min_{\mathbf{W}, \mathbf{B}, \mathbf{C}}, \tag{4.6}$$

for the object with continuous output,

$$\sum_{p=1}^{M} \left\{ \sum_{j=1}^{m} \left[\mu^{d_j}(\mathbf{X}_p, \mathbf{W}, \mathbf{B}, \mathbf{C}) - \mu_p^{d_j}(y) \right]^2 \right\} = \min_{\mathbf{W}, \mathbf{B}, \mathbf{C}}, \tag{4.7}$$

for the object with discrete output, where

$$\mu_p^{d_j} = \begin{cases} 1, & \textit{if}\ \ d_j = d_p \\ 0, & \textit{if}\ \ d_j \neq d_p \end{cases}.$$

To solve these optimization problems it is appropriate to use a hybrid genetic and neuro approach.

4.3 Genetic Algorithm for Rules Extraction

The chromosome describing desired parameter matrix (Table 4.1), we define by the row shown in Fig. 4.1, where r_{jp} is the code of IF-THEN rule with number jp, $p=\overline{1,k_j}$, $j=\overline{1,m}$.

The operation of chromosomes crossover is defined in Fig. 4.2. It consists of exchanging parts of chromosomes in each rule r_{jp} ($j=\overline{1,m}$) and rules weights vector. The total number of exchange points makes $\overline{k_1}+\overline{k_2}+...+\overline{k_m}+1$: one for each rule and one for rules weights vector.

The operation of mutation (Mu) consists in random change (with some probability) of chromosome elements:

$$Mu(w_{jp}) = RANDOM([0,1]) ,$$

$$Mu\ (b_i^{jp}) = RANDOM([\underline{x}_i, \overline{x}_i]) ,$$

$$Mu\ (c_i^{jp}) = RANDOM([\underline{c}_i^{jp}, \overline{c}_i^{jp}]) ,$$

where $RANDOM([\underline{x},\overline{x}])$ is the operation of finding random number which is uniformly distributed on the interval $[\underline{x},\overline{x}]$.

If rules weights can take values 1 (rule available) or 0 (rule not available), then weights mutation can take place by way of random choice of 1 or 0.

Fitness function of chromosomes - solutions is calculated on the basis of (4.6) and (4.7) criteria.

If $P(t)$ - parent chromosomes, and $C(t)$ - offspring chromosomes on t-th iteration then genetic procedure of optimization will be carried out according to the following algorithm [18, 19]:

begin
  t:=0 ;
  assign initial value $P(t)$;
  estimate $P(t)$ using criteria (4.6) and (4.7);
    while (**not** condition for completion) **do**
      Crossover $P(t)$ to obtain $C(t)$;
      Estimate $C(t)$ using criteria (4.6) and (4.7);
      Choose $P(t+1)$ from $P(t)$ and $C(t)$;
      t:=t+1 ;
    end
end

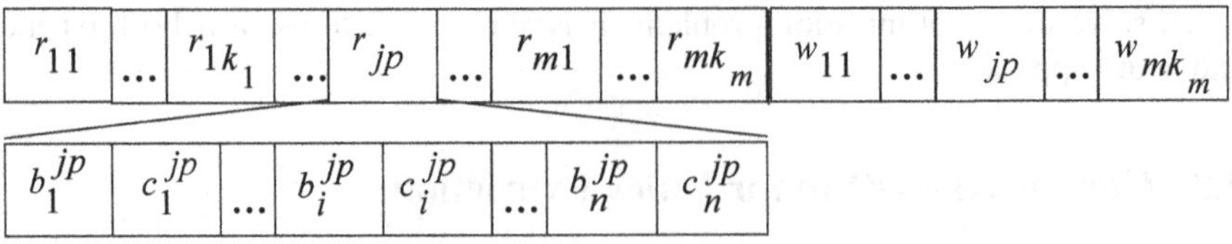

Fig. 4.1. Coding of parameter matrix

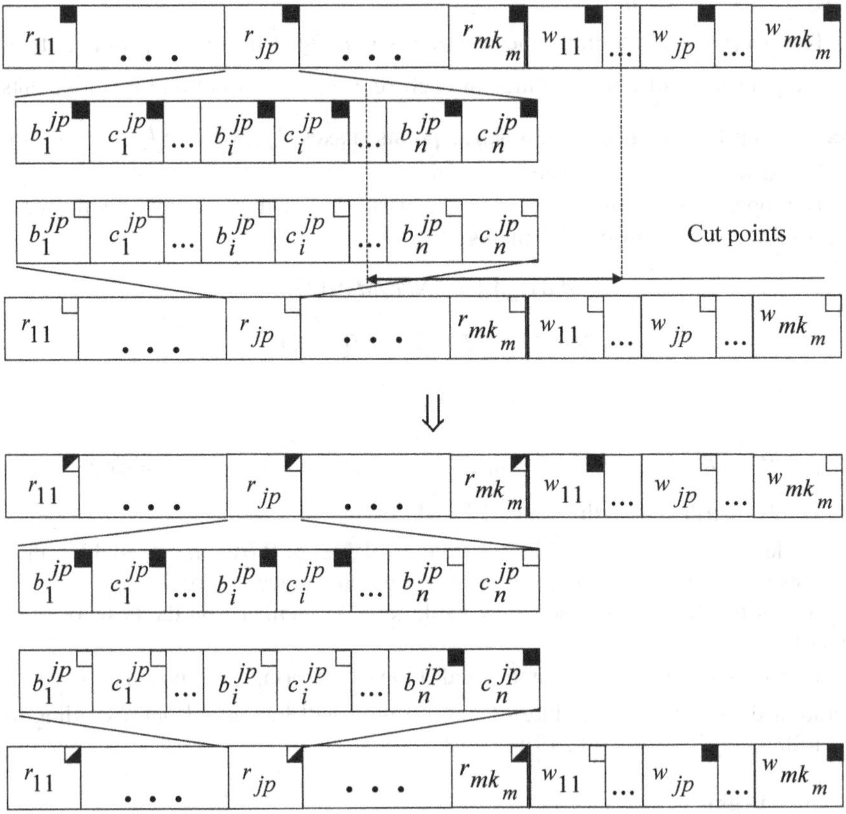

Fig. 4.2. Crossover operation (■, □ - parents symbols, ◪, ◩ - offspring symbols)

4.4 Neuro-fuzzy Network for Rules Extraction from Data

Let us impose limitations on the knowledge base (4.2) volume in the following form:

$$q_1 \le \overline{q_1}\,,\; q_2 \le \overline{q_2}\,,\; ...,\; q_n \le \overline{q_n}\,,$$

where $\overline{q}_i$ is the maximum permissible total number of fuzzy terms describing a variable x_i, $i=\overline{1,n}$;

This allows embedding system (4.2) into the special neuro-fuzzy network, which is able to extract knowledge [7, 17]. The neuro-fuzzy network for knowledge extraction is shown in Fig. 4.3, and the nodes are presented in Table 3.1.

As is seen from Fig. 4.3 the neuro-fuzzy network has the following structure:

layer 1 for object identification inputs (the number of nodes is equal to n),

layer 2 for fuzzy terms used in knowledge base (the number of nodes is equal to $\overline{q}_1+\overline{q}_2+...+\overline{q}_n$),

layer 3 for strings-conjunctions (the number of nodes is equal to $\overline{q}_1 \cdot \overline{q}_2 \cdot ... \cdot \overline{q}_n$),

layer 4 for fuzzy rules making classes (the layer is fully connected, the number of nodes is equal to the number of output classes m),

layer 5 for a defuzzification operation.

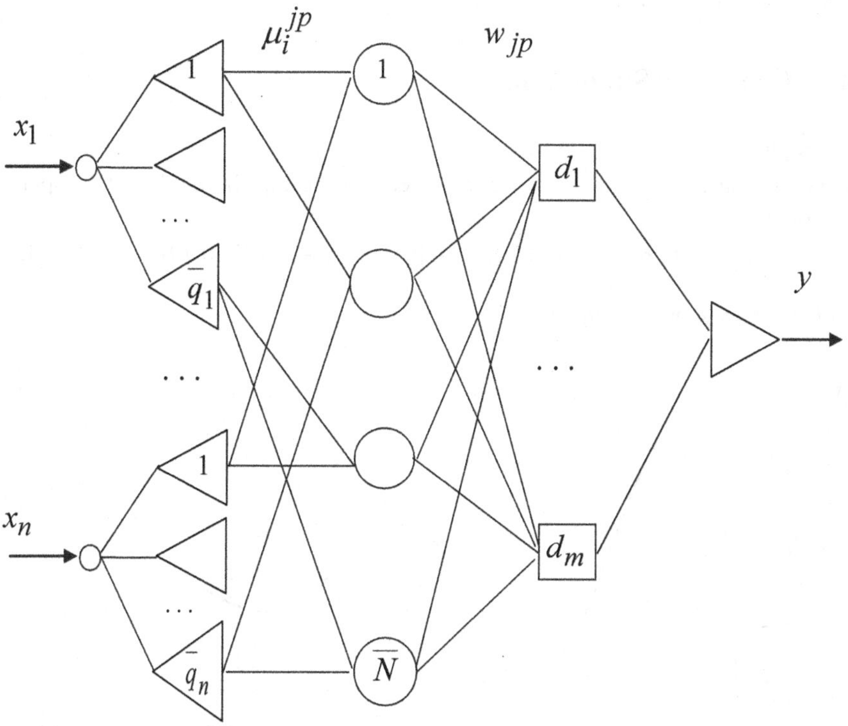

Fig. 4.3. Neuro-fuzzy network for knowledge extraction

To train the parameters of the neuro-fuzzy network, the recurrent relations

$$w_{jp}(t+1) = w_{jp}(t) - \eta \frac{\partial E_t}{\partial w_{jp}(t)},$$

$$c_i^{jp}(t+1) = c_i^{jp}(t) - \eta \frac{\partial E_t}{\partial c_i^{jp}(t)}, \quad b_i^{jp}(t+1) = b_i^{jp}(t) - \eta \frac{\partial E_t}{\partial b_i^{jp}(t)}$$

are used which minimize the criterion

$$E_t = \frac{1}{2}(\hat{y}_t - y_t)^2,$$

applied in the neural network theory, where $\hat{y}_t$ (y_t) are experimental and model outputs of the object at the t-th step of training;

$w_{jp}(t)$, $c_i^{jp}(t)$, $b_i^{jp}(t)$ are rules weights and parameters for the fuzzy terms membership functions at the t-th step of training;

η is a parameter of training [20].

The partial derivatives appearing in recurrent relations can be obtained according to the results from Section 3.3.

4.5 Computer Simulations

Example 1

Experimental data about the object was generated using the model "one input – one output"

$$y = f(x) = e^{\frac{-x}{4}} \cdot \sin(\tfrac{\pi}{2} x), \ x \in [0, 10], \ y \in [-0.47, 0.79], \tag{4.8}$$

which is represented in Fig. 4.4.

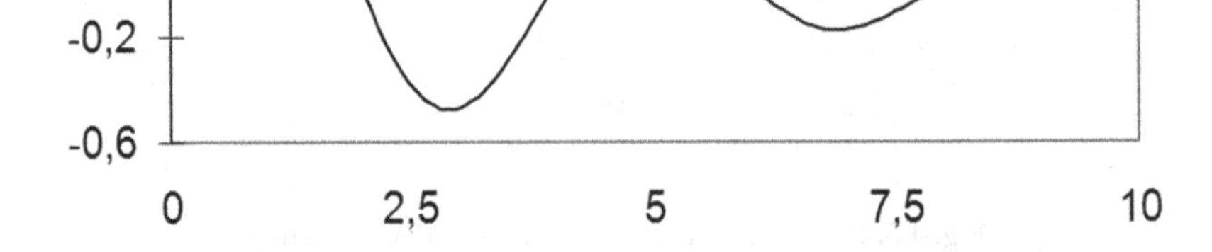

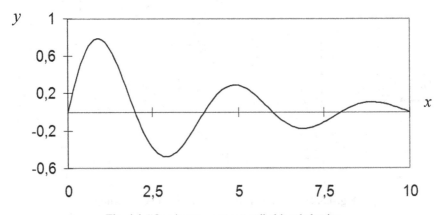

Fig. 4.4. "One input – one output" object behavior

The object output was divided into seven classes:

$$y \in \underbrace{[-0.47,-0.30)}_{d_1} \cup \underbrace{[-0.30,-0.05)}_{d_2} \cup \underbrace{[-0.05,0.15)}_{d_3} \cup \underbrace{[0.15,0.30)}_{d_4} \cup$$
$$\cup \underbrace{[0.30,0.45)}_{d_5} \cup \underbrace{[0.45,0.65)}_{d_6} \cup \underbrace{[0.65,0.78]}_{d_7}$$

The goal was to synthesize 5 rules for every class describing the object (4.8).

Rules weights were accepted as equal to 0 and 1. As the result of using the genetic and neuro algorithm of optimization we obtained the parameters matrix represented in Table 4.2.

Table 4.2. Rules parameters matrix

IF x				THEN y
Genetic algorithm		Neuro-fuzzy network		
Term parameters (b, c)	Weight	Term parameters (b, c)	Weight	
(2.85, 0.96)	1	(2.81, 1.12)	1	d_1
(2.77, 1.05)	1	(2.72, 0.70)	1	
(2.90, 0.88)	1	(2.93, 0.85)	1	
(0.25, 0.85)	0	(0.13, 0.64)	0	
(2.88, 1.24)	1	(2.81, 1.17)	1	
(6.85, 1.94)	1	(6.11, 1.13)	1	d_2
(8.74, 1.26)	1	(3.71, 0.25)	0	
(8.91, 2.17)	1	(6.91, 2.05)	1	
(6.92, 1.83)	1	(6.83, 0.72)	1	
(0.93, 1.21)	0	(1.13, 0.92)	0	
(0.06, 0.74)	1	(0.13, 0.87)	1	d_3
(8.91, 2.53)	0	(9.10, 1.25)	0	
(9.72, 2.12)	1	(8.62, 2.20)	1	
(9.90, 1.30)	1	(9.92, 1.12)	1	
(8.25, 1.15)	0	(8.7, 1.33)	1	
(4.85, 0.11)	1	(4.91, 0.21)	1	d_4
(5.33, 1.72)	1	(5.20, 1.50)	1	
(5.10, 1.08)	1	(5.01, 0.90)	1	
(6.54, 0.70)	0	(5.12, 0.83)	0	
(9.48, 2.31)	0	(9.17, 1.19)	0	
(2.00, 0.94)	0	(2.13, 0.72)	0	d_7
(0.64, 2.46)	0	(0.70, 1.25)	0	
(0.88, 0.76)	1	(0.92, 0.70)	1	
(1.25, 0.67)	0	(0.93, 1.12)	0	
(0.97, 2.18)	1	(1.01, 1.90)	1	

After linguistic interpretation the genetically generated rules look like this:

IF $x =$ *about* 2.8 THEN $y \in d_1$

IF $x =$ *about* 6.9 OR $x =$ *about* 8.8 THEN $y \in d_2$

IF $x =$ *about* 0 OR $x =$ *about* 10 THEN $y \in d_3$

IF $x =$ *about* 5 THEN $y \in d_4$

IF $x =$ *about* 0.9 THEN $y \in d_7$

Rules specified using neural adjustment after linguistic interpretation look like this:

IF $x =$ *about* 2.8 THEN $y \in d_1$

IF $x =$ *about* 6.9 THEN $y \in d_2$

IF $x =$ *about* 0 OR $x =$ *about* 8.8 OR $x =$ *about* 10 THEN $y \in d_3$

IF $x =$ *about* 5 THEN $y \in d_4$

IF $x =$ *about* 0.9 THEN $y \in d_7$

The model derived according to synthesized rules in comparison with the target one is shown in Fig. 4.5, 4.6.

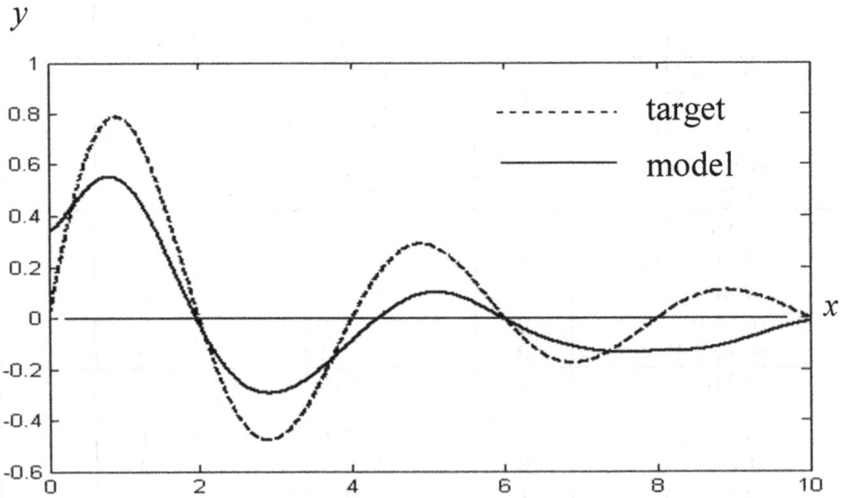

Fig. 4.5. Comparison of the genetically synthesized linguistic model with the standard

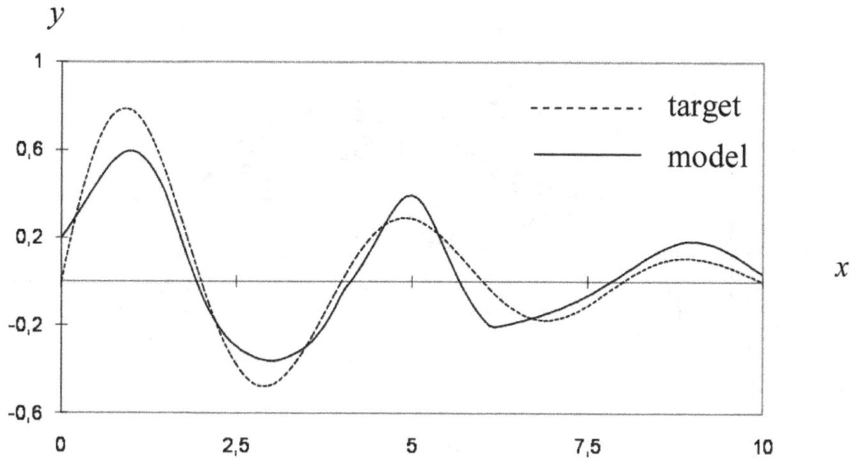

Fig. 4.6. Comparison of the linguistic model specified using neural adjustment with the standard

Further increase of linguistic model precision is possible on the account of its fine tuning.

Example 2

Experimental data about the object was generated using the model "two inputs – one output":

$$y = f(x_1, x_2) = \frac{1}{10}(2z-0.9)\ (7z-1)\ (17z-19)\ (15z-2)\,, \tag{4.9}$$

where $z = \dfrac{(x_1-3.0)^2+(x_2-3.0)^2}{40}$,

which is represented in Fig. 4.7.

The object output was divided into five classes:

$$y \in \underbrace{[-5.08,-4.50)}_{d_1} \cup \underbrace{[-4.50,-3.0)}_{d_2} \cup \underbrace{[-3.0,-0.5)}_{d_3} \cup \underbrace{[-0.5,0)}_{d_4} \cup \underbrace{[0,0.855)}_{d_5}\,.$$

The goal was to synthesize 20 rules for every class describing the object (4.9). Rules weights were accepted as equal to 0 and 1. As the result of using the genetic and neuro algorithm of optimization we obtained the parameters matrix represented in Table 4.3.

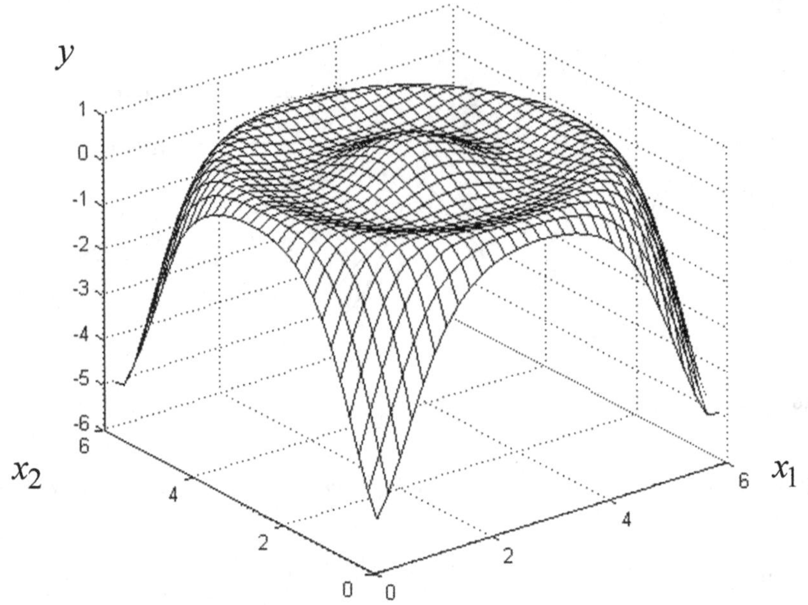

Fig. 4.7. "Two inputs – one output" object behaviour

Table 4.3. Rules parameters (b , c) matrix

Genetic algorithm			Neuro-fuzzy network			d
x_1	x_2	weight	x_1	x_2	weight	
(0.05, 0.12)	(1.10, 0.99)	1	(0.15, 0.08)	(1.16, 0.83)	1	d_1
(0.39, 0.98)	(0.02, 0.17)	1	(0.32, 0.75)	(0.09, 0.06)	1	
(4.83, 0.86)	(0.20, 0.11)	1	(4.72, 1.14)	(0.18, 0.09)	1	
(5.99, 0.15)	(1.33, 0.84)	1	(5.97, 0.12)	(1.48, 1.17)	1	
(0.20, 0.15)	(5.08, 0.92)	1	(0.17, 0.09)	(5.62, 0.79)	1	
(0.77, 0.96)	(5.92, 0.14)	1	(0.92, 0.81)	(5.99, 0.06)	1	
(5.95, 0.17)	(4.91, 0.83)	1	(5.85, 0.10)	(4.69, 0.72)	1	
(4.93, 1.36)	(5.90, 0.17)	1	(5.24, 1.17)	(5.99, 0.07)	1	
(0.08, 0.12)	(0.16, 0.08)	1	(0.04, 0.06)	(0.05, 0.11)	1	d_2
(5.99, 0.20)	(0.19, 0.18)	1	(5.98, 0.11)	(0.17, 0.04)	1	
(0.13, 0.17)	(5.92, 0.12)	1	(0.10, 0.09)	(5.97, 0.08)	1	
(5.97, 0.11)	(5.90, 0.20)	1	(5.87, 0.09)	(6.00, 0.10)	1	
(0.44, 0.96)	(0.87, 0.91)	1	(0.56, 1.17)	(1.28, 0.99)	0	
(4.06, 0.52)	(0.03, 0.08)	1	(5.88, 0.14)	(0.12, 0.14)	1	
(0.58, 1.07)	(5.71, 1.20)	1	(0.82, 1.34)	(5.86, 0.92)	0	
(4.91, 0.78)	(1.48, 0.77)	1	(5.32, 0.89)	(1.54, 0.65)	0	

Table 4.3. *(continued)*

(0.09, 0.15)	(2.04, 0.56)	1	(0.10, 0.12)	(2.17, 0.45)	1	d_3
(3.65, 0.74)	(1.52, 0.73)	1	(0.44, 0.96)	(0.87, 0.91)	1	
(5.91, 0.08)	(3.71, 0.67)	1	(1.86, 0.37)	(0.16, 0.09)	1	
(0.16, 0.07)	(3.94, 0.64)	1	(4.06, 0.52)	(0.03, 0.08)	1	
(0.04, 0.20)	(3.05, 0.86)	1	(4.91, 0.78)	(1.48, 0.77)	1	
(4.88, 0.84)	(5.32, 0.98)	1	(5.94, 0.09)	(2.11, 0.56)	1	
(3.02, 0.77)	(5.94, 0.13)	1	(0.06, 0.15)	(3.67, 0.39)	1	
(5.91, 0.34)	(0.12, 0.19)	0	(0.58, 1.07)	(5.71, 1.20)	1	
(5.34, 0.76)	(4.18, 0.56)	0	(5.96, 0.04)	(3.94, 0.65)	1	
(0.16, 0.25)	(3.44, 0.95)	0	(5.17, 0.88)	(4.98, 0.70)	1	
(4.97, 0.56)	(5.11, 0.93)	0	(2.02, 0.60)	(5.99, 0.06)	1	
(3.22, 0.91)	(5.99, 0.32)	0	(3.74, 0.49)	(5.87, 0.09)	1	
(0.22, 1.17)	(3.07, 0.85)	1	(0.16, 0.09)	(2.86, 0.59)	1	d_4
(1.25, 0.93)	(1.96, 0.53)	1	(1.07, 1.15)	(2.25, 0.35)	1	
(2.17, 0.75)	(0.74, 0.72)	1	(1.96, 0.54)	(0.37, 0.88)	1	
(3.00, 0.92)	(0.04, 0.26)	1	(3.04, 0.79)	(0.09, 0.16)	1	
(1.08, 0.54)	(3.45, 0.65)	1	(1.06, 0.94)	(3.75, 0.49)	1	
(5.93, 0.18)	(2.16, 0.78)	1	(4.07, 0.52)	(0.42, 0.30)	1	
(1.85, 0.46)	(0.06, 0.15)	1	(1.92, 0.33)	(1.96, 0.51)	1	
(3.03, 0.88)	(2.03, 0.47)	1	(2.96, 0.81)	(2.40, 0.38)	1	
(5.92, 0.20)	(2.34, 0.67)	1	(3.61, 0.42)	(2.08, 0.44)	1	
(2.03, 0.68)	(3.00, 0.91)	1	(4.75, 0.79)	(1.96, 0.50)	1	
(5.99, 0.08)	(2.92, 0.79)	1	(2.17, 0.38)	(3.08, 0.72)	1	
(1.98, 0.93)	(5.74, 1.17)	1	(3.81, 0.54)	(2.99, 0.85)	1	
(3.81, 0.69)	(3.66, 0.61)	1	(5.96, 0.11)	(3.06, 0.69)	1	
(4.82, 1.45)	(3.52, 0.93)	1	(1.77, 0.42)	(3.68, 0.47)	1	
(2.26, 0.74)	(4.65, 1.14)	1	(3.07, 0.68)	(4.05, 0.32)	1	
(3.67, 0.81)	(5.86, 0.26)	1	(3.91, 0.53)	(3.89, 0.37)	1	
(4.55, 1.34)	(3.22, 0.96)	0	(4.78, 1.15)	(3.61, 0.45)	1	
(1.87, 0.72)	(5.08, 0.33)	0	(2.18, 0.39)	(5.67, 0.95)	1	
(3.77, 0.21)	(4.26, 1.91)	0	(3.65, 0.47)	(4.86, 0.71)	1	
(3.08, 0.83)	(5.07, 2.36)	0	(2.97, 0.75)	(5.96, 0.11)	1	
(3.68, 1.31)	(4.78, 1.56)	1	(0.26, 0.81)	(3.02, 0.75)	1	d_5
(2.97, 0.93)	(0.52, 0.09)	1	(3.02, 0.70)	(0.56, 0.15)	1	
(2.92, 0.55)	(3.02, 0.98)	1	(2.96, 0.64)	(3.09, 0.66)	1	
(5.64, 0.97)	(3.00, 1.17)	1	(5.41, 0.79)	(3.03, 0.82)	1	
(3.02, 1.26)	(5.44, 0.97)	1	(3.06, 0.67)	(5.56, 1.13)	1	
(2.33, 0.85)	(2.07, 0.46)	1	(2.17, 1.68)	(1.74, 0.61)	0	
(3.92, 1.45)	(1.89, 0.92)	1	(3.12, 2.65)	(1.28, 1.12)	0	
(3.90, 1.58)	(3.02, 0.77)	1	(3.18, 0.54)	(3.00, 0.38)	0	
(1.82, 0.23)	(3.48, 0.82)	1	(1.89, 0.74)	(3.91, 0.60)	0	
(3.06, 1.72)	(4.01, 2.12)	1	(3.00, 2.16)	(4.871, 0.53)	0	

The generated rules after linguistic interpretation are presented in Table 4.4, where the parameters of fuzzy terms for variables x_1 and x_2 evaluation are interpreted as follows: *about* 0 – *Low* (*L*), *about* 0.5 – *higher than Low* (*hL*), *about* 1.5 – *lower than Average* (*lA*), *about* 3 – *Average* (*A*), *about* 4.5 – *higher than Average* (*hA*), *about* 5.5 – *lower than High* (*lH*), *about* 6 – *High* (*H*).

Table 4.4. Fuzzy knowledge base

Genetic algorithm		Neuro-fuzzy network		d
x_1	x_2	x_1	x_2	
L	*hL*	*L*	*hL*	d_1
hL	*L*	*hL*	*L*	
lH	*L*	*lH*	*L*	
H	*hL*	*H*	*hL*	
L	*lH*	*L*	*lH*	
hL	*H*	*hL*	*H*	
H	*lH*	*H*	*lH*	
lH	*H*	*lH*	*H*	
L	*L*	*L*	*L*	d_2
H	*L*	*H*	*L*	
L	*H*	*L*	*H*	
H	*H*	*H*	*H*	
hL	*hL*			
hA	*L*			
hL	*lH*			
lH	*hL*			
L	*lA*	*L*	*lA*	d_3
hA	*hL*	*hL*	*hL*	
H	*hA*	*lA*	*L*	
L	*hA*	*hA*	*L*	
L	*A*	*lH*	*hL*	
lH	*lH*	*H*	*lA*	
A	*H*	*L*	*hA*	
		hL	*lH*	
		H	*hA*	
		lH	*lH*	
		lA	*H*	
		hA	*H*	

Genetic algorithm		Neuro-fuzzy network		d
x_1	x_2	x_1	x_2	
hL	*A*	*L*	*A*	d_4
hL	*lA*	*hL*	*lA*	
lA	*hL*	*lA*	*hL*	
A	*L*	*A*	*L*	
hL	*hA*	*hL*	*hA*	
H	*lA*	*hA*	*hL*	
lA	*L*	*lA*	*lA*	
A	*lA*	*A*	*lA*	
lH	*lA*	*hA*	*lA*	
lA	*A*	*lH*	*lA*	
H	*A*	*lA*	*A*	
lA	*H*	*hA*	*A*	
hA	*hA*	*H*	*A*	
lH	*hA*	*lA*	*hA*	
lA	*lH*	*A*	*hA*	
hA	*H*	*hA*	*hA*	
		lH	*hA*	
		lA	*lH*	
		hA	*lH*	
		A	*H*	
hA	*lH*	*hL*	*A*	d_5
A	*hL*	*A*	*hL*	
A	*A*	*A*	*A*	
lH	*A*	*lH*	*A*	
A	*lH*	*A*	*lH*	
lA	*lA*			
hA	*lA*			
hA	*A*			
lA	*hA*			
A	*hA*			

The model of the object derived according to synthesized rules is shown in Fig. 4.8, 4.9.

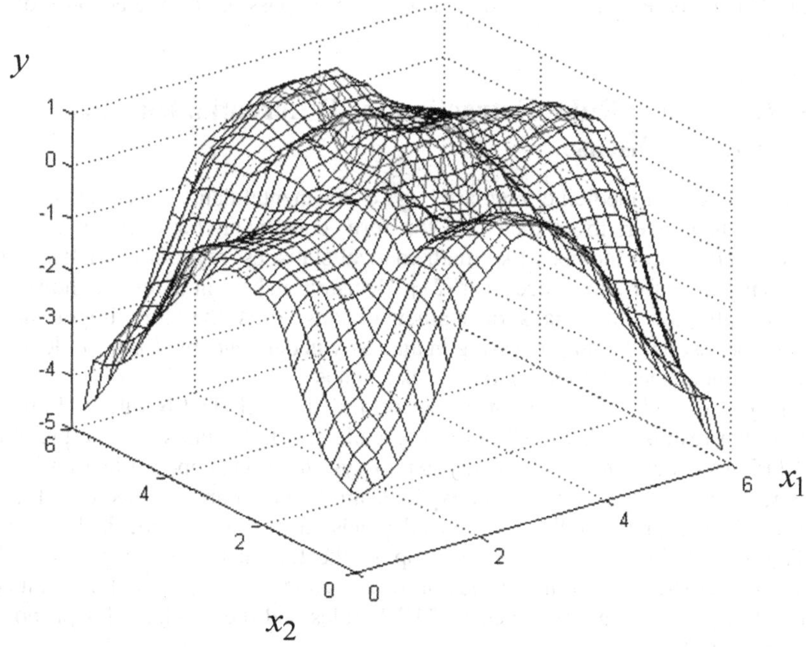

Fig. 4.8. Linguistic model synthesized using the genetic algorithm

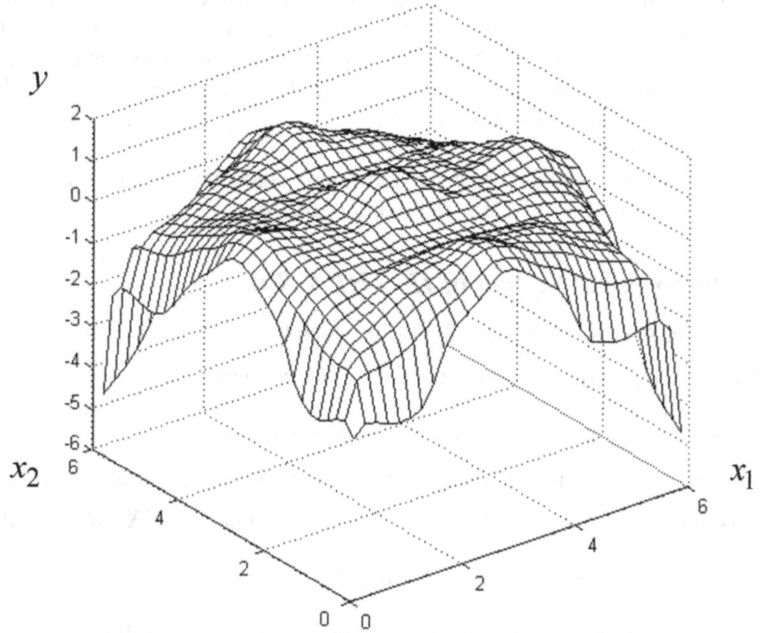

Fig. 4.9. Linguistic model specified using neural adjustment

Further increase of linguistic model precision is possible on the account of its fine tuning.

4.6 Example 3: Rules Extraction for Differential Diagnosis of Heart Disease

In a lot of areas of medicine there are huge experimental data collections and it is necessary to convert these data into the form convenient for decision making. Several well-known methods like mathematical statistics, regression analyses etc. are usually used for data processing [21]. Decision makers in medicine, however, are typically not statisticians or mathematicians. It is therefore important to present the results of data processing in an easily understandable form for decision makers without special mathematical backgrounds.

Fuzzy information granulation in the form of fuzzy IF-THEN rules [1] allows making the results of data analysis easily understandable and well interpretable. But during the development of fuzzy expert systems it is supposed that an initial knowledge base is generated by an expert from the given area of medicine [2, 3]. That is why the quality of these systems depends on the skill of a medical expert.

The aim of this section is (1) to propose the formal procedure of fuzzy IF-THEN rules extraction from histories of diseases and (2) to compare the results of medical diagnosis using extracted IF-THEN rules and the similar rules proposed by an expert [3].

A specific feature of fuzzy rules bases for medical diagnosis consists of their hierarchical character. In this section we propose the formal procedure for extraction of a hierarchical system of fuzzy rules for medical diagnosis from real histories of diseases. The suggested procedure is based on the optimal solution growing from a set of primary IF-THEN rules variants using the genetic cross-over, mutation and selection operations [18, 19]. The neural approach is used for adaptive correction of the diagnostic rules by pruning redundant membership functions and rules.

The efficiency of proposed genetic and neuro algorithms is illustrated by an example of ischemia heart disease (IHD) diagnosis [3].

4.6.1 Hierarchical System of IF-THEN Rules

Let us consider the object (3.30) - (3.32) for which the following is known:

- intervals of inputs (parameters of the patient state) change $x_i \in [\underline{x_i}, \overline{x_i}]$, $i = \overline{1,n}$,

- classes of decisions d_j $(j = \overline{1,m})$ (types of diagnoses),

- training data (histories of diseases) in the form of M pairs of experimental data “parameters of patient state - type of diagnose” $\{\mathbf{X}_p, d_p\}$, where $\mathbf{X}_p = \{x_1^p, x_2^p, ..., x_n^p\}$ - input vector in p-th pair, $p = \overline{1,M}$.

It is necessary to transfer the available training data into the following systems of the fuzzy IF-THEN rules:

1) for the instrumental danger y depending on parameters $\{x_2, x_3, x_4, x_5, x_{10}, x_{11}\}$:

$$\text{IF}\left[\;(x_2 = a_2^{j1}) \text{ AND } (x_3 = a_3^{j1}) \text{ AND} \ldots (x_{11} = a_{11}^{j1})\;\right] \text{(with weight } w_{j1}^{y})$$

...

$$\text{OR}\left[\;\left(x_2 = a_2^{jk_j}\right) \text{ AND } \left(x_3 = a_3^{jk_j}\right) \text{ AND} \ldots \left(x_{11} = a_{11}^{jk_j}\right)\;\right] \text{(with weight } w_{jk_j}^{y}),$$

$$\text{THEN } \; y \in y_j \text{, for all } j = \overline{1,5}; \tag{4.10}$$

2)) for the biochemical danger z depending on parameters $\{x_6, x_7, x_8, x_9, x_{12}\}$:

$$\text{IF}\left[\;(x_6 = a_6^{j1}) \text{ AND } (x_7 = a_7^{j1}) \text{ AND} \ldots (x_{12} = a_{12}^{j1})\;\right] \text{(with weight } w_{j1}^{z})$$

...

$$\text{OR}\left[\;\left(x_6 = a_6^{jk_j}\right) \text{ AND } \left(x_7 = a_7^{jk_j}\right) \text{ AND} \ldots \left(x_{12} = a_{12}^{jk_j}\right)\;\right] \text{(with weight } w_{jk_j}^{z}),$$

$$\text{THEN } z \in z_j \text{, for all } j = \overline{1,5}; \tag{4.11}$$

3) *for the danger of IHD d depending on parameters* $\{x_1, y, z\}$*:*

$$\text{IF}\left[\;(x_1 = a_1^{j1}) \text{ AND } (y = a_y^{j1}) \text{ AND } (z = a_z^{j1})\;\right] \text{(with weight } w_{j1})$$

...

$$\text{OR}\left[\;\left(x_1 = a_1^{jk_j}\right) \text{ AND } \left(y = a_y^{jk_j}\right) \text{ AND } \left(z = a_z^{jk_j}\right)\;\right] \text{(with weight } w_{jk_j}),$$

$$\text{THEN } d \in d_j \text{, for all } j = \overline{1,m}, \tag{4.12}$$

where a_i^{jp} is the linguistic term for the estimation of variable x_i in the row with number $p = \overline{1, k_j}$,

a_y^{jp} (a_z^{jp}) is the linguistic term for the estimation of variable *y (z)* in the row with number $p = \overline{1, k_j}$, and it is supposed that term a_y^{jp} (a_z^{jp}) should be chosen from estimates y_j (z_j), $j = \overline{1,5}$;

k_j is the number of conjunction rows corresponding to the classes d_j, y_j, z_j;

w_{jp}^{y}, w_{jp}^{z}, w_{jp} the weights of the expressions with number jp in (4.10) - (4.12).

4.6.2 Hierarchical System of Parameter Matrices

The problem of fuzzy IF-THEN rules (4.10) - (4.12) extraction can be considered as finding three matrices presented in Tables 4.5 - 4.7. Each element (b_i^{jp} , c_i^{jp}) of these matrices corresponds to the membership function parameters and can be interpreted as a fuzzy term (low, average, high, etc.). Each element a_y^{jp} (a_z^{jp}) in Table 4.7 is chosen from the decision classes y_j (z_j) in Table 4.5, 4.6.

Table 4.5. Matrix of IF-THEN rules parameters for model (3.31)

Rule	IF			Weight	THEN
№	x_2	...	x_{11}		y
11	(b_2^{11}, c_2^{11})		$(b_{11}^{11}, c_{11}^{11})$	w_{11}^{y}	
...	...		...	...	y_1
$1\,k_1$	$(b_2^{1k_1}, c_2^{1k_1})$		$(b_{11}^{1k_1}, c_{11}^{1k_1})$	$w_{1k_1}^{y}$	
...	...		...	...	...
51	(b_2^{51}, c_2^{51})		$(b_{11}^{51}, c_{11}^{51})$	w_{51}^{y}	
...	...		...	...	y_5
$5\,k_5$	$(b_2^{5k_5}, c_2^{5k_5})$		$(b_{11}^{5k_5}, c_{11}^{5k_5})$	$w_{5k_5}^{y}$	

Table 4.6. Matrix of IF-THEN rules parameters for model (3.32)

Rule	IF			Weight	THEN
№	x_6	...	x_{12}		z
11	(b_6^{11}, c_6^{11})		$(b_{12}^{11}, c_{12}^{11})$	w_{11}^{z}	
...	...		...	...	z_1
$1\,k_1$	$(b_6^{1k_1}, c_6^{1k_1})$		$(b_{12}^{1k_1}, c_{12}^{1k_1})$	$w_{1k_1}^{z}$	
...	...		...	...	...
51	(b_6^{51}, c_6^{51})		$(b_{12}^{51}, c_{12}^{51})$	w_{51}^{z}	
...	...		...	...	z_5
$5\,k_5$	$\left(b_6^{5k_5}, c_6^{5k_5}\right)$		$\left(b_{12}^{5k_5}, c_{12}^{5k_5}\right)$	$w_{5k_5}^{z}$	

Table 4.7. Matrix of IF-THEN rules parameters for model (3.30)

Rule	IF			Weight	THEN
№	x_1	y	z		d
11	(b_1^{11}, c_1^{11})	a_y^{11}	a_z^{11}	w_{11}	
...	...	...	...	...	d_1
$1\,k_1$	$(b_1^{1k_1}, c_1^{1k_1})$	$a_y^{1k_1}$	$a_z^{1k_1}$	w_{1k_1}	
...	...	...	...	...	...
$m\,1$	(b_1^{m1}, c_1^{m1})	a_y^{m1}	a_z^{m1}	w_{m1}	
...	...	...	...	...	d_m
$m\,k_m$	$(b_1^{mk_m}, c_1^{mk_m})$	$a_y^{mk_m}$	$a_z^{mk_m}$	w_{mk_m}	

4.6.3 Computer Experiment

The total number of patients with IHD in our study was 65. The aim of computer experiment was to generate three rules for each class of decision (*y*-, *z*-, *d*-) according to the models (3.30) - (3.32). The results of this optimization problem solving using genetic and neuro algorithm are presented in Tables 4.8 - 4.13. According to these tables it is easy to make interpretation of each pairs of parameters using fuzzy terms: *L – Low, lA – lower than Average, A – Average, hA – higher than Average, H – High.* For example, the pairs (176.5, 87.8), (256.1, 25.1), (368.3, 49.8) correspond to the membership functions shown in Fig. 4.10, which can be interpreted as *lower than Average* (*lA*), *Average* (*A*), *High* (*H*).

After linguistic interpretation we can describe the optimal solutions (Tables 4.8 - 4.13) in the form of fuzzy IF-THEN rules matrices (Tables 4.14 - 4.16), where

GA – genetic algorithm;
NN – neuro-fuzzy network.

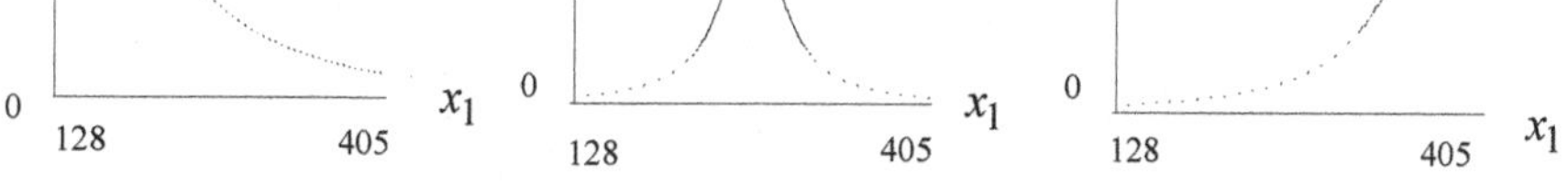

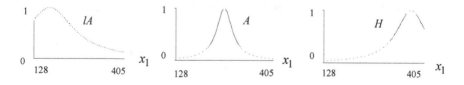

Fig. 4.10. Example of linguistic interpretation

Table 4.8. Parameters of rules for model (3.31) synthesized using the genetic algorithm

x_2	x_3	x_4	x_5	x_{10}	x_{11}	y
(366.22, 83.44)	(941.93, 251.67)	(3.22, 5.24)	(0.43, 0.08)	(34.28, 11.42)	(275.50, 535.50)	
(176.48, 206.91)	(667.20, 120.90)	(1.84, 5.63)	(0.25, 0.02)	(17.79, 41.88)	(298.45, 135.26)	L
(145.31, 50.27)	(109.43, 1350.49)	(0.81, 0.41)	(0.09, 0.11)	(24.23, 3.10)	(65.13, 21.18)	
(368.30, 102.18)	(955.80, 842.19)	(1.31, 2.48)	(0.17, 0.15)	(11.42, 12.05)	(251.02, 7.03)	
(256.11, 90.71)	(128.85, 408.26)	(2.14, 0.46)	(0.32, 0.59)	(40.57, 25.13)	(179.88, 160.36)	lA
(128.00, 48.30)	(92.78, 180.36)	(0.60, 0.58)	(0.10, 0.05)	(7.40, 3.86)	(199.77, 52.74)	
(184.79, 350.26)	(914.18, 1942.50)	(2.41, 5.78)	(0.23, 0.18)	(26.33, 18.37)	(227.31, 229.50)	
(130.77, 80.12)	(808.73, 224.63)	(0.62, 2.60)	(0.40, 0.23)	(8.91, 10.84)	(140.10, 200.05)	A
(162.63, 45.64)	(306.45, 1406.27)	(0.66, 0.39)	(0.12, 0.35)	(8.41, 4.69)	(290.80, 150.46)	
(315.67, 50.92)	(123.30, 917.02)	(0.88, 5.78)	(0.28, 0.27)	(33.53, 7.18)	(191.35, 688.50)	
(188.94, 346.25)	(142.73, 268.38)	(1.89, 2.05)	(0.36, 0.07)	(8.91, 9.04)	(325.23, 116.83)	hA
(128.00, 74.17)	(645.00, 138.73)	(0.76, 0.49)	(0.10, 0.03)	(8.49, 16.79)	(208.95, 10.25)	
(202.79, 120.62)	(597.83, 340.36)	(1.47, 1.42)	(0.11, 0.35)	(16.53, 8.17)	(185.23, 137.25)	
(290.74, 80.56)	(434.10, 380.95)	(1.06, 7.02)	(0.46, 0.21)	(39.90, 18.37)	(277.80, 155.48)	H
(128.00, 60.04)	(114.98, 570.30)	(0.61, 0.78)	(0.09, 0.08)	(7.74, 5.28)	(46.00, 40.34)	

Table 4.9. Parameters of rules for model (3.31) specified using the neuro-fuzzy network

x_2	x_3	x_4	x_5	x_{10}	x_{11}	w	y
(330.21, 207.75)	(539.55, 260.85)	(1.76, 5.78)	(0.26, 0.03)	(18.12, 6.20)	(75.84, 688.50)	0.98	
(314.28, 42.24)	(114.98, 238.65)	(0.77, 5.78)	(0.20, 0.35)	(33.20, 25.13)	(59.77, 535.50)	0.51	L
(205.56, 623.25)	(711.60, 185.93)	(3.64, 5.78)	(0.35, 0.06)	(26.83, 25.46)	(114.09, 688.50)	0.99	
(179.25, 484.75)	(575.63, 2497.50)	(2.68, 0.82)	(0.46, 0.11)	(38.30, 4.69)	(216.60, 688.50)	0.54	
(206.95, 346.25)	(950.25, 1387.50)	(3.49, 0.83)	(0.19, 0.35)	(40.73, 8.29)	(176.82, 229.50)	0.59	lA
(397.38, 623.25)	(1197.23, 1942.50)	(1.37, 7.43)	(0.20, 1.06)	(37.22, 7.04)	(205.12, 688.50)	0.88	
(140.47, 58.86)	(797.63, 1942.50)	(3.19, 0.59)	(0.49, 0.59)	(23.40, 8.38)	(296.16, 69.62)	0.51	
(215.95, 195.29)	(794.85, 8.33)	(3.26, 7.43)	(0.15, 0.59)	(19.21, 41.88)	(259.44, 231.03)	0.70	A
(299.74, 346.25)	(1086.23, 1387.50)	(1.81, 5.78)	(0.22, 0.04)	(24.65, 41.88)	(301.51, 95.63)	0.95	
(226.34, 484.75)	(395.25, 832.50)	(1.59, 2.48)	(0.49, 0.10)	(38.14, 8.12)	(155.40, 74.21)	0.50	
(200.71, 346.25)	(425.78, 2497.50)	(2.14, 2.48)	(0.27, 0.12)	(38.89, 7.79)	(62.07, 65.79)	0.53	hA
(202.10, 74.79)	(1039.05, 563.33)	(0.90, 0.54)	(0.26, 0.20)	(14.10, 25.13)	(332.88, 387.09)	0.97	
(321.21, 623.25)	(148.28, 122.10)	(0.81, 0.43)	(0.33, 0.59)	(36.88, 5.53)	(262.50, 229.50)	0.50	
(146.70, 46.40)	(1061.25 230.33)	(1.44, 4.13)	(0.17, 0.09)	(11.59, 3.10)	(86.55, 45.14)	0.50	H
(232.57, 346.25)	(237.08, 740.93)	(2.49, 4.21)	(0.53, 0.06)	(20.72, 58.63)	(152.34, 382.50)	1.00	

Table 4.10. Parameters of rules for model (3.32) synthesized using the genetic algorithm

x_6	x_7	x_8	x_9	x_{12}	z
(50.32, 26.25) (49.71, 9.16) (35.09, 8.75)	(20.56, 11.28) (22.53, 4.17) (22.84, 2.75)	(13.41, 4.45) (15.47, 41.13) (4.42, 0.57)	(4.50, 5.06) (3.82, 0.72) (1.01, 3.58)	(21.92, 30.85) (16.59, 28.14) (3.90, 10.52)	L
(62.31, 7.80) (61.70, 15.03) (35.01, 8.75)	(26.91, 20.48) (20.87, 3.37) (11.90, 8.51)	(15.88, 25.16) (24.69, 12.04) (3.66, 5.28)	(2.33, 7.95) (2.75, 5.98) (1.01, 1.05)	(23.56, 41.17) (24.74, 27.44) (4.29, 7.20)	lA
(49.10, 6.11) (65.38, 12.34) (56.45, 9.72)	(28.09, 39.38) (27.74, 21.88) (15.71, 4.76)	(16.94, 27.06) (7.30, 12.65) (3.66, 5.88)	(5.32, 1.41) (3.80, 5.00) (2.48, 0.88)	(21.85, 15.22) (20.60, 5.17) (4.10, 3.24)	A
(58.64, 43.75) (47.35, 20.85) (34.66, 78.75)	(16.84, 8.90) (22.36, 5.03) (11.90, 4.55)	(4.60, 4.18) (5.95, 1.03) (5.07, 13.18)	(4.71, 6.27) (3.77, 8.16) (1.00, 0.93)	(24.94, 15.88) (7.91, 11.07) (3.97, 5.43)	hA
(58.72, 26.25) (34.57, 8.75) (34.57, 6.28)	(28.83, 30.63) (15.27, 20.15) (11.90, 4.74)	(24.40, 9.47) (9.24, 22.94) (3.84, 16.32)	(5.32, 10.29) (4.88, 9.84) (1.01, 4.40)	(16.79, 6.29) (6.67, 30.15) (18.76, 3.65)	H

Table 4.11. Parameters of rules for model (3.32) specified using the neuro-fuzzy network

x_6	x_7	x_8	x_9	x_{12}	w	z
(52.79, 43.75) (55.68, 78.75) (63.90, 78.75)	(27.96, 4.38) (24.02, 2.71) (12.82, 30.63)	(9.59, 17.63) (8.12, 52.88) (19.29, 17.63)	(4.77, 3.40) (1.72, 3.53) (2.75, 0.88)	(29.48, 3.48) (10.61, 33.53) (16.46, 5.65)	0.57 0.98 0.69	L
(47.54, 78.75) (39.84, 8.75) (35.90, 44.63)	(20.08, 21.88) (17.24, 30.63) (24.11, 21.88)	(25.28, 29.38) (11.77, 52.88) (23.05, 41.13)	(2.30, 5.88) (2.42, 1.18) (4.30, 3.53)	(21.85, 32.88) (25.86, 46.03) (17.64, 32.88)	0.97 1.00 0.99	lA
(56.73, 43.75) (40.98, 78.75) (56.29, 6.74)	(15.49, 4.33) (28.00, 13.13) (28.35, 9.23)	(17.41, 52.88) (26.63, 5.76) (8.12, 41.13)	(4.45, 10.58) (1.95, 5.88) (4.63, 1.32)	(7.38, 5.33) (20.86, 6.58) (25.33, 19.73)	0.61 0.78 0.93	A
(68.19, 26.25) (37.48, 8.75) (61.10, 78.75)	(13.08, 4.38) (14.96, 30.63) (27.13, 3.98)	(19.87, 29.38) (5.72, 5.88) (6.07, 52.88)	(4.78, 8.23) (2.54, 5.88) (2.36, 3.53)	(26.91, 6.58) (20.27, 32.88) (26.52, 6.05)	0.60 0.70 0.50	hA
(66.18, 43.75) (44.91, 8.75) (49.73, 61.25)	(20.30, 30.63) (26.12, 39.38) (16.41, 39.38)	(21.87, 29.38) (3.78, 5.88) (13.59, 52.88)	(2.35, 3.53) (4.81, 10.58) (1.99, 10.58)	(19.94, 59.18) (18.43, 32.88) (27.37, 32.88)	0.80 1.00 1.00	H

Table 4.12. Parameters of rules for model (3.30) synthesized using the genetic algorithm

x_1	y	z	d
(38.56, 25.19)	H	L	d_1
(54.83, 40.26)	A	H	
(31.07, 10.04)	H	H	
(55.30, 6.74)	hA	A	d_2
(51.25, 10.57)	lA	H	
(31.00, 4.36)	A	lA	
(55.91, 12.11)	lA	A	d_3
(49.83, 4.67)	lA	lA	
(34.38, 5.12)	lA	lA	
(56.04, 12.20)	L	A	d_4
(31.14, 37.21)	lA	hA	
(32.01, 4.23)	L	L	
(42.34, 11.45)	L	lA	d_5
(46.80, 5.17)	hA	hA	
(32.96, 4.82)	L	hA	
(33.30, 6.31)	A	hA	d_6
(45.78, 16.70)	hA	hA	
(31.07, 4.48)	L	lA	

Table 4.13. Parameters of rules for model (3.30) specified using the neuro-fuzzy network

x_1	y	z	w	d
(38.90, 60.75)	hA	L	0.93	d_1
(31.47, 33.75)	L	H	0.70	
(51.05, 33.75)	hA	H	0.70	
(57.46, 19.85)	H	A	0.50	d_2
(45.92, 9.79)	A	H	0.99	
(50.04, 33.75)	L	hA	0.50	
(51.52, 60.75)	A	A	1.00	d_3
(48.15, 33.75)	A	hA	0.70	
(52.40, 33.75)	A	hA	0.50	
(52.06, 6.62)	lA	A	0.50	d_4
(40.38, 47.25)	A	lA	0.83	
(42.00, 20.25)	lA	L	0.50	
(57.53, 47.25)	lA	hA	0.72	d_5
(34.85, 60.75)	H	lA	0.50	
(44.16, 33.75)	lA	lA	0.97	
(36.54, 60.75)	L	hA	1.00	d_6
(44.84, 20.25)	H	lA	0.60	
(31.47, 35.91)	lA	hA	1.00	

Table 4.14. Fuzzy knowledge base for the instrumental danger y

x_2	x_3	x_4	x_5	x_{10}	x_{11}	y
GA / NN	GA / NN	GA / NN	GA / NN	GA / NN	GA / NN	
hA	hA / A	hA / lA	hA / lA	hA / lA	hA / L	
lA / hA	A / lA	A / L	lA	lA / hA	hA / L	L
L / lA	L / A	lA / H	L / A	A	lA	
hA / lA	hA / A	lA / hA	lA / hA	lA / H	hA / A	
A / lA	L / hA	A / H	A / lA	H	A	lA
L / H	L / H	L / lA	L / lA	L / H	A	
lA / L	hA	A / hA	lA / hA	A	A / hA	
L / lA	hA	L / hA	hA / lA	L / lA	lA / hA	A
lA / A	lA / H	L / lA	lA	L / A	hA	
hA / lA	L / lA	lA	A / hA	hA / H	A / lA	
lA	L / lA	A	A	L / H	H / L	hA
L lA	A / hA	L	L / lA	L / lA	A / H	
lA / hA	A / L	lA / L	L / A	lA / hA	A / hA	
A / L	lA / hA	lA	hA / lA	H / L	hA / L	H
L / A	L / lA	L / A	L / H	L / lA	L / lA	

Table 4.15. Fuzzy knowledge base for the biochemical danger z

x_6	x_7	x_8	x_9	x_{12}	z
GA / NN	GA / NN	GA / NN	GA / NN	GA / NN	
A	A / H	A / lA	hA	hA / H	
A	A / hA	A / lA	A / lA	A / lA	L
L / hA	hA / L	L / hA	L / lA	L / A	
hA / lA	hA / A	A / H	lA	hA	
hA / lA	A / lA	H / lA	lA	hA	lA
L	L / hA	L / hA	L / hA	L / A	
A	H / lA	A	H / hA	hA / lA	
H / lA	H	lA / H	A / lA	hA	A
hA / A	lA / H	L / lA	lA / hA	L / hA	
hA	lA / L	L / A	hA	hA	
lA	A / lA	lA / L	A / lA	lA / A	hA
L hA	L / hA	lA / L	L / lA	L / hA	
hA / H	H / A	H / hA	H / lA	A	
L / lA	lA / hA	lA / L	hA	lA / A	H
L / A	L / lA	L / A	L / lA	A / H	

Table 4.16. Fuzzy knowledge base for IHD danger d

x_1		y		z		d
GA /	NN	GA /	NN	GA /	NN	
lA		H /	hA	L		
H /	L	A /	L	H		d_1
L /	hA	H /	hA	H		
H		hA /	H	A		
hA /	A	lA /	A	H		d_2
L /	hA	A /	L	lA /	hA	
H /	hA	lA /	A	A		
hA /	A	lA /	A	lA /	hA	d_3
lA /	hA	lA /	A	lA /	hA	
H /	hA	L /	lA	A		
L /	lA	lA /	A	hA /	lA	d_4
L /	A	L /	lA	L		
A /	H	L /	lA	lA /	hA	
A /	lA	hA /	H	hA /	lA	d_5
lA /	A	L /	lA	hA /	lA	
lA		A /	L	hA		
A		hA /	H	hA /	lA	d_6
L		L /	lA	lA /	hA	

4.6.4 *Comparison of the Expert and Extracted from Histories of Diseases IF-THEN Rules*

Comparison of the expert [3] and extracted from the real histories of diseases IF-THEN rules is presented in Tables 4.17 – 4.19. As can be seen

- fuzzy terms marked by (!) fully coincide;
- instead of terms marked by (+) the adjacent terms were extracted;
- instead of terms marked by (-), the terms which are too far from the expert ones were extracted.

No coincidences of the terms are due to the parameters c- of membership

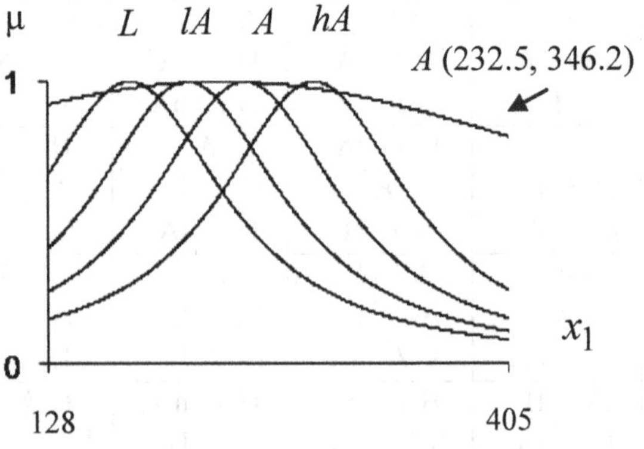

Fig. 4.11. Comparison of fuzzy terms

functions compression-extension. For example, the pair (232.5, 346.2) in the first column of Table 4.9, to which term *Average (A)* corresponds in Fig. 4.11, can be presented by a term set: *L – Low, lA – lower than Average, A – Average, hA – higher than Average.* If some expert rule contains the term from this set, then this rule is not at variance with the rule extracted from data.

Table 4.17. Comparison of the extracted and expert rules for instrumental danger y

Number of the extracted rule in Table 4.14	Expert rules						y
	x_2	x_3	x_4	x_5	x_{10}	x_{11}	
Rule 3	H (-)	H (-)	H (!)	L (-)	H (-)	H (-)	L
Rule 1	H (+)	hA (+)	H (-)	lA (!)	H (-)	H (-)	
Rule 2	hA (!)	H (-)	hA (-)	L (+)	H (+)	H (-)	
Rule 3	hA (+)	hA (+)	H (-)	lA (!)	H (!)	hA (+)	lA
Rule 2	H (-)	H (+)	hA (+)	A (+)	H (!)	H (-)	
Rule 1	hA (-)	hA (+)	H (+)	lA (-)	hA (+)	hA (+)	
Rule 3	A (!)	A (-)	A (+)	A (+)	A (!)	A (+)	A
Rule 2	hA (-)	hA (!)	A (+)	lA (!)	hA (-)	A (+)	
Rule 1	A (-)	hA (!)	hA (!)	A (+)	hA (+)	hA (!)	
Rule 1	lA (!)	A (+)	lA (!)	hA (!)	lA (-)	lA (!)	hA
Rule 2	lA (!)	lA (!)	A (!)	A (!)	L (-)	lA (+)	
Rule 3	A (+)	lA (-)	lA (+)	hA (-)	lA (!)	A (-)	
Rule 1	L (-)	L (!)	L (!)	hA (+)	L (-)	L (-)	H
Rule 3	lA (+)	L (+)	lA (+)	H (!)	L (+)	lA (!)	
Rule 2	L (!)	lA (-)	lA (!)	hA (-)	L (!)	L (!)	

Table 4.18. Comparison of the extracted and expert rules for biochemical danger z

Number of the extracted rule in Table 4.15	Expert rules					z
	x_6	x_7	x_8	x_9	x_{12}	
Rule 1	H (-)	H (!)	H (-)	H (+)	H (!)	L
Rule 2	hA (+)	H (+)	hA (-)	hA (-)	hA (-)	
Rule 3	H (+)	hA (-)	H (+)	A (+)	hA (+)	
Rule 1	hA (-)	hA (+)	A (-)	A (+)	hA (!)	lA
Rule 2	A (+)	hA (-)	A (+)	hA (-)	H (+)	
Rule 3	A (-)	H (+)	hA (!)	hA (!)	hA (+)	

Table 4.18. *(continued)*

Rule 1 Rule 3 Rule 2	A (!) hA (+) hA (-)	A (+) hA (+) A (-)	A (!) A (+) hA (+)	hA (!) A (+) hA (-)	hA (-) A (+) A (+)	A
Rule 2 Rule 1 Rule 3	lA (!) hA (!) L (-)	A (+) lA (+) A (+)	lA (+) A (!) A (-)	A (+) lA (-) lA (!)	A (!) lA (-) A (+)	hA
Rule 1 Rule 2 Rule 3	L (-) lA (!) L (-)	L (-) L (-) lA (!)	L (-) lA (+) lA (+)	L (+) L (-) L (+)	lA (+) L (-) lA (-)	H

Table 4.19. Comparison of the extracted and expert rules for IHD danger d

Number of the extracted rule in Table 4.16	Expert rules			d
	x_1	y	z	
Rule 1 Rule 2 Rule 3	L (+) lA (+) lA (-)	L (-) L (!) lA (-)	L (!) lA (-) H (!)	d_1
Rule 3 Rule 2 Rule 1	lA (-) lA (+) lA (-)	lA (+) A (!) lA (-)	lA (-) lA (-) A (!)	d_2
Rule 2 Rule 3 Rule 1	lA (+) hA (!) A (+)	A (!) hA (+) hA (+)	A (+) lA (-) A (!)	d_3
Rule 3 Rule 2 Rule 1	A (!) hA (-) hA (!)	hA (-) A (!) lA (!)	hA (-) hA (-) hA (+)	d_4
Rule 1 Rule 3 Rule 2	H (!) hA (+) hA (-)	A (+) hA (-) H (!)	A (+) H (-) hA (-)	d_5
Rule 2 Rule 3 Rule 1	H (-) H (-) H (-)	H (!) hA (-) A (-)	H (-) hA (!) hA (!)	d_6

4.6.5 Comparison of the Results of Medical Diagnosis

The separate aim of our study was to compare the results of medical diagnosis obtained by formally extracted IF-THEN rules (using a genetic and neuro algorithm) and the same rules proposed by a medical expert in the field of ischemia heart disease [3]. The fragment of data sample is presented in Table 4.20.

Comparison of diagnoses for 65 patients shows the following (See Table 4.21). As a result of the genetic algorithm operation, there are full coincidences of all types of diagnoses for 54 patients. In 9 cases we can observe decisions on a boundary between classes of diagnoses (these cases are marked by *). In 2 cases the results of computer decision were too far from the real medical doctor diagnosis (these cases are marked by **). After neural correction of diagnostic rules there are full coincidences of all types of diagnoses for 57 patients. In 8 cases we can observe decisions on a boundary between classes of diagnoses (these cases are marked by *).

These results (obtained by extracted IF-THEN rules) are close enough to similar results obtained by the fuzzy expert system described in [3]. Future quality improvement of extracted fuzzy IF-THEN rules can be reached by increasing the number of tuning parameters.

The number of unknown parameters in our computer experiment was 486, and for the optimization problem solving we spent about 3 hours (Intel Core 2 Duo P7350 2.0 GHz).

Table 4.20. Comparison of the diagnosis results

	IF-THEN rules		
	Expert	Extracted from histories of diseases	
		Genetic algorithm	Neuro-fuzzy network
Full coincidences of all types of diagnoses	56	54	57
Decisions on a boundary between classes of diagnoses (*)	8	9	8
Computer decision is too far from the real medical doctor diagnosis (**)	1	2	0

Table 4.21. Fragment of the data sample and diagnosis results

№	Patient state parameters												Diagnosis			
	x_1	x_2	x_3	x_4	x_5	x_6	x_7	x_8	x_9	x_{10}	x_{11}	x_{12}	$\hat{d}$	d_e	d_G	d_N
1	324	980	2.8	0.12	34.2	266	50.07	22.76	8.05	3.7	19.3	31	d1	d1	d1	d1
2	330	900	2.9	0.14	29.7	242	56.52	24.33	9.02	4.1	21.0	36	d1	d1	d1	d1
3	260	800	2.3	0.18	28.5	194	51.73	25.62	8.53	4.2	23.8	39	d2	d2	d2	d2
4	272	867	2.5	0.28	28.7	198	59.31	28.44	8.53	4.0	19.4	42	d2	d2	d2	d3*
5	287	491	2.2	0.24	25.3	156	52.77	21.61	8.53	3.5	20.5	48	d3	d3	d3	d3
6	175	507	2.4	0.25	22.4	172	60.70	26.14	10.40	3.9	26.1	53	d3	d3	d3	d3

Table 4.21. *(continued)*

7	247	728	2.0	0.34	26.5	144	62.06	26.14	5.55	2.3	22.9	45	d4	d4	d4	d4
8	231	768	1.5	0.36	20.0	158	62.77	23.01	6.83	2.5	23.8	52	d4	d4	d5*	d4
9	151	610	1.3	0.42	19.8	104	54.49	23.91	5.55	2.4	25.7	32	d5	d5	d5	d5
10	177	542	1.6	0.48	21.7	120	62.06	26.14	5.55	2.3	28.1	45	d5	d6*	d6*	d6*
11	128	349	1.4	0.48	13.9	92	67.03	24.46	5.20	1.9	30.2	38	d6	d6	d6	d6
12	145	304	1.2	0.56	14.4	74	64.15	25.62	7.11	2.6	25.5	38	d6	d6	d6	d6
13	327	930	2.2	0.24	35.4	347	59.31	25.62	7.56	3.3	18.9	40	d1	d2*	d1	d1
14	348	952	1.8	0.20	34.2	352	34.48	20.79	9.56	5.7	21.6	38	d1	d1	d1	d1
15	307	800	1.9	0.21	30.1	304	57.90	25.08	6.83	2.9	19.3	34	d2	d4**	d2	d1*
16	284	738	2.0	0.26	29.7	339	62.06	25.08	8.53	3.4	20.4	48	d2	d2	d2	d2
17	174	600	1.7	0.32	27.2	312	55.18	24.46	8.56	3.8	22.0	35	d3	d3	d1**	d3
18	229	515	2.1	0.30	22.4	300	61.34	22.20	6.83	2.4	23.4	49	d3	d4*	d3	d4*
19	265	421	2.0	0.26	17.7	258	60.07	22.76	4.08	1.8	23.8	58	d4	d4	d4	d4
20	330	650	1.5	0.25	20.3	244	69.49	25.08	6.83	2.5	22.0	49	d4	d4	d4	d4
21	187	475	1.4	0.34	21.4	204	60.39	23.31	5.55	2.1	22.7	48	d5	d5	d5	d5
22	224	400	1.5	0.39	20.4	215	55.18	21.05	7.11	2.7	22.5	42	d5	d5	d5	d5
23	195	100	1.2	0.48	22.6	191	60.70	21.61	7.52	2.7	25.9	32	d6	d6	d5*	d6
24	192	292	1.3	0.45	19.2	188	62.77	23.70	5.55	1.6	24.4	51	d6	d6	d6	d6
25	347	952	2.9	0.10	35.7	298	62.40	23.70	12.50	4.3	19.6	36	d1	d1	d1	d1
26	314	902	3.2	0.14	33.5	287	59.40	24.20	10.50	4.2	18.8	48	d1	d1	d1	d1
27	352	875	3.2	0.16	38.2	322	52.30	22.70	9.50	3.9	19.0	42	d1	d1	d1	d1
28	323	1040	2.7	0.20	30.4	290	59.60	25.20	8.80	3.2	18.2	40	d1	d2*	d1	d2*
29	377	988	2.9	0.09	32.5	275	60.40	24.30	10.20	3.4	17.7	41	d1	d1	d1	d1
30	309	932	3.2	0.15	31.5	312	60.80	25.40	9.40	4.4	18.5	34	d1	d1	d1	d1
31	279	1056	2.7	0.09	33.4	334	59.90	21.30	8.80	3.7	18.7	52	d1	d1	d1	d1
32	376	895	2.7	0.18	30.4	312	61.50	23.60	9.50	3.6	20.1	44	d2	d2	d2	d2
33	304	929	2.6	0.22	32.5	346	58.20	25.10	10.70	3.8	19.2	46	d2	d2	d2	d2
34	292	904	2.2	0.24	29.3	290	56.00	27.90	10.10	4.0	18.5	46	d2	d2	d1*	d2
35	276	885	2.4	0.25	27.8	226	61.40	29.40	11.20	3.6	20.8	42	d2	d2	d2	d2
36	311	930	2.7	0.19	25.6	249	62.50	23.80	9.80	2.9	21.0	31	d2	d1*	d2	d2
37	335	992	2.4	0.22	24.6	255	61.60	24.70	9.90	3.3	20.3	44	d2	d2	d2	d2
38	346	873	2.3	0.18	28.7	267	57.70	22.50	10.60	3.7	18.8	47	d2	d2	d1*	d2
39	288	804	2.4	0.27	20.9	275	60.00	22.20	11.50	3.5	19.5	48	d3	d3	d1**	d3
40	316	875	2.1	0.31	22.5	302	61.40	24.00	9.30	2.8	21.2	50	d3	d4*	d3	d4*
41	292	774	2.0	0.28	26.7	277	62.50	25.90	8.80	3.0	22.5	51	d3	d4*	d3	d3
42	315	766	2.2	0.22	21.4	265	53.70	26.20	8.70	2.7	20.5	54	d3	d4*	d3	d2*
43	300	865	2.1	0.25	21.9	303	59.40	25.80	9.30	3.5	21.4	40	d3	d3	d3	d2*
44	270	777	2.1	0.28	22.3	316	61.00	26.10	9.70	4.1	21.3	36	d3	d3	d3	d3
45	275	859	2.3	0.30	24.0	295	62.50	27.00	9.60	4.2	22.5	34	d3	d3	d3	d3
46	261	776	1.7	0.36	20.4	204	65.00	22.50	8.40	2.7	23.8	52	d4	d4	d5*	d4
47	258	785	1.5	0.36	19.8	225	62.70	23.80	7.60	2.5	24.0	41	d4	d4	d5*	d4
48	290	845	1.8	0.39	18.7	268	57.10	24.00	7.20	2.5	22.5	53	d4	d4	d4	d4
49	203	723	2.0	0.40	17.1	209	58.50	23.70	6.20	2.8	24.7	39	d4	d4	d4	d4
50	244	802	1.7	0.35	18.5	212	62.00	25.30	6.30	3.0	24.9	45	d4	d4	d5*	d4
51	233	795	1.9	0.39	17.4	251	57.90	24.90	5.20	2.4	23.5	46	d4	d4	d4	d4
52	262	805	1.8	0.38	19.2	244	57.90	24.50	7.70	2.2	22.1	54	d4	d4	d4	d4
53	245	595	1.3	0.44	16.5	204	64.20	26.40	5.60	2.1	24.7	51	d5	d5	d5	d5
54	209	772	1.5	0.45	14.7	195	60.20	27.80	5.90	2.4	25.0	40	d5	d5	d5	d5
55	198	621	1.4	0.42	12.2	225	58.80	25.20	6.10	2.6	24.5	42	d5	d5	d5	d5
56	245	523	1.5	0.39	14.1	207	57.50	23.30	6.50	2.2	26.9	44	d5	d5	d5	d5

Table 4.21. *(continued*

57	237	652	1.6	0.45	11.9	262	63.70	24.70	6.40	2.1	24.2	50	d5	d5	d5	d5
58	202	744	1.3	0.45	12.3	226	61.80	25.70	5.70	2.4	22.6	56	d5	d5	d5	d5
59	247	723	1.2	0.38	10.4	230	62.50	26.90	5.60	2.3	25.8	51	d5	d5	d6*	d5
60	192	516	1.1	0.52	9.9	200	60.10	22.70	5.50	2.0	22.9	48	d6	d6	d6	d6
61	188	446	1.2	0.48	9.5	212	59.00	23.50	5.20	2.4	26.7	39	d6	d6	d6	d6
62	212	406	0.9	0.56	8.2	225	61.70	26.00	5.30	1.9	29.4	49	d6	d6	d6	d6
63	247	527	0.7	0.51	7.4	197	62.60	27.40	5.10	2.0	28.5	45	d6	d6	d6	d6
64	206	448	0.8	0.55	7.4	188	57.40	22.10	6.30	2.1	30.1	44	d6	d6	d6	d6
65	228	512	1.0	0.52	7.8	204	53.90	25.60	5.40	2.3	29.5	42	d6	d6	d6	d6

$\hat{d}$ - diagnosis obtained by medical doctor.

d_e - computer diagnosis obtained by the expert IF-THEN rules.

d_G - computer diagnosis obtained by the genetically grown rules.

d_N - computer diagnosis specified using the neural network.

References

1. Zadeh, L.: Toward a Theory of Fuzzy Information Granulation and its Centrality in Human Reasoning and Fuzzy Logic. Fuzzy Sets and Systems 90, 111–127 (1997)
2. Zimmermann, H.-J.: Fuzzy Set Theory and Its Application. Kluwer, Dordrecht (1991)
3. Rotshtein, A.: Design and Tuning of Fuzzy Rule-Based Systems for Medical Diagnosis. In: Teodorescu, N.-H., Kandel, A. (eds.) Fuzzy and Neuro-Fuzzy Systems in Medicine, pp. 243–289. CRC Press (1998)
4. Whitley, D., Starkweather, T., Bogart, C.: Genetic algorithms and neural networks: optimizing connection and connectivity. Parallel Computing 14, 347–361 (1990)
5. Shann, I.J., Fu, H.C.: A fuzzy neural network for rule acquiring on fuzzy control systems. Fuzzy Sets and Systems 71, 345–357 (1995)
6. Tsukimoto, H.: Extracting Rules From Trained Neural Networks. IEEE Trans. on Neural Networks 11(2), 377–389 (2000)
7. Chakraborty, D., Pal, N.R.: Integrated Feature Analysis and Fuzzy Rule-Based System Identification in a Neuro-Fuzzy Paradigm. IEEE Transactions on Systems, Man and Cybernetics – Part B: Cybernetics 31(3), 391–400 (2001)
8. Nomura, H., et al.: A self-tuning method of fuzzy reasoning by genetic algorithm. In: Proc. of the Int'l Fuzzy Systems and Intelligent Control Conf., pp. 236–245 (1992)
9. Yuan, Y., Zhuang, H.: A genetic algorithm for generating fuzzy classification rules. Fuzzy Sets and Systems 25, 41–55 (1996)
10. Ishigami, H., Fukuda, T., Shibata, T., Arai, F.: Structure optimization of fuzzy neural network by genetic algorithm. Fuzzy Sets and Systems 71(3), 257–264 (1995)
11. Ishigami, H., Fukuda, J., Shibata, T.: Automatic Fuzzy Tuning and Its Applications. In: Sanchez, E., Zadeh, L., Shibata, T. (eds.) Advances in Fuzzy Systems, vol. 7, pp. 49–69 (2001)
12. Suand, H.T., Chang, M.C.: Application of neural networks incorporated with real-valued genetic algorithms in knowledge acquisition. Fuzzy Sets and Systems 112(1), 85–97 (2000)

13. Oh, S.K., Pedricz, W., Park, H.S.: Rule–based multi–FNN identification with the aid of evolutionary fuzzy granulation. Klnowledge Based Systems 17(1), 1–13 (2004)
14. Rotshtein, A., Katel'nikov, D.: Identification of Non-linear Objects by Fuzzy Knowledge Bases. Cybernetics and Systems Analysis 34(5), 676–683 (1998)
15. Rotshtein, A., Mityushkin, Y.: Extraction of Fuzzy Knowledge Bases from Experimental Data by Genetic Algorithms. Cybernetics and Systems Analysis 37(4), 501–508 (2001)
16. Rotshtein, A.P., Posner, M., Rakytyanska, H.: Fuzzy IF-THEN Rules Extraction for Medical Diagnosis Using Genetic Algorithm. WSEAS Transactions on Systems Journal 3(2), 995–1001 (2004)
17. Rotshtein, A., Mityushkin, Y.: Neurolinguistic Identification of Nonlinear Dependencies. Cybernetics and Systems Analysis 36(2), 179–187 (2000)
18. Gen, M., Cheng, R.: Genetic Algorithms and Engineering Design, p. 352c. John Wiley & Sons, New York (1997)
19. Cordon, O., Herrera, F., Hoffmann, F., Magdalena, L.: Genetic Fuzzy Systems. Evolutionary Tuning and Learning of Fuzzy Knowledge Bases. World Scientific, New York (2001)
20. Tsypkin, Y.Z.: Information Theory of Identification, p. 320. Nauka, Moscow (1984) (in Russian)
21. Hartmann, K.: Planning of Experiment, p. 552. Mir, Moscow (1977) (in Russian)

Chapter 5
Inverse Inference Based on Fuzzy Relational Equations

Application of a fuzzy methodology in system failure engineering encompasses the fault diagnosis problem [1, 2]. According to Cai [1] by fault we mean a system state which deviates from the desired system state. The task of fault diagnosing may include detecting whether a fault has occurred, diagnosing where the fault occurred, determining the type of fault, assessing the fault damage, and reconfiguring the system to accommodate the fault. Fault diagnosis partially answers one of the basic issues in system failure engineering: why does it fail.

Cause and effect analysis is an important part of fault diagnosis [3, 4]. Obviously, various symptoms of a system during its operation are essential to implement tasks of fault diagnosis. However, vague symptoms frequently emerge [5, 6].

Fuzzy abduction is a promising approach to fault detection [7, 8]. Simulation of the cause-effect connections is done by way of interpreting Zadeh's compositional rule of inference which connects input and output variables of an object (causes and effects) using a fuzzy relational matrix [9]. The problem of inputs restoration and identification is formulated in the form of inverse fuzzy logical inference and requires solution of a system of fuzzy relational equations [10]. In this case some renewal of causes takes place according to observable effects. Thus fault causes diagnosis implies (1) fuzzy relations construction and (2) fuzzy relational equations solution.

Precise relationships between causes and effects are rarely documented in the literature. To determine the fuzzy relational matrix, either linguistically documented or statistically acquired from databases assessments are used [11, 12]. However, experts often establish the cause-effect connections using the comparisons like "*Cause A has obvious advantage in comparison with cause B while effect C is occurring*". Such paired comparisons can be used for fuzzy relational matrix construction. In this case, effects can be considered as fuzzy sets given on the universal set of causes. The definition of membership degree is accomplished on the basis of expert information regarding cause paired comparisons with the help of a 9-mark Saaty's scale [13].

The insufficient use of the inverse logical inference is stipulated through the lack of effective algorithms for solving fuzzy relational equations. In this chapter, the search for a system solution amounts to the solution of an optimization

A.P. Rotshtein et al.: Fuzzy Evidence in Identif., Forecast. and Diagn., STUDFUZZ 275, pp. 149–162.
springerlink.com

problem. It has to be emphasized that the identification of a fuzzy relational equation solution is a complex optimization task with many local minima. Another difficulty comes from the exponential growth of the solution search space with the increase in the number of causes and effects considered in the diagnosis process. Generally, this problem is identified as being among *NP*-hard ones [14 – 16].

Genetic programming [17] provides a way to solve such complex optimization problems. We suggest some procedures of numerical solution of the fuzzy relational equations using genetic algorithms. The procedures envisage the optimal solution growing from a set of primary variants using genetic cross-over, mutation and selection operations. To serve the illustration of the procedures and genetic algorithm effectiveness study we present an example of technical diagnosis.

This chapter is written using original work materials [18 – 20].

5.1 Fuzzy Relational Equations in Diagnostic Problems

The diagnosis object is treated as a black box with n inputs and m outputs:

$\mathbf{X}=(x_1, x_2, ..., x_n)$ is the set of inputs;

$\mathbf{Y}=(y_1, y_2, ..., y_m)$ is the set of outputs.

Simulation of the cause-effect "input-output" connections is done by way of interpreting Zadeh's compositional rule of inference [9]

$$\mathbf{B} = \mathbf{A} \circ \mathbf{R}, \tag{5.1}$$

where:

$\mathbf{A} = (a_1, a_2, ..., a_n)$ is the fuzzy causes vector with elements $a_i \in [0, 1]$, interpreted as some significance measures of x_i causes;

$\mathbf{B} = (b_1, b_2, ..., b_m)$ is the fuzzy effects vector with elements $b_j \in [0, 1]$, interpreted as some significance measures of y_j effects;

$\mathbf{R}$ is the fuzzy relational matrix with elements r_{ij}, $i = \overline{1,n}$, $j = \overline{1,m}$, where r_{ij} is the number in the range of [0,1] characterizing the degree to which cause x_i influences upon the rise of effect y_j;

$\circ$ is the operation of *max-min* composition [9].

The diagnostic problem is set in the following way. According to the known matrix **R** and fuzzy effects vector **B**, it is necessary to find some fuzzy causes vector **A**. It is suggested that matrix **R** and fuzzy effects vector **B** are formed on the basis of expert assessments, for example, by way of Saaty's paired comparisons [13].

Finding vector **A** amounts to the solution of the fuzzy relational equations:

$$\begin{aligned} b_1 &= (a_1 \wedge r_{11}) \vee (a_2 \wedge r_{21}) ... \vee (a_n \wedge r_{n1}) \\ b_2 &= (a_1 \wedge r_{12}) \vee (a_2 \wedge r_{22}) ... \vee (a_n \wedge r_{n2}) \\ & ... \quad ... \quad ... \quad ... \\ b_m &= (a_1 \wedge r_{1m}) \vee (a_2 \wedge r_{2m}) ... \vee (a_n \wedge r_{nm}) , \end{aligned} \tag{5.2}$$

which is derived from relation (5.1). Taking into account the fact that operations $\vee$ and $\wedge$ are replaced by *max* and *min* in fuzzy set theory [9], system (5.2) is rewritten in the form

$$b_j = \max_{i=1,n}(\min(a_i, r_{ij})), \quad j = \overline{1,m} . \tag{5.3}$$

5.2 Solving Fuzzy Relational Equations as an Optimization Problem

The problem of solving fuzzy relational equations (5.3) is formulated as follows. Vector $\mathbf{A} = (a_1, a_2, ..., a_n)$ should be found which satisfies limitations of

$$a_i \in [0, 1] , \quad i = \overline{1,n} ,$$

and also provides the least distance between expert and analytical measures of effects significances, that is between the left and the right parts of each system equation (5.3):

$$F(\mathbf{A}) = \sum_{j=1}^{m} [b_j - \max_{i=\overline{1,n}}(\min(a_i, r_{ij}))]^2 = \min_{\mathbf{A}} . \tag{5.4}$$

In the general case, system (5.3) can have no unique solution but rather a set of them. Therefore, according to [8] we find the fuzzy relational equations (5.3) solution in the form of intervals:

$$a_i = [\underline{a}_i, \bar{a}_i] \subset [0, 1] , \quad i = \overline{1,n} , \tag{5.5}$$

where $\underline{a}_i$ ($\bar{a}_i$) is the lower (upper) bound of cause x_i significance measure.

Formation of intervals (5.5) is accomplished by way of solving a multiple optimization problem (5.4) and it begins with the search for its null solution. As the null solution of optimization problem (5.4) we designate $\mathbf{A}^{(0)} = (a_1^{(0)}, a_2^{(0)}, ..., a_n^{(0)})$, where $a_i^{(0)} \in [\underline{a}_i, \bar{a}_i]$, $i = \overline{1,n}$.

The upper bound ($\bar{a}_i$) is found in range $[a_i^{(0)}, 1]$, and the lower bound ($\underline{a}_i$) - in range $[0, a_i^{(0)}]$.

Let $\mathbf{A}^{(k)} = (a_1^{(k)}, a_2^{(k)}, ..., a_n^{(k)})$ be some k-th solution of optimization problem (5.4), that is $F(\mathbf{A}^{(k)}) = F(\mathbf{A}^{(0)})$, since for the all values of parameter a_i in the range $[\underline{a}_i, \overline{a}_i]$ we have the same value of criterion (5.4). While searching for upper bounds ($\overline{a}_i$) it is suggested that $a_i^{(k)} \geq a_i^{(k-1)}$, and while searching for lower bounds ($\underline{a}_i$) it is suggested that $a_i^{(k)} \leq a_i^{(k-1)}$ (Fig. 5.1).

The definition of the upper (lower) bounds follows the rule: if $\mathbf{A}^{(k)} \neq \mathbf{A}^{(k-1)}$, then $\overline{a}_i$ ($\underline{a}_i$)=$a_i^{(k)}$, $i = \overline{1,n}$. If $\mathbf{A}^{(k)} = \mathbf{A}^{(k-1)}$, then the search is stopped.

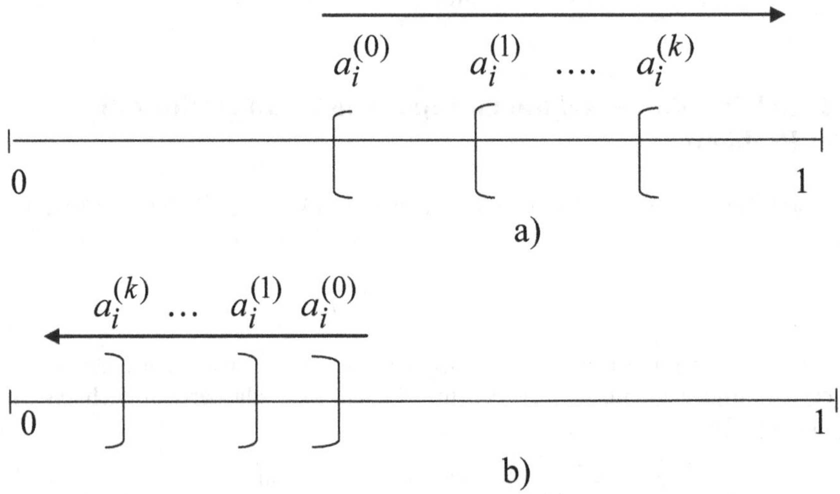

Fig. 5.1. Search for upper (a) and lower (b) bounds of the interval

5.3 Genetic Algorithm for Solving Fuzzy Relational Equations

To realize the genetic algorithm for solving the optimization problem (5.4) it is necessary to define the following main notions and operations [17]: chromosome – coded solution variant; gene – solution element; population – original set of solution variants; fitness function – criterion of variants selection; cross over – operation of providing variants-offsprings from variants-parents; mutation – random change of chromosome elements

Let $P(t)$ be chromosome-parents, and $C(t)$ – chromosome-offsprings of the t-th iteration. The general structure of the genetic algorithm will have this form:

```
begin
    t:=0;
    set the initial set P(t);
    assess P(t) using fitness function;
        while (no condition of completion) do
            generate C(t) by way of crossing over P(t);
            perform mutation C(t);
            assess C(t) using fitness function;
            select P(t+1) from P(t) and C(t);
            t:=t+1;
        end;
end.
```

We define the chromosome as the vector-line of binary solution codes a_i, $i = \overline{1,n}$. The number of bits g_i for variable a_i coding is defined according to formula

$$2^{g_i - 1} < (\overline{d}_i - \underline{d}_i) \cdot 10^q \leq 2^{g_i} - 1,$$

where $[\underline{d}_i, \overline{d}_i]$ is the range of changes of variable a_i;

q is the required precision, that is the number of digits after the decimal point in a_i solution.

For example, if there is a solution in the range [0,1] with set precision q=3, and

$$a_1 = 0.593, \quad a_2 = 0.814, \quad a_3 = 0.141, \quad a_4 = 0.970, \quad a_5 = 0.300,$$

then the following chromosome will fit this solution

$v =$					
	1001010001	1100101110	0010001101	1111001010	0100101100

The chromosomes of the initial population will be defined by

$$a_i = \text{RANDOM}([0,1]), \quad i = \overline{1,n},$$

where RANDOM([0,1]) denotes a random number within the interval [0,1].

We choose a fitness function as the negative of criterion (5.4) so that:

$$f(v) = -F(v). \tag{5.6}$$

Thus, the higher the degree of adaptability of the chromosome to perform the criteria of optimization, the greater is the fitness function.

Let p_c be some cross-over factor, that is the share of the offsprings of the each iteration performed, and let K be the population dimension. Then it is necessary to select $\frac{K \cdot p_c}{2}$ pairs of chromosome-parents at the each iteration.

Selection of chromosome-parents for the cross-over operation should not be performed randomly. The greater the fitness function of some chromosome the greater is the probability for the given chromosome to yield offsprings. The probability of selection p_k, corresponding to chromosome v_k, $k = \overline{1,K}$, is calculated according to formula [17]:

$$p_k = \frac{f(v_k) - \min_{j=1,K}[f(v_j)]}{\sum_{k=1}^{K}\left(f(v_k) - \min_{j=1,K}[f(v_j)]\right)}, \qquad \sum_{k=1}^{K} p_k = 1. \tag{5.7}$$

Using these probabilities we define chromosome-parents in the following way. Let us mark row p_k on the horizontal axis (Fig.5.2), and generate uniform random number z on interval [0,1].

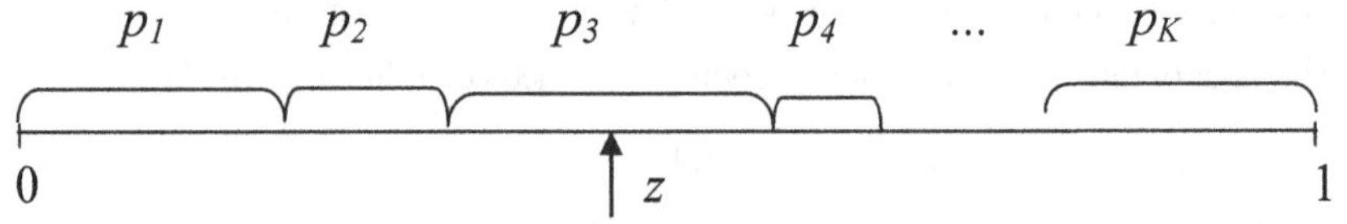

Fig. 5.2. Selection of chromosome-parent

We select chromosome v_k as the parent, this chromosome corresponding to subinterval p_k, within which number z finds itself. For example, in Fig.5.2 generated number z defines chromosome v_3 as the parent.

Selection of the second chromosome-parent is carried out in similar way.

The cross-over operation is shown in Fig. 5.3. It is carried out by way of exchanging genes inside each variable a_i, $i = \overline{1,n}$. The cross-over points shown as dotted lines are selected randomly.

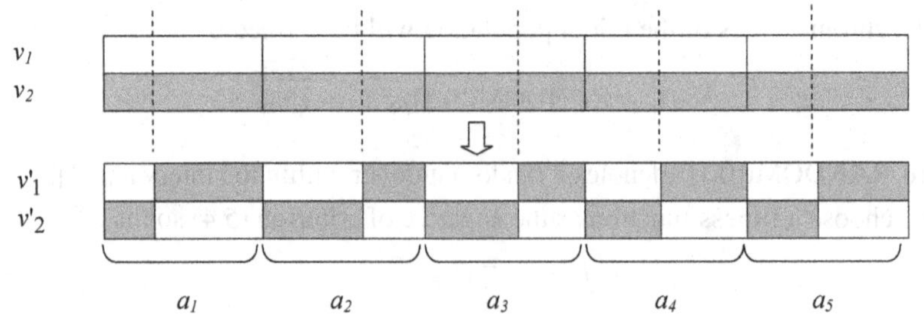

Fig. 5.3. Example of cross-over operation performance with n=5.

The mutation operation (Mu) implies random change (with some probability p_m) of chromosome elements

$$Mu(a_i) = RANDOM\ ([\underline{d}_i, \overline{d}_i]).$$

So as this solution is coded by binary line then the mutation is reduced to inversion of some separate bits.

While performing the genetic algorithm the dimension of the population stays constant at K. That is why after cross-over and mutation operations it is necessary to remove $K \cdot p_c$ chromosomes having the fitness function of the worst significance from the obtained population.

5.4 Example 4: Car Engine Diagnosis

Let us consider the algorithm performance having the recourse to the example of the car engine faults causes diagnosis.

Engine fault effects are: y_1 – engine power insufficiency; y_2 – difficulties with engine starting; y_3 – smoky exhaust; y_4 – oil pressure too low.

Fault causes to be identified: x_1 – wear out of crank gear; x_2 – valve timing gear wear out; x_3 – carburetor fault; x_4 – battery fault; x_5 – oil pump fault.

Let the expert matrix of fuzzy relations has this form:

R =

	y_1	y_2	y_3	y_4
x_1	0.8	0.4	0.8	0.3
x_2	0.7	0.3	0.6	0.1
x_3	0.9	0.7	0.3	0.1
x_4	0.1	0.9	0.1	0.1
x_5	0.5	0.6	0.4	0.9

As the result of the examination the expert defined the following measures of significance for the engine fault criteria:

$$b_1 = 0.8,\ \ b_2 = 0.6,\ \ b_3 = 0.2,\ \ b_4 = 0.1.$$

It means that engine power insufficiency and difficulties in starting the engine with the smoky exhaust and oil pressure being normal were identified.

The system of fuzzy logical equations in this case will appear in this form:

$$b_1 = (a_1 \wedge 0.8) \vee (a_2 \wedge 0.7) \vee (a_3 \wedge 0.9) \vee (a_4 \wedge 0.1) \vee (a_5 \wedge 0.5),$$

$$b_2 = (a_1 \wedge 0.4) \vee (a_2 \wedge 0.3) \vee (a_3 \wedge 0.7) \vee (a_4 \wedge 0.9) \vee (a_5 \wedge 0.6),$$

$$b_3 = (a_1 \wedge 0.8) \vee (a_2 \wedge 0.6) \vee (a_3 \wedge 0.3) \vee (a_4 \wedge 0.1) \vee (a_5 \wedge 0.4),$$

$$b_4 = (a_1 \wedge 0.3) \vee (a_2 \wedge 0.1) \vee (a_3 \wedge 0.1) \vee (a_4 \wedge 0.1) \vee (a_5 \wedge 0.9). \qquad (5.8)$$

5.4.1 Genetic Search for the Null Solution

To implement the genetic algorithm the initial population was formed as consisting of seven solutions:

v_1 =(0.343, 0.257, 0.489, 0.136, 0.967),

v_2 =(0.345, 0.415, 0.848, 0.724, 0.261),

v_3 =(0.536, 0.134, 0.677, 0.869, 0.880),

v_4 =(0.791, 0.010, 0.411, 0.245, 0.279),

v_5 =(0.665, 0.315, 0.631, 0.199, 0.456),

v_6 =(0.400, 0.652, 0.943, 0.673, 0.551),

v_7 =(0.622, 0.284, 0.992, 0.933, 0.072);

Fitness functions of these solutions found using formulae (5.4) and (5.6) made up this representation:

$f(v_1)$= - 0.770; $f(v_2)$= - 0.104; $f(v_3)$= - 0.809; $f(v_4)$= - 0.425;
$f(v_5)$= - 0.362; $f(v_6)$= - 0.383; $f(v_7)$= - 0.318.

Let the cross-over probability be p_c =0.3. So that $\frac{K \cdot p_c}{2} = \frac{7 \cdot 0.3}{2} \approx 1$, then one pair of chromosomes must be selected to realize the cross-over operation. On the basis of formula (5.7) the probability distribution for chromosome selection is given by:

p_1 =0.01558; p_2 =0.28306; p_3 =0.00000; p_4 =0.15409;
p_5 =0.17936; p_6 =0.17082; p_7 =0.19705.

Let us assume that for the chromosomes-parents two random numbers z_1 =0.18320 and z_2 =0.50780 were generated. Then according to the algorithm of chromosome-parents selection, chromosomes v_2 and v_5 must be subjected to the cross-over.

To realize cross-over operation, we used 5 points of exchange which were generated randomly in range [1,10] what corresponds to solutions a_i representation using 10 digits. These random numbers made up 4, 3, 5, 4, 1 and defined points of chromosome exchange shown in Fig. 5.4, where v_2, v_5 are chromosomes-parents, v'_2, v'_5 are chromosomes-offsprings.

The same figure depicts mutation operation which implied inversion of the 49-th gene of v'_5 chromosome-offspring. The mutation ratio (p_m) was set at the level of 0.01.

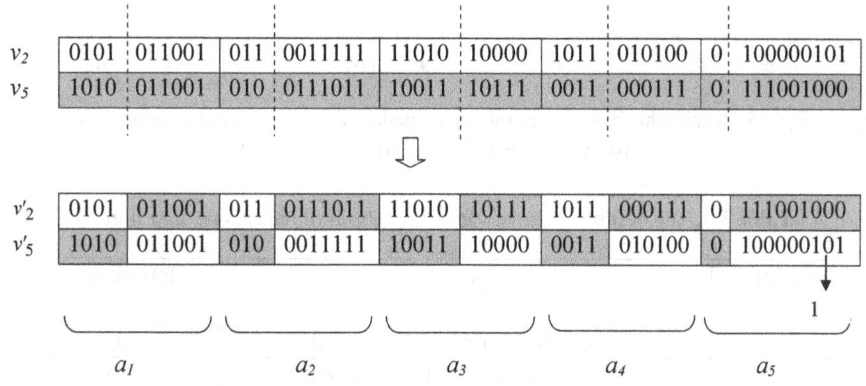

Fig. 5.4. Cross-over and mutation operations performance

Fitness function of chromosomes-offsprings

v'_2=(0.345,0.443,0.855,0.711,0.456), v'_5=(0.665,0.287,0.624,0.212,0.263)

made up this representation:

$$f(v'_2) = -0.201, \quad f(v'_5) = -0.275.$$

To maintain the same total population, let us exempt 2 chromosomes with the worst fitness functions, that is v_1 and v_3. Then, the new population will include also chromosomes: v_2, v_4, v_5, v_6, v_7, v'_2, v'_5.

This operation completes one iteration of the genetic algorithm.

Sequential application of genetic cross-over, mutation and selection operations to the initial set of variants provides for the growth of the fitness function of the solutions being obtained. The dynamics of change of the optimization criterion (*F*) relative to the number of iterations (*N*) are shown in Fig. 5.5. Table 5.1 shows the list of chromosomes which were the best in carrying out some definite iterations of the genetic algorithm.

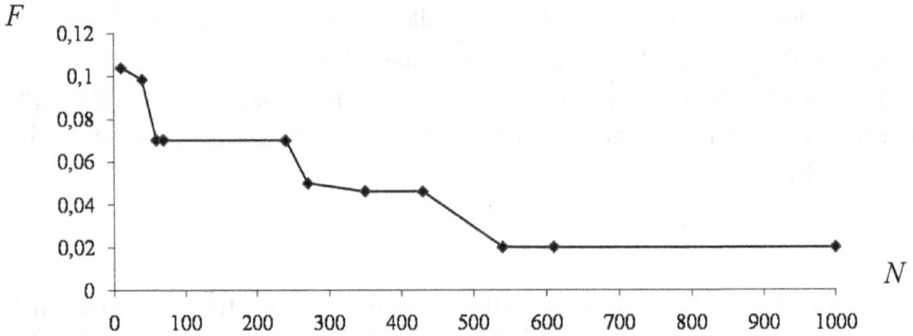

Fig. 5.5. Relationship between optimization criterion (F) and number of iterations (N) (crossover $p_c = 0.3$, mutation $p_m = 0.01$)

Table 5.1. The best solutions of various iterations of the genetic algorithm

Iteration number	Solution	Optimization criterion
1	(0.345,0.415,0.848,0.724,0.261)	0.10390
4	(0.345,0.415,0.848,0.660,0.261)	0.09853
6	(0.320,0.031,0.784,0.724,0.256)	0.07003
7	(0.320,0.031,0.785,0.724,0.256)	0.07000
24	(0.320,0.287,0.789,0.724,0.256)	0.06990
27	(0.256,0.287,0.789,0.724,0.256)	0.04983
35	(0.256,0.287,0.789,0.708,0.256)	0.04612
43	(0.256,0.279,0.797,0.708,0.256)	0.04601
54	(0.075,0.162,0.808,0.256,0.024)	0.02006
61	(0.075,0.162,0.800,0.256,0.024)	0.02000
100	(0.075,0.162,0.800,0.256,0.024)	0.02000

It is seen from Fig. 5.5 and Table 5.1 that as the number of iterations increases, the resulting optimization criterion decreases, finally converging to 0.02000 approximately.

5.4.2 Genetic Search for the Complete Solution Set

The obtained null solution

$$\mathbf{A}^{(0)} = (a_1^{(0)} = 0.075, a_2^{(0)} = 0.162, a_3^{(0)} = 0.800, a_4^{(0)} = 0.256, a_5^{(0)} = 0.024)$$

allows us to arrange for the genetic search for a_i variables significances intervals. Search dynamics of upper and lower solutions bounds is depicted in Tables 5.2 and 5.3.

Table 5.2. Genetic search for upper bounds of intervals

Iteration	$\bar{a}_1$	$\bar{a}_2$	$\bar{a}_3$	$\bar{a}_4$	$\bar{a}_5$
1	0.075	0.162	0.800	0.256	0.024
10	0.084	0.178	0.800	0.359	0.054
20	0.088	0.251	0.800	0.420	0.078
30	0.095	0.275	0.800	0.573	0.095
40	0.098	0.287	0.800	0.585	0.100
50	0.099	0.298	0.800	0.640	0.100
60	0.100	0.299	0.800	0.688	0.100
70	0.100	0.300	0.800	0.695	0.100
80	0.100	0.300	0.800	0.699	0.100
90	0.100	0.300	0.800	0.700	0.100
100	0.100	0.300	0.800	0.700	0.100

Table 5.3. Genetic search for lower bounds of intervals

Iteration	$\underline{a}_1$	$\underline{a}_2$	$\underline{a}_3$	$\underline{a}_4$	$\underline{a}_5$
1	0.075	0.162	0.800	0.256	0.024
10	0.039	0.151	0.800	0.173	0.010
20	0.022	0.085	0.800	0.159	0.004
30	0.017	0.043	0.800	0.084	0.001
40	0.011	0.028	0.800	0.047	0.000
50	0.005	0.016	0.800	0.038	0.000
60	0.001	0.008	0.800	0.020	0.000
70	0.000	0.003	0.800	0.007	0.000
80	0.000	0.000	0.800	0.002	0.000
90	0.000	0.000	0.800	0.000	0.000
100	0.000	0.000	0.800	0.000	0.000

These tables show that the fuzzy relational equations (5.8) solution can be represented in the form of intervals:

$$a_1 \in [0, 0.1];\ a_2 \in [0, 0.3];\ a_3 = 0.8;\ a_4 \in [0, 0.7];\ a_5 \in [0, 0.1].$$

5.4.3 Solution Interpretation and Model Testing

The resulting solution allows the analyst to make the following conclusions. The cause of the observed engine state should be located and identified as the carburetor fault (x_3), so that the measure of significance of this fault is maximal. In addition, the observed state can be the effect of the battery fault (x_4), since the significance measure of this cause is sufficiently high. Insufficient wear-out of the valve timing gear (x_2) can also tell on engine proper functioning, the significance measure of which is indicative of the cause. Crank gear (x_1) and oil pump (x_5) functionate properly and have practically no influence on engine fault, so that the significance measures of the given causes are small.

The genetic algorithm's accuracy was measured using the 250 test instances, obtained from the expert at car engine fault diagnosis domain. The goal was to identify one of five possible car engine fault causes $x_1 \div x_5$ by observed effects $y_1 \div y_4$. The 250 cases acted as input events to test the fuzzy model diagnosis results, and the percentage of correct predictions was recorded. The results are shown in Fig. 5.6 which shows the relationship between average accuracy of the fault causes diagnosis and number of iterations. The fault causes diagnosis obtained an accuracy rate of 95.2 % after 1000 iterations (120 seconds on Intel Core 2 Duo P7350 2.0 GHz).

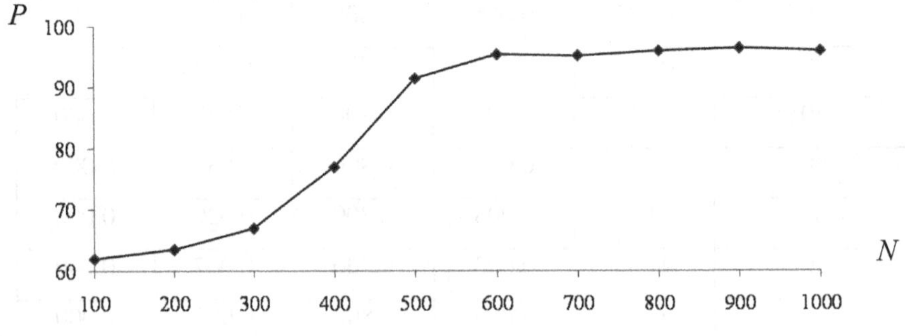

Fig. 5.6. Relationship between average accuracy (P) and number of iterations (N) (crossover $p_c = 0.3$, mutation $p_m = 0.01$)

5.4.4 Assessment of Genetic Algorithm Effectiveness

Dependence of the number of iterations, which is necessary to obtain an optimal solution, on the population volume (*K*), frequency of cross-over (p_c) and mutation (p_m) was studied in the course of computer experiment. Dependence of optimization criterion (*F*) on iteration number (*N*) under conditions of various parameters of the genetic algorithm is shown in Fig. 5.7.

It was determined that a population volume of *K*=7 is sufficient for fuzzy logical equations system (5.8) resolution. To exclude hitting the local minimum the experiment was carried out for large values of p_c and p_m. Fig. 5.7,a shows that under conditions of p_c =0.6 and p_m =0.02 about 100 iterations were required to grow an optimal solution.

To cut time losses in unpromising fields studies, some parameters of the main genetic operations were experimentally selected. Fig. 5.7,b shows that setting of cross-over frequency at the level of 0.3 allowed to cut the number of iterations on the average to 75. Reduction of the mutation number to 0.01 allowed to cut iteration number to 50 (Fig. 5.7,c).

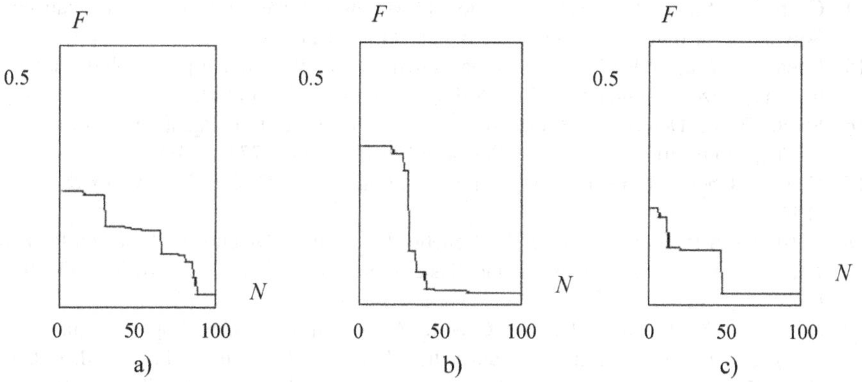

Fig. 5.7. Dependence of optimization criterion (*F*) on iteration number (*N*) under conditions of various parameters of genetic algorithm a) *K*=7, p_c = 0.6, p_m = 0.02; b) *K*=7, p_c = 0.3, p_m = 0.02; c) *K*=7, p_c = 0.3, p_m = 0.01.

References

1. Cai, K.-Y.: Introduction to Fuzzy Reliability. Kluwer Academic Publishers (1996)
2. Onisawa, T., Kacprzyk, J. (eds.): Reliability and Safety under Fuzziness, p. 390. Physica – Verlag, Germany (1995)
3. Abramovici, M., Breuer, M.A.: Fault diagnosis based on effect-cause analysis. In: Proc. 17th IEEE Design Automation Conf., pp. 69–76. ACM Press, NY (1980)

4. Isermann, R.: Supervision, fault-detection and fault-diagnosis methods — An introduction. Control Engineering Practice 5(5), 639–652 (1997)
5. Bowles, J.B., Pelaez, C.E.: Application of Fuzzy Logic to Reliability Engineering. IEEE Trans. J. 83(3), 435–449 (1995)
6. Fagarasan, I., Ploixand, S., Gentil, S.: Causal fault detection and isolation based on a set-membership approach. Automatica 40(12), 2099–2110 (2004)
7. Peng, Y., Reggia, J.A.: Abductive Inference Models for Diagnostic Problem Solving. Springer, New York (1990)
8. Terano, T., Asai, K., Sugeno, M. (eds.): Applied Fuzzy Systems. Omsya, Tokyo (1989); Mir, Moscow (1993) (in Russian)
9. Zadeh, L.: The Concept of Linguistic Variable and It's Application to Approximate Decision Making. Mir, Moscow (1976)
10. Di Nola, A., Sessa, S., Pedrycz, W., Sancez, E.: Fuzzy Relational Equations and Their Applications to Knowledge Engineering. Kluwer, Boston (1989)
11. Mordeson, J.N., Malik, D.S., Cheng, S.-C.: Fuzzy Mathematics in Medicine, p. 257. Phisica-Verlag, Heidelberg (2000)
12. Adlassnig, K.P., Kolarz, G.: CADIAG–2: computer–assisted medical diagnosis using fuzzy subsets. In: Gupta, M., Sanchez, E. (eds.) Approximate Reasoning in Decision Analysis, pp. 219–247. North Holland Publ. Comp. (1982)
13. Rotshtein, A.: Modification of Saaty Method for the Construction of Fuzzy Set Membership Functions. In: FUZZY 1997 – International Conference "Fuzzy Logic and Its Applications", Zichron, Israel, pp. 125–130 (1997)
14. Chen, L., Wang, P.P.: Fuzzy Relation Equations (I): the General and Specialized Solving Algorithms. Soft Computing 6, 428–435 (2002)
15. Chen, L., Wang, P.P.: Fuzzy Relation Equations (II): the Branch-point-solutions and the Categorized Minimal Solutions. Soft Computing 11, 33–40 (2007)
16. Shieh, B.-S.: Deriving minimal solutions for fuzzy relation equations with max-product composition. Information Sciences 178(19), 3766–3774 (2008)
17. Gen, M., Cheng, R.: Genetic Algorithms and Engineering Design. John Wiley & Sons (1997)
18. Rotshtein, A.P., Rakytyanska, H.B.: Genetic algorithm for Diagnostics based on Fuzzy Relations. Journal of Computer and Systems Sciences International 40(5), 793–798 (2001)
19. Rotshtein, A., Rakytyanska, H.: Genetic Algorithm for Fuzzy Logical Equations Solving in Diagnostic Expert Systems. In: Monostori, L., Váncza, J., Ali, M. (eds.) IEA/AIE 2001. LNCS (LNAI), vol. 2070, pp. 349–358. Springer, Heidelberg (2001)
20. Rotshtein, A., Posner, M., Rakytyanska, H.: Cause and Effect Analysis by Fuzzy Relational Equations and a Genetic Algorithm. Reliability Engineering & System Safety 91(9), 1095–1107 (2006)

Chapter 6
Inverse Inference with Fuzzy Relations Tuning

Diagnosis, i.e. determination of the identity of the observed phenomena, is the most important stage of decision making in different domains of human activity: medicine, engineering, economics, military affairs, and others. In the case of the diagnosis of problems where physical mechanisms are not well known due to high complexity and nonlinearity, a fuzzy relational model may be useful. A fuzzy relational model for simulating cause and effect connections in diagnosing problems has been introduced by Sanchez [1, 2]. A model for diagnosis can be built on the basis of Zadeh's compositional rule of inference [3], in which the fuzzy matrix of "causes-effects" relations serves as the support of the diagnostic information. In this case, the problem of diagnosis amounts to solving fuzzy relational equations.

Inverse problem resolution is of interest to both exact methods and approximate ones. The complete bibliographical notes are presented in [4]. Analytically exact methods for fuzzy relational equations on various lattices and with different kinds of composition laws for fuzzy relations are given in [4 – 8]. There exist tasks in which approximate solutions instead of exact ones are reasonable [9]. Solvability and approximate solvability conditions of fuzzy relational equations are considered in [10 – 14]. In the general case, an optimization environment is a convenient tool for decomposing fuzzy relations. Solving fuzzy relational equations by neural networks is described in [15, 16]. The use of genetic optimization for decomposition of fuzzy relations is proposed in [17].

The necessary condition of diagnostic problem solving is to ascertain the cause-effect relationship. A general methodological scheme envisages structure determination, parameter identification and model validation [18, 19]. An approach of integrated genetic and gradient-based learning in construction of fuzzy relational models is proposed in [20]. An approach of identification of fuzzy relational models by fuzzy neural networks is proposed in [21, 22].

In those cases, when domain experts are involved in developing fuzzy models, construction of the cause-effect connections can be considered as rough tuning of the fuzzy relational model [23]. The observed (output) and diagnosed (input) parameters of a system are considered as linguistic variables [3]. Fuzzy terms, e.g.

A.P. Rotshtein et al.: Fuzzy Evidence in Identif., Forecast. and Diagn., STUDFUZZ 275, pp. 163–192.
springerlink.com

"temperature rise", "pressure drop" etc., associated with causes and effects are used for these variables evaluation. The use of the expert relational matrix cannot guarantee the coincidence of theoretical results of diagnosis and real data. In other words, the "quality" of the model strongly depends on the "quality" of the expert forming the diagnostic matrix. In addition, the problem of solving fuzzy relational equations is still relevant – as of yet there does not exist a satisfactory answer for computing a complete solution set [4].

In chapter 5, a pure expert system using a genetic algorithm [24, 25] as a tool to solve the diagnosis problem was proposed. In this chapter, we propose an approach for building fuzzy systems of diagnosis, which enables solving fuzzy relational equations together with design and tuning of fuzzy relations on the basis of expert and experimental information [26, 27]. The essence of tuning consists of the selection of such membership functions of the fuzzy terms for the input and output variables (causes and effects) and such "causes-effects" fuzzy relations, which provide minimal difference between theoretical and experimental results of diagnosis.

To overcome the *NP*-hardness, chapter 5 used the ideology of genetic optimization [24, 25], which quickly established the domain of global minimum of the discrepancy between the left and right sides of the system of equations followed by a fine adjustment of the solution by search methods available. The genetic algorithm uses all the available experimental information for the optimization, i.e., operates off-line and becomes toilful and inefficient if new experimental data are obtained, i.e., in the on-line mode. The process of diagnosis should be augmented by a hybrid genetic and neuro approach to designing adaptive diagnostic systems [28]. The essence of the approach is in constructing and training a special neuro-fuzzy network isomorphic to the diagnostic equations, which allows on-line correction of decisions.

This chapter is written using original work materials [26 – 28].

6.1 Diagnostic Approximator Based on Fuzzy Relations

The diagnosis object is treated as a black box with n inputs and m outputs (Fig. 6.1). Outputs of the object are associated with the observed effects (symptoms). Inputs correspond to the causes of the observed effects (diagnoses). The problem of diagnosis consists of restoration and identification of the causes (inputs) through the observed effects (outputs). Inputs and outputs can be considered as linguistic variables given on the corresponding universal sets. Fuzzy terms are used for these linguistic variables evaluation.

We shall denote:

$\{x_1, x_2, ..., x_n\}$ is the set of input parameters, $x_i \in [\underline{x}_i, \overline{x}_i]$, $i = \overline{1,n}$;

$\{y_1, y_2, ..., y_m\}$ is the set of output parameters, $y_j \in [\underline{y}_j, \overline{y}_j]$, $j = \overline{1,m}$;

$\{c_{i1}, c_{i2}, ..., c_{ik_i}\}$ is the set of linguistic terms for parameter x_i evaluation, $i = \overline{1,n}$;

$\{e_{j1}, e_{j2}, ..., e_{jq_j}\}$ is the set of linguistic terms for parameter y_j evaluation, $j = \overline{1,m}$.

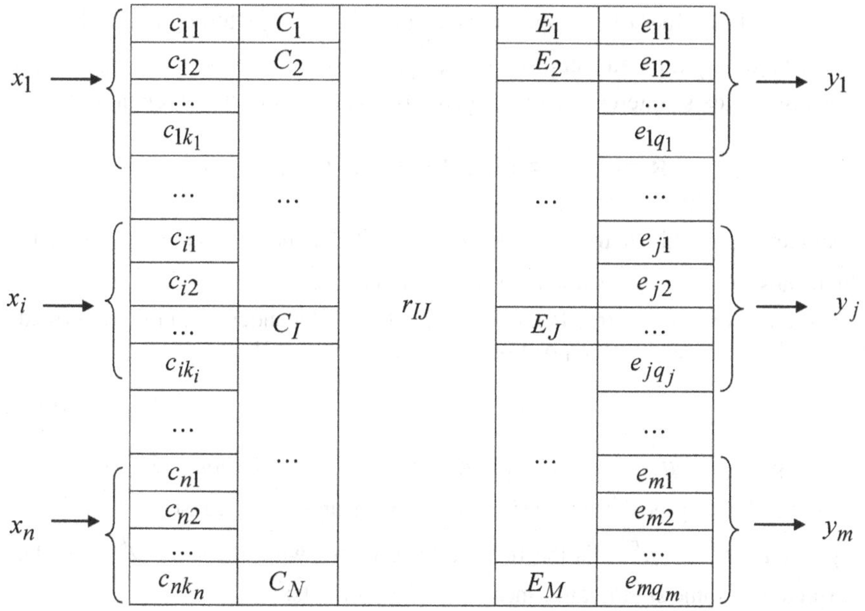

Fig. 6.1. The object of diagnosis

Each term-assessment is described with the help of a fuzzy set:

$$c_{il} = \{(x_i, \mu^{c_{il}}(x_i))\}, \quad i = \overline{1,n}, \; l = \overline{1,k_i};$$

$$e_{jp} = \{(y_j, \mu^{e_{jp}}(y_j))\}, \quad j = \overline{1,m}, \; p = \overline{1,q_j}.$$

where $\mu^{c_{il}}(x_i)$ is a membership function of variable x_i to the term-assessment c_{il}, $i = \overline{1,n}$, $l = \overline{1,k_i}$;

$\mu^{e_{jp}}(y_j)$ is a membership function of variable y_j to the term-assessment e_{jp}, $j = \overline{1,m}$, $p = \overline{1,q_j}$.

We shall redenote the set of input and output terms-assessments in the following way:

$\{C_1, C_2, ..., C_N\} = \{c_{11}, c_{12}, ..., c_{1k_1}, ..., c_{n1}, c_{n2}, ..., c_{nk_n}\}$ is the set of terms for input parameters evaluation, where $N = k_1 + k_2 + ... + k_n$;

$\{E_1, E_2, ..., E_M\} = \{e_{11}, e_{12}, ..., e_{1q_1}, ..., e_{m1}, e_{m2}, ..., e_{mq_m}\}$ is the set of terms for output parameters evaluation, where $M = q_1 + q_2 + ... + q_m$.

Set $\{C_I, I = \overline{1,N}\}$ is called fuzzy causes (diagnoses), and set $\{E_J, J = \overline{1,M}\}$ is called fuzzy effects (symptoms).

The diagnostic problem is set in the following way: it is necessary to restore and identify the values of the input parameters $(x_1^*, x_2^*, ..., x_n^*)$ through the values of the observed output parameters $(y_1^*, y_2^*, ..., y_m^*)$.

"Causes-effects" interconnection is given by the matrix of fuzzy relations

$$\mathbf{R} \subseteq C_I \times E_J = [\, r_{IJ},\; I = \overline{1,N},\; J = \overline{1,M}\,].$$

An element of this matrix is a number $r_{IJ} \in [0, 1]$, characterizing the degree to which cause C_I influences upon the rise of effect E_J.

In the presence of matrix **R** the "causes-effects" dependency can be described with the help of Zadeh's compositional rule of inference [3]

$$\boldsymbol{\mu}^E = \boldsymbol{\mu}^C \circ \mathbf{R}, \tag{6.1}$$

where $\boldsymbol{\mu}^C = (\mu^{C_1}, \mu^{C_2}, ..., \mu^{C_N})$ is the fuzzy causes vector with elements $\mu^{C_I} \in [0, 1]$, interpreted as some significance measures of C_I causes;

$\boldsymbol{\mu}^E = (\mu^{E_1}, \mu^{E_2}, ..., \mu^{E_M})$ is the fuzzy effects vector with elements $\mu^{E_J} \in [0, 1]$, interpreted as some significance measures of E_J effects;

$\circ$ is the operation of *max-min* composition [3].

Finding vector $\boldsymbol{\mu}^C$ amounts to the solution of the fuzzy relational equations:

$$\begin{aligned}
\mu^{E_1} &= (\mu^{C_1} \wedge r_{11}) \vee (\mu^{C_2} \wedge r_{21}) ... \vee (\mu^{C_N} \wedge r_{N1}) \\
\mu^{E_2} &= (\mu^{C_1} \wedge r_{12}) \vee (\mu^{C_2} \wedge r_{22}) ... \vee (\mu^{C_N} \wedge r_{N2}) \\
& \dots \qquad \dots \qquad \dots \qquad \dots \\
\mu^{E_M} &= (\mu^{C_1} \wedge r_{1M}) \vee (\mu^{C_2} \wedge r_{2M}) ... \vee (\mu^{C_N} \wedge r_{NM}),
\end{aligned} \tag{6.2}$$

which is derived from relation (6.1). Taking into account the fact that operations $\vee$ and $\wedge$ are replaced by *max* and *min* in fuzzy set theory [3], system (6.2) is rewritten in the form:

$$\mu^{E_J} = \max_{I=\overline{1,N}}(\min(\mu^{C_I}, r_{IJ})),\; J = \overline{1,M}. \tag{6.3}$$

In order to translate the specific values of the input and output variables into the measures of the causes and effects significances it is necessary to define a membership function of fuzzy terms C_I and E_J, $I = \overline{1,N}$, $J = \overline{1,M}$. We use a bell-shaped membership function model of variable u to arbitrary term T in the form:

$$\mu^T(u) = \frac{1}{1 + \left(\dfrac{u - \beta}{\sigma}\right)^2}, \tag{6.4}$$

where β is a coordinate of function maximum, $\mu^T(\beta)=1$; σ is a parameter of concentration-extension (Fig. 6.2).

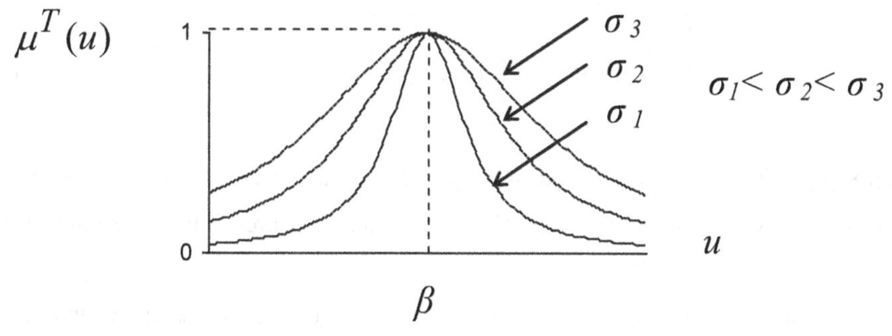

Fig. 6.2. Model of the bell-shaped membership function

This function was determined in [23] and was used for nonlinear dependencies identification by fuzzy IF-THEN rules [29, 30].

Correlations (6.3) and (6.4) define the generalized fuzzy model of diagnosis as follows:

$$\boldsymbol{\mu}^E(\mathbf{Y}, \mathbf{B}_E, \boldsymbol{\Omega}_E) = F_R(\mathbf{X}, \mathbf{R}, \mathbf{B}_C, \boldsymbol{\Omega}_C), \tag{6.5}$$

where $\mathbf{B}_C = (\beta^{C_1}, \beta^{C_2}, ..., \beta^{C_N})$ and $\boldsymbol{\Omega}_C = (\sigma^{C_1}, \sigma^{C_2}, ..., \sigma^{C_N})$ are the vectors of β- and σ- parameters for fuzzy causes C_1, C_2,..., C_N membership functions;

$\mathbf{B}_E = (\beta^{E_1}, \beta^{E_2}, ..., \beta^{E_M})$ and $\boldsymbol{\Omega}_E = (\sigma^{E_1}, \sigma^{E_2}, ..., \sigma^{E_M})$ are the vectors of β- and σ- parameters for fuzzy effects E_1, E_2,..., E_M membership functions;

F_R is the operator of inputs-outputs connection, corresponding to formulae (6.3), (6.4).

6.2 Optimization Problem for Fuzzy Relations Based Inverse Inference

Following the approach proposed in [24, 25], the problem of solving fuzzy relational equations (6.3) is formulated as follows. Fuzzy causes vector $\boldsymbol{\mu}^C = (\mu^{C_1}, \mu^{C_2}, ..., \mu^{C_N})$ should be found which satisfies the constraints $\mu^{C_I} \in [0, 1]$, $I = \overline{1,N}$, and also provides the least distance between observed and model measures of effects significances, that is between the left and the right parts of each system equation (6.3):

$$\sum_{J=1}^{M} [\mu^{E_J} - \max_{I=\overline{1,N}}(\min(\mu^{C_I}, r_{IJ}))]^2 = \min_{\boldsymbol{\mu}^C}. \tag{6.6}$$

Following [4], in the general case, system (6.3) has a solution set $S(\mathbf{R},\boldsymbol{\mu}^E)$, which is completely characterized by the unique greatest solution $\overline{\boldsymbol{\mu}}^C$ and the set of lower solutions $S^*(\mathbf{R},\boldsymbol{\mu}^E)=\{\underline{\boldsymbol{\mu}}_l^C,\ l=\overline{1,T}\}$:

$$S(\mathbf{R},\boldsymbol{\mu}^E) = \bigcup_{\underline{\boldsymbol{\mu}}_l^C \in S^*} \left[\underline{\boldsymbol{\mu}}_l^C, \overline{\boldsymbol{\mu}}^C\right]. \tag{6.7}$$

Here $\overline{\boldsymbol{\mu}}^C=(\overline{\mu}^{C_1},\overline{\mu}^{C_2},...,\overline{\mu}^{C_N})$ and $\underline{\boldsymbol{\mu}}_l^C=(\underline{\mu}_l^{C_1},\underline{\mu}_l^{C_2},...,\underline{\mu}_l^{C_N})$ are the vectors of the upper and lower bounds of causes C_I significance measures, where the union is taken over all $\underline{\boldsymbol{\mu}}_l^C \in S^*(\mathbf{R},\boldsymbol{\mu}^E)$.

Following [24, 25], formation of intervals (6.7) is accomplished by way of solving a multiple optimization problem (6.6) and it begins with the search for its null solution. As the null solution of optimization problem (6.6) we designate $\boldsymbol{\mu}_0^C=(\mu_0^{C_1},\mu_0^{C_2},...,\mu_0^{C_N})$, where $\mu_0^{C_I}\le\overline{\mu}^{C_I}$, $I=\overline{1,N}$. The upper bound ($\overline{\mu}^{C_I}$) is found in the range $\left[\mu_0^{C_I},1\right]$. The lower bound ($\underline{\mu}_l^{C_I}$) for $l=1$ is found in the range $\left[0,\mu_0^{C_I}\right]$, and for $l>1$ – in the range $\left[0,\overline{\mu}^{C_I}\right]$, where the minimal solutions $\underline{\mu}_k^C$, $k<l$, are excluded from the search space.

Let $\boldsymbol{\mu}^C(t)=(\mu^{C_1}(t),\mu^{C_2}(t),...,\mu^{C_N}(t))$ be some t-th solution of optimization problem (6.6), that is $F(\boldsymbol{\mu}^C(t))=F(\boldsymbol{\mu}_0^C)$, since for all $\boldsymbol{\mu}^C\in S(\mathbf{R},\boldsymbol{\mu}^E)$ we have the same value of criterion (6.6). While searching for upper bounds ($\overline{\mu}^{C_I}$) it is suggested that $\mu^{C_I}(t)\ge\mu^{C_I}(t-1)$, and while searching for lower bounds ($\underline{\mu}_l^{C_I}$) it is suggested that $\mu^{C_I}(t)\le\mu^{C_I}(t-1)$ (Fig. 6.3).

The definition of the upper (lower) bounds follows the rule: if $\boldsymbol{\mu}^C(t)\ne\boldsymbol{\mu}^C(t-1)$, then $\overline{\mu}^{C_I}$ ($\underline{\mu}_l^{C_I}$)$=\mu^{C_I}(t)$, $I=\overline{1,N}$. If $\boldsymbol{\mu}^C(t)=\boldsymbol{\mu}^C(t-1)$, then the search for the interval solution $\left[\underline{\boldsymbol{\mu}}_l^C,\overline{\boldsymbol{\mu}}^C\right]$ is stopped. Formation of intervals (6.7) will go on until the condition $\underline{\boldsymbol{\mu}}_l^C\ne\underline{\boldsymbol{\mu}}_k^C$, $k<l$, has been satisfied.

The hybrid genetic and neuro approach is proposed for solving optimization problem (6.6).

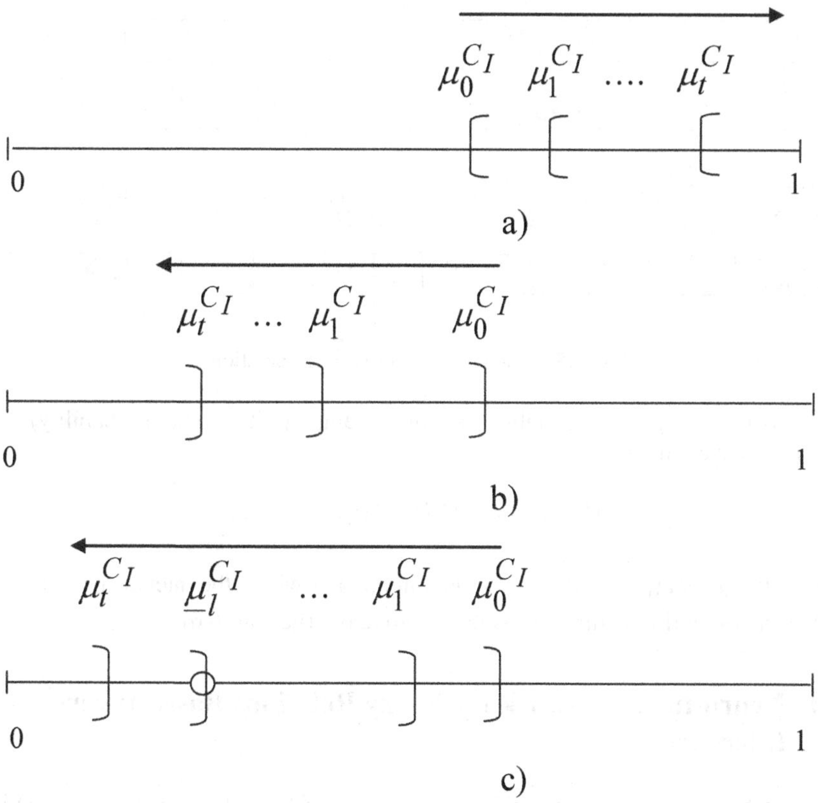

Fig. 6.3. Search for the upper (a) and lower bounds of the intervals for $l = 1$ (b) and $l > 1$ (c)

6.3 Genetic Algorithm for Fuzzy Relations Based Inverse Inference

The chromosome needed in the genetic algorithm for solving the optimization problem (6.6) is defined as the vector-line of binary codes of the lower and upper bounds of the solutions μ^{C_l} , $l = \overline{1,N}$ (Fig. 6.4) [31].

$\underline{\mu}^{C_1}$	$\underline{\mu}^{C_2}$	...	$\underline{\mu}^{C_N}$	$\overline{\mu}^{C_1}$	$\overline{\mu}^{C_2}$	...	$\overline{\mu}^{C_N}$

Fig. 6.4. Structure of the chromosome

The crossover operation is defined in Fig. 6.5, and is carried out by way of exchanging genes inside each solution μ^{C_l} . The points of cross-over shown in dotted lines are selected randomly. Upper symbols (1 and 2) in the vectors of parameters correspond to the first and second chromosomes-parents.

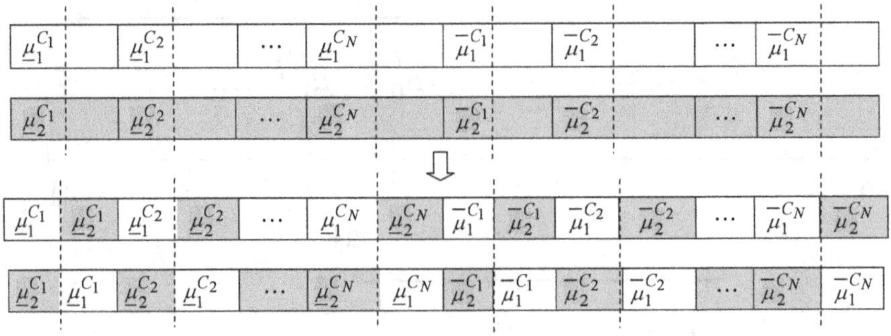

Fig. 6.5. Structure of the crossover operation

A mutation operation implies random change (with some probability) of chromosome elements

$$Mu\,(\mu^{C_I}) = RANDOM([\underline{\mu}^{C_I}, \overline{\mu}^{C_I}]),$$

where $RANDOM([\underline{x}, \overline{x}])$ denotes a random number within the interval $[\underline{x}, \overline{x}]$.

We choose a fitness function as the negative of criterion (6.6).

6.4 Neuro-fuzzy Network for Fuzzy Relations Based Inverse Inference

A neuro-fuzzy network isomorphic to the system of fuzzy logic equations (6.3) is presented in Fig. 6.6. Table 3.1 shows elements of the neuro-fuzzy network [28].

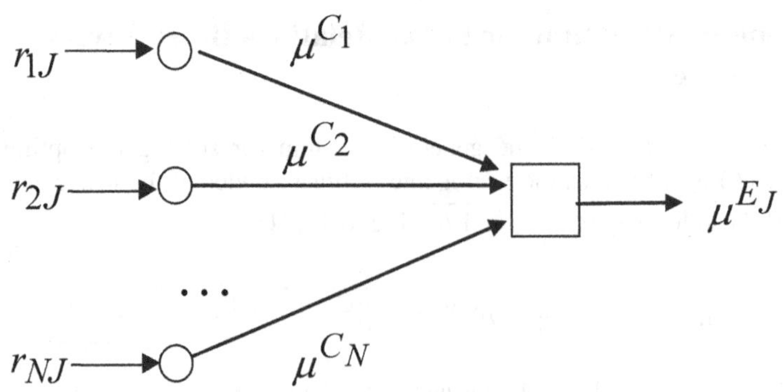

Fig. 6.6. Neuro-fuzzy model of diagnostic equations

The network is designed so that the adjusted weights of arcs are the unknown significance measures of $C_1, C_2, ..., C_N$ causes.

Network inputs are elements of the matrix of fuzzy relations. As follows from the system of fuzzy logic equations (6.3), the fuzzy relation r_{IJ} is the significance measure of the effect μ^{E_J} provided that the significance measure μ^{C_I} is equal to unity, and the significance measures of other causes are equal to zero, i.e. $r_{IJ} = \mu^{E_J}$ ($\mu^{C_I} =1$, $\mu^{C_K} =0$), $K = \overline{1,N}$, $K \neq I$. At the network outputs, actual significance measures of the effects $\max\limits_{I=1,N}[\min(\mu^{C_I}, r_{IJ})]$ obtained with allowance for the actual weights of arcs μ^{C_I} are united.

Thus, the problem of solving the system of fuzzy logic equations (6.3) is reduced to the problem of training of a neuro fuzzy network (see Fig. 6.6) with the use of points

$$(r_{1J}, r_{2J}, ..., r_{NJ}, \mu^{E_J}), \quad J = \overline{1,M}.$$

To train the parameters of the neuro-fuzzy network, the recurrent relations:

$$\mu^{C_I}(t+1) = \mu^{C_I}(t) - \eta \frac{\partial \varepsilon_t}{\partial \mu^{C_I}(t)}, \tag{6.8}$$

that minimize the criterion

$$\varepsilon_t = \frac{1}{2}(\hat{\boldsymbol{\mu}}^E(t) - \boldsymbol{\mu}^E(t))^2, \tag{6.9}$$

applied in the neural network theory, where

$\hat{\boldsymbol{\mu}}^E(t)$ and $\boldsymbol{\mu}^E(t)$ are the experimental and the model fuzzy effects vectors at the t-th step of training;

$\mu^{C_I}(t)$ are the significance measures of causes C_I at the t-th step of training;

η is a parameter of training, which can be selected according to the results from [32].

The partial derivatives appearing in recurrent relations (6.8) characterize the sensitivity of the error (ε_t) to variations in parameters of the neuro-fuzzy network and can be calculated as follows:

$$\frac{\partial \varepsilon_t}{\partial \mu^{C_I}} = \sum_{J=1}^{M} \left[\frac{\partial \varepsilon_t}{\partial \mu^{E_J}} \cdot \frac{\partial \mu^{E_J}}{\partial \mu^{C_I}} \right].$$

Since determining the element "fuzzy output" from Table 3.1 involves the *min* and *max* fuzzy-logic operations, the relations for training are obtained using finite differences.

6.5 Expert Method of Fuzzy Relations Construction

To obtain matrix **R** between causes $C_1, C_2, ..., C_N$ and effects $E_1, E_2, ..., E_M$, included in correlation (6.1), we shall use the method of membership functions construction proposed in [33] on the basis of the 9-mark scale of Saaty's paired comparisons [34].

We consider an effect E_J as a fuzzy set, which is given on the universal set of causes as follows:

$$E_J = \left\{ \frac{r_{1J}}{C_1}, \frac{r_{2J}}{C_2}, \ldots, \frac{r_{NJ}}{C_N} \right\}, \quad J = \overline{1,M}, \qquad (6.10)$$

where r_{1J}, r_{2J},..., r_{NJ} represent the degrees of membership of causes $C_1, C_2, \ldots, C_N$ to fuzzy set E_J, and correspond to the J-th column of the fuzzy relational matrix.

Following [33], to obtain membership degrees r_{IJ}, included in (6.10), it is necessary to form the matrix of paired comparisons for each effect E_J, which reflects the influence of causes $C_1, C_2, \ldots, C_N$ upon the rise of effect E_J, $J = \overline{1,M}$.

For an effect E_J the matrix of paired comparisons looks as follows:

$$\mathbf{A}_J = \begin{array}{c} \\ C_1 \\ C_2 \\ \vdots \\ C_N \end{array} \begin{array}{c} \begin{array}{cccc} C_1 & C_2 & \cdots & C_N \end{array} \\ \begin{bmatrix} a_{11}^J & a_{12}^J & \ldots & a_{1N}^J \\ a_{21}^J & a_{22}^J & \ldots & a_{2N}^J \\ \ldots & \ldots & \ldots & \ldots \\ a_{N1}^J & a_{N2}^J & \ldots & a_{NN}^J \end{bmatrix} \end{array}, \quad J = \overline{1,M}, \qquad (6.11)$$

where the element a_{IK}^J is evaluated by an expert according to the 9-mark Saaty's scale:

1 — if cause C_K has *no advantage* over cause C_I;

3 — if C_K has *a weak advantage* over C_I;

5 — if C_K has *an essential advantage* over C_I;

7 — if C_K has *an obvious advantage* over C_I;

9— if C_K has *an absolute advantage* over C_I.

Values of 2, 4, 6, 8 correspond to *intermediate* comparative assessments

In accordance with [33], we assume that matrix (6.11) has the following properties:

- elements placed symmetrically relative to the main diagonal are connected by correlation $a_{IK}^J = 1/a_{KI}^J$;
- transitivity, i. e., $a_{IL}^J a_{LK}^J = a_{IK}^J$;
- diagonality, i.e., $a_{II}^J = 1$, $I = \overline{1,N}$, as the consequence from symmetry and transitivity.

These properties allow us to define all elements of matrix (6.11) by using elements of only a single row. If the L-th row is known, i. e. the elements a_{LK}^J, $K=\overline{1,N}$, then an arbitrary element a_{IK}^J is defined as follows:

$$a_{IK}^J = \frac{a_{LK}^J}{a_{LI}^J}, \qquad I,K,L=\overline{1,N}, \quad J=\overline{1,M}.$$

After defining matrix (6.11), the degrees of membership needed for constructing fuzzy set (6.10) are calculated by formula [33]:

$$r_{IJ} = \frac{1}{a_{I1}^J + a_{I2}^J + \ldots + a_{IN}^J}, \quad I=\overline{1,N}, \; J=\overline{1,M}. \tag{6.12}$$

Obtained membership degrees (6.12) are to be normalized by way of dividing into the highest degree of membership.

6.6 Problem of Fuzzy Relations Tuning

It is assumed that the training data which is given in the form of L pairs of experimental data is known:

$$\langle \hat{\mathbf{X}}_p, \hat{\mathbf{Y}}_p \rangle, \; p=\overline{1,L},$$

where $\hat{\mathbf{X}}_p = (\hat{x}_1^p, \hat{x}_2^p, ..., \hat{x}_n^p)$ and $\hat{\mathbf{Y}}_p = (\hat{y}_1^p, \hat{y}_2^p, ..., \hat{y}_m^p)$ are the vectors of the values of the input and output variables in the experiment number p.

Let $\mathbf{\Lambda} = (\lambda_1, \lambda_2, .., \lambda_M)$ be the vector of concentration parameters for fuzzy sets of effects (6.10), such as:

$$\mathbf{R} = \begin{bmatrix} (r_{11})^{\lambda_1} & (r_{12})^{\lambda_2} & \ldots & (r_{1M})^{\lambda_M} \\ (r_{21})^{\lambda_1} & (r_{22})^{\lambda_2} & \ldots & (r_{2M})^{\lambda_M} \\ \ldots & \ldots & \ldots & \ldots \\ (r_{N1})^{\lambda_1} & (r_{N2})^{\lambda_2} & \ldots & (r_{NM})^{\lambda_M} \end{bmatrix}.$$

The essence of tuning of the fuzzy model (6.5) consists of finding such null solutions $\boldsymbol{\mu}_0^C(\hat{x}_1^p, \hat{x}_2^p, ..., \hat{x}_n^p)$ of the inverse problem, which minimize criterion (6.6) for all the points of the training data:

$$\sum_{p=1}^{L} [F_R(\boldsymbol{\mu}_0^C(\hat{x}_1^p, \hat{x}_2^p, ..., \hat{x}_n^p)) - \hat{\boldsymbol{\mu}}^E(\hat{y}_1^p, \hat{y}_2^p, ..., \hat{y}_m^p)]^2 = min.$$

In other words, the essence of tuning of the fuzzy model (6.5) consists of finding such a vector of concentration parameters $\mathbf{\Lambda}$ and such vectors of membership functions parameters $\mathbf{B}_C$, $\mathbf{\Omega}_C$, $\mathbf{B}_E$, $\mathbf{\Omega}_E$, which provide the least distance between model and experimental fuzzy effects vectors:

$$\sum_{p=1}^{L}[F_R(\hat{\mathbf{X}}_p,\mathbf{\Lambda},\mathbf{B}_C,\mathbf{\Omega}_C)-\hat{\boldsymbol{\mu}}^E(\hat{\mathbf{Y}}_p,\mathbf{B}_E,\mathbf{\Omega}_E)]^2=\min_{\mathbf{\Lambda},\mathbf{B}_C,\mathbf{\Omega}_C,\mathbf{B}_E,\mathbf{\Omega}_E}. \qquad (6.13)$$

6.7 Genetic Algorithm of Fuzzy Relations Tuning

The chromosome needed in the genetic algorithm for solving the optimization problem (6.13) is defined as the vector-line of binary codes of parameters $\mathbf{\Lambda}$, $\mathbf{B}_C$, $\mathbf{\Omega}_C$, $\mathbf{B}_E$, $\mathbf{\Omega}_E$ (Fig. 6.7) [31].

$\mathbf{\Lambda}$	$\mathbf{B}_C$	$\mathbf{\Omega}_C$	$\mathbf{B}_E$	$\mathbf{\Omega}_E$

Fig. 6.7. Structure of the chromosome

The crossover operation is defined in Fig. 6.8, and is carried out by way of exchanging genes inside the vector of concentration parameters ($\mathbf{\Lambda}$) and each of the vectors of membership functions parameters $\mathbf{B}_C$, $\mathbf{\Omega}_C$, $\mathbf{B}_E$, $\mathbf{\Omega}_E$. The points of cross-over shown in dotted lines are selected randomly. Upper symbols (1 and 2) in the vectors of parameters correspond to the first and second chromosomes-parents.

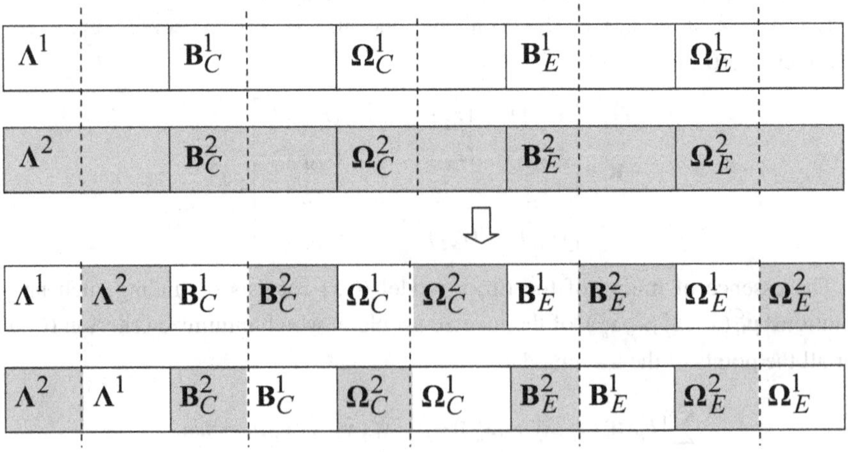

Fig. 6.8. Structure of the crossover operation

A mutation operation implies random change (with some probability) of chromosome elements:

$$Mu\left(\beta^{C_I}\right) = RANDOM\left(\left[\underline{\beta}^{C_I}, \overline{\beta}^{C_I}\right]\right);\; Mu\left(\sigma^{C_I}\right) = RANDOM\left(\left[\underline{\sigma}^{C_I}, \overline{\sigma}^{C_I}\right]\right);$$

$$Mu\left(\beta^{E_J}\right) = RANDOM\left(\left[\underline{\beta}^{E_J}, \overline{\beta}^{E_J}\right]\right);\; Mu\left(\sigma^{E_J}\right) = RANDOM\left(\left[\underline{\sigma}^{E_J}, \overline{\sigma}^{E_J}\right]\right);$$

$$Mu\,(\lambda_J) = RANDOM\,([\underline{\lambda}_J, \overline{\lambda}_J]),$$

where $RANDOM([\underline{x}, \overline{x}])$ denotes a random number within the interval $[\underline{x}, \overline{x}]$.

We choose criterion (6.13) with the negative sign as the fitness function; that is, the higher the degree of adaptability of the chromosome to perform the criterion of optimization the greater is the fitness function.

6.8 Adaptive Tuning of Fuzzy Relations

The neuro-fuzzy model of the object of diagnostics (6.5) is represented in Fig. 6.9, and the nodes are in Table. 3.1. The neuro-fuzzy model is obtained by embedding the matrix of fuzzy relations into the neural network so that the weights of arcs subject to tuning are fuzzy relations and membership functions for causes and effects fuzzy terms [28, 30].

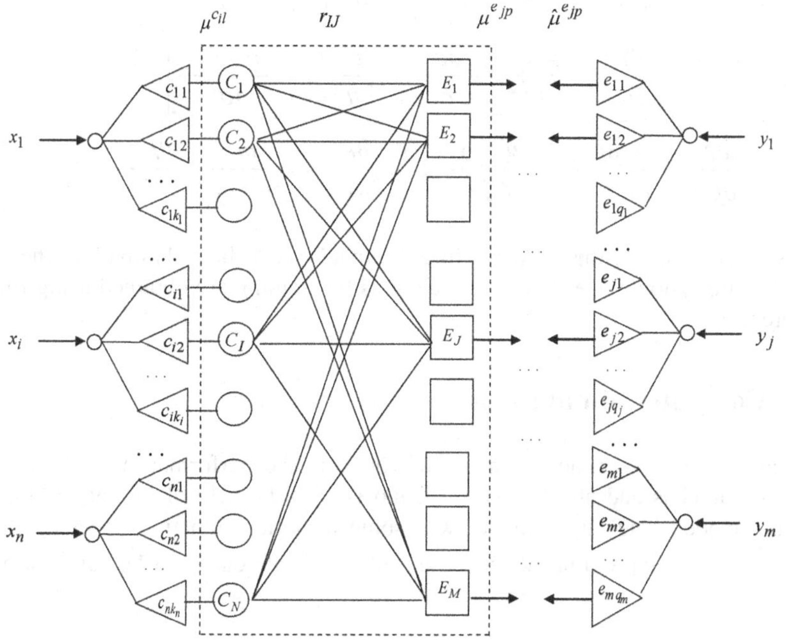

Fig. 6.9. Neuro-fuzzy model of the object of diagnostics

To train the parameters of the neuro-fuzzy network, the recurrent relations:

$$r_{IJ}(t+1)=r_{IJ}(t)-\eta\frac{\partial\varepsilon_t}{\partial r_{IJ}(t)};$$

$$\beta^{c_{il}}(t+1)=\beta^{c_{il}}(t)-\eta\frac{\partial\varepsilon_t}{\partial\beta^{c_{il}}(t)};\quad \sigma^{c_{il}}(t+1)=\sigma^{c_{il}}(t)-\eta\frac{\partial\varepsilon_t}{\partial\sigma^{c_{il}}(t)};$$

$$\beta^{e_{jp}}(t+1)=\beta^{e_{jp}}(t)-\eta\frac{\partial\varepsilon_t}{\partial\beta^{e_{jp}}(t)};\quad \sigma^{e_{jp}}(t+1)=\sigma^{e_{jp}}(t)-\eta\frac{\partial\varepsilon_t}{\partial\sigma^{e_{jp}}(t)},\qquad (6.14)$$

minimizing criterion (6.9) are used, where

$r_{IJ}(t)$ are fuzzy relations at the t-th step of training;

$\beta^{c_{il}}(t)$, $\sigma^{c_{il}}(t)$, $\beta^{e_{jp}}(t)$, $\sigma^{e_{jp}}(t)$ are the parameters of the membership functions for causes and effects fuzzy terms at the t-th step of training.

The partial derivatives appearing in recurrent relations (6.14) characterize the sensitivity of the error (ε_t) to variations in parameters of the neuro-fuzzy network and can be calculated as follows:

$$\frac{\partial\varepsilon_t}{\partial r_{IJ}}=\frac{\partial\varepsilon_t}{\partial\mu^{E_J}(X)}\cdot\frac{\partial\mu^{E_J}(X)}{\partial r_{IJ}};$$

$$\frac{\partial\varepsilon_t}{\partial\beta^{c_{il}}}=\sum_{j=1}^{m}\sum_{p=1}^{q_j}\left[\frac{\partial\varepsilon_t}{\partial\mu^{e_{jp}}(x_i)}\cdot\frac{\partial\mu^{e_{jp}}(x_i)}{\partial\mu^{c_{il}}(x_i)}\cdot\frac{\partial\mu^{c_{il}}(x_i)}{\partial\beta^{c_{il}}}\right];$$

$$\frac{\partial\varepsilon_t}{\partial\sigma^{c_{il}}}=\sum_{j=1}^{m}\sum_{p=1}^{q_j}\left[\frac{\partial\varepsilon_t}{\partial\mu^{e_{jp}}(x_i)}\cdot\frac{\partial\mu^{e_{jp}}(x_i)}{\partial\mu^{c_{il}}(x_i)}\cdot\frac{\partial\mu^{c_{il}}(x_i)}{\partial\sigma^{c_{il}}}\right];$$

$$\frac{\partial\varepsilon_t}{\partial\beta^{e_{jp}}}=\frac{\partial\varepsilon_t}{\partial\mu^{e_{jp}}(y_j)}\cdot\frac{\partial\mu^{e_{jp}}(y_j)}{\partial\beta^{e_{jp}}};\quad \frac{\partial\varepsilon_t}{\partial\sigma^{e_{jp}}}=\frac{\partial\varepsilon_t}{\partial\mu^{e_{jp}}(y_j)}\cdot\frac{\partial\mu^{e_{jp}}(y_j)}{\partial\sigma^{e_{jp}}}.$$

Since determining the element "fuzzy output" (see Table 3.1) involves the *min* and *max* fuzzy-logic operations, the relations for training are obtained using finite differences.

6.9 Computer Simulations

The aim of the experiment consists of checking the performance of the above proposed models and algorithms of diagnosis with the help of the target "input-output" model. The target model was some analytical function $y=f(x)$. This function was approximated by the rule of inference (6.1), and served simultaneously as training and testing data generator. The input values (x) restored for each output (y) were compared with the target values.

The target model is given by the formula:

$$y = \frac{(1.8x+0.8)(5x-1.1)(4x-2.9)(3x-2.1)(9.5x-9.5)(3x-0.05)+20}{80},$$

which is represented in Fig. 6.10 together with the fuzzy terms of causes C_1=*low (L)*, C_2=*lower than average (lA)*, C_3=*average (A)*, C_4=*higher than average (hA)*, C_5=*lower than high*, C_6=*high (H)* and effects E_1=*lower than average (lA)*, E_2=*average (A)*, E_3=*higher than average (hA)*, E_4=*high (H)*.

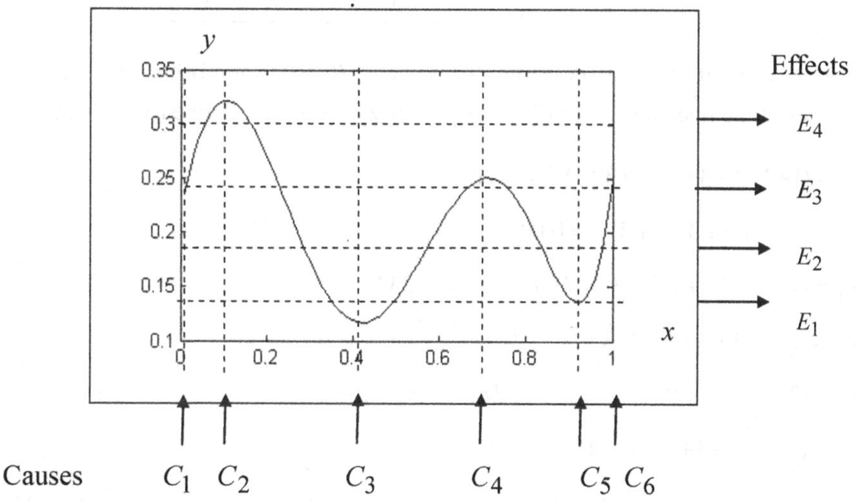

Fig. 6.10. "Input-output" model-generator

A fuzzy relational matrix was formed on the basis of expert assessments. For example, the procedure of fuzzy relations construction for effect E_1 consists of the following. Cause C_2 is the least important for effect E_1, so that the visual difference between the output values $y = E_1$ and $y(x = C_2)$, i.e. $|E_1 - y(x = C_2)|$, is maximal. Therefore, we start forming the matrix of paired comparisons $\mathbf{A}_1$ (6.11) from the 2nd row. This row is formed by an expert and contains the assessments, which define the degree of advantage of the rest causes C_K,

$K = \overline{1,6}$, over C_2. The advantage of cause C_K over cause C_2 is defined by the fact, how much the distance $|E_1 - y(x = C_K)|$ is less than the distance $|E_1 - y(x = C_2)|$. Matrix $\mathbf{A}_1$ (6.11) is defined by the known 2nd row as follows:

$\mathbf{A}_1 =$

	C_1	C_2	C_3	C_4	C_5	C_6
C_1	1	1/3	3	1	8/3	1
C_2	3	1	9	3	8	3
C_3	1/3	1/9	1	1/3	8/9	1/3
C_4	1	1/3	3	1	8/3	1
C_5	3/8	1/8	9/8	3/8	1	3/8
C_6	1	1/3	3	1	8/3	1

Matrix $\mathbf{A}_1$ allows us to construct fuzzy set E_1 (6.10) using formula (6.12). The degrees of membership r_{I1} of causes C_I to fuzzy set E_1 are defined as follows:

$r_{11} = (1+1/3+3+1+8/3+1)^{-1} = 0.11$;

$r_{21} = (3+1+9+3+8+3)^{-1} = 0.04$;

$r_{31} = (1/3+1/9+1+1/3+8/9+1/3)^{-1} = 0.33$;

$r_{41} = (1+1/3+3+1+8/3+1)^{-1} = 0.11$;

$r_{51} = (3/8+1/8+9/8+3/8+1+3/8)^{-1} = 0.30$;

$r_{61} = (1+1/3+3+1+8/3+1)^{-1} = 0.11$.

The obtained membership degrees should be normalized, i.e. r_{11}=0.11/0.33 $\approx$ 0.33; r_{21}=0.04/0.33 $\approx$ 0.12; r_{31}=0.33/0.33=1.00; r_{41}=0.11/0.33 $\approx$ 0.33; r_{51}=0.30/0.33 $\approx$ 0.91; r_{61}=0.11/0.33 $\approx$ 0.33.

Thus, fuzzy set E_1, whose elements correspond to the 1st column of the fuzzy relational matrix, takes the form:

$$E_1 = \left\{ \frac{0.33}{C_1}, \frac{0.12}{C_2}, \frac{1.00}{C_3}, \frac{0.33}{C_4}, \frac{0.91}{C_5}, \frac{0.33}{C_6} \right\}.$$

The resulting expert fuzzy relational matrix takes the form:

$\mathbf{R} =$

	E_1	E_2	E_3	E_4
C_1	0.33	1.00	0.67	0.21
C_2	0.12	0.10	0.33	1.00
C_3	1.00	0.23	0.11	0.11
C_4	0.33	0.33	1.00	0.21
C_5	0.91	0.77	0.22	0.34
C_6	0.33	0.90	0.67	0.21

The results of the fuzzy model tuning are given in Tables 6.1, 6.2.

Table 6.1. Parameters of the membership functions for the causes fuzzy terms before (after) tuning

Fuzzy terms	Parameters (β -, σ -)		
	Before tuning	Genetic algorithm	Neural net
C_1	(0, 0.17)	(0, 0.114)	(0, 0.114)
C_2	(0.1, 0.17)	(0.091, 0.121)	(0.091, 0.121)
C_3	(0.4, 0.17)	(0.430, 0.115)	(0.446, 0.115)
C_4	(0.7, 0.17)	(0.703, 0.100)	(0.711, 0.118)
C_5	(0.9, 0.17)	(0.919, 0.112)	(0.919, 0.112)
C_6	(1.0, 0.08)	(1.0, 0.041)	(1.0, 0.041)

Table 6.2. Parameters of the membership functions for the effects fuzzy terms before (after) tuning

Fuzzy terms	Parameters (β -, σ -)		
	Before tuning	Genetic algorithm	Neural net
E_1	(0.15, 0.05)	(0.171, 0.032)	(0.172, 0.037)
E_2	(0.2, 0.05)	(0.209, 0.040)	(0.209, 0.040)
E_3	(0.25, 0.05)	(0.257, 0.039)	(0.259, 0.041)
E_4	(0.3, 0.05)	(0.350, 0.037)	(0.352, 0.040)

Fuzzy relational equations after tuning take the form:

$$\mu^{E_1} = (\mu^{C_1} \wedge 0.27) \vee (\mu^{C_2} \wedge 0.13) \vee (\mu^{C_3} \wedge 0.97) \vee (\mu^{C_4} \wedge 0.20) \vee (\mu^{C_5} \wedge 0.86) \vee (\mu^{C_6} \wedge 0.21)$$
$$\mu^{E_2} = (\mu^{C_1} \wedge 0.93) \vee (\mu^{C_2} \wedge 0.09) \vee (\mu^{C_3} \wedge 0.28) \vee (\mu^{C_4} \wedge 0.44) \vee (\mu^{C_5} \wedge 0.75) \vee (\mu^{C_6} \wedge 0.82)$$
$$\mu^{E_3} = (\mu^{C_1} \wedge 0.63) \vee (\mu^{C_2} \wedge 0.41) \vee (\mu^{C_3} \wedge 0.15) \vee (\mu^{C_4} \wedge 0.95) \vee (\mu^{C_5} \wedge 0.26) \vee (\mu^{C_6} \wedge 0.67)$$
$$\mu^{E_4} = (\mu^{C_1} \wedge 0.12) \vee (\mu^{C_2} \wedge 0.88) \vee (\mu^{C_3} \wedge 0.07) \vee (\mu^{C_4} \wedge 0.08) \vee (\mu^{C_5} \wedge 0.32) \vee (\mu^{C_6} \wedge 0.12) \quad (6.15)$$

The results of solving the problem of inverse inference before and after tuning are shown in Fig. 6.11 and 6.12. The same figure depicts the membership functions of the fuzzy terms for the causes and effects before and after tuning.

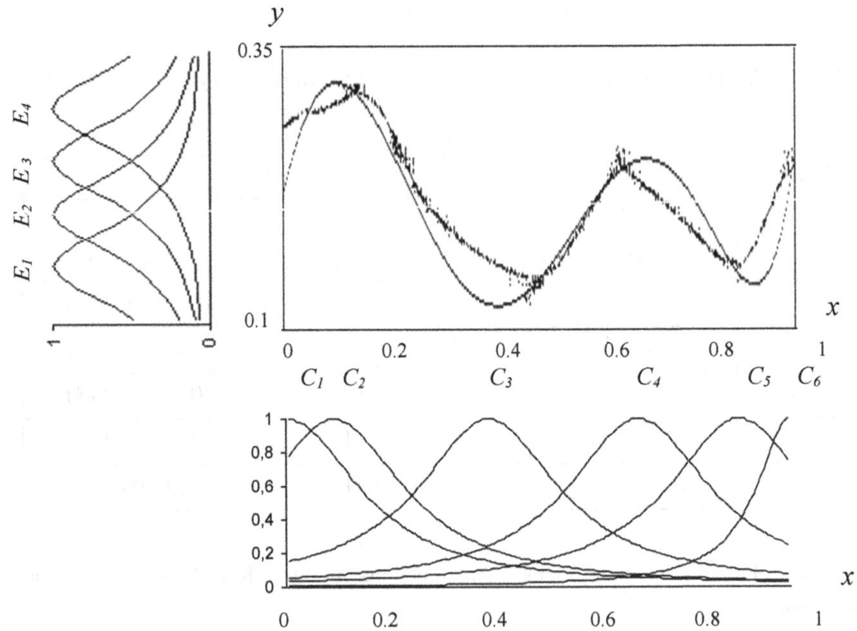

Fig. 6.11. Solution to the problem of inverse fuzzy inference before tuning

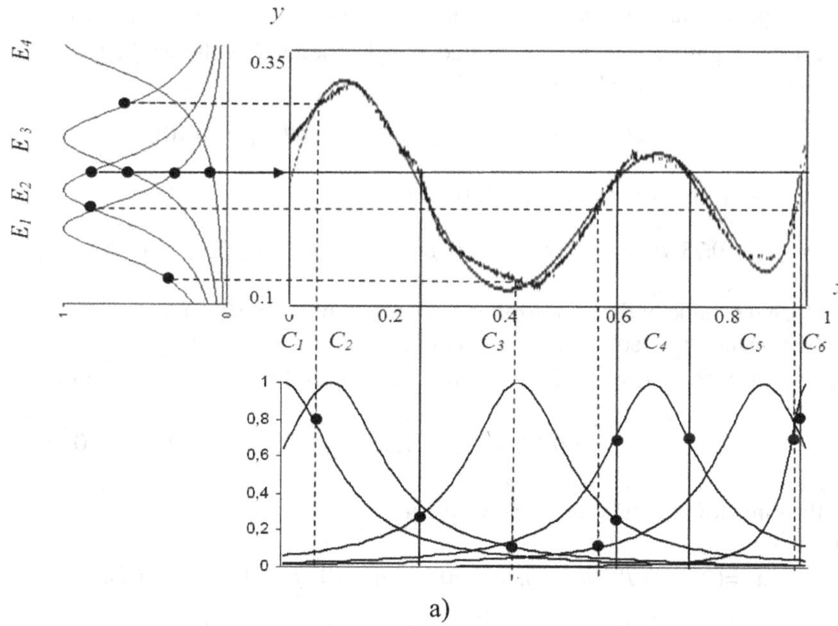

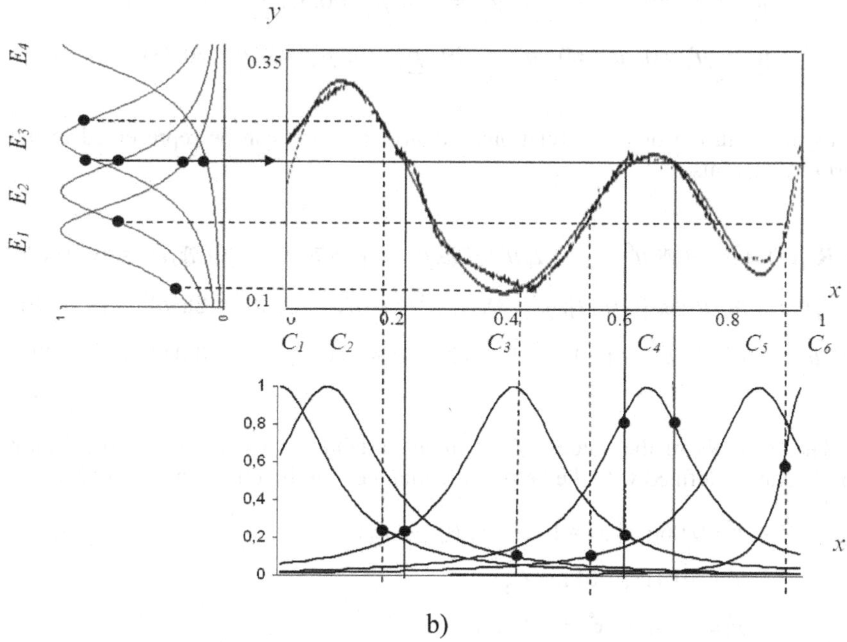

Fig. 6.12. Solution to the problem of inverse fuzzy inference after tuning:(a) y^*=0.23; (b) y^*=0.24

Let a specific value of the output variable consists of y^*=0.23. The measures of the effects significances for this value can be defined with the help of the membership functions in Fig. 6.12,a:

$$\boldsymbol{\mu}^E(y^*)=(\mu^{E_1}=0.29;\ \mu^{E_2}=0.78;\ \mu^{E_3}=0.67;\ \mu^{E_4}=0.10).$$

The genetic algorithm yields a null solution

$$\boldsymbol{\mu}_0^C=(\mu_0^{C_1}=0.78, \mu_0^{C_2}=0.10, \mu_0^{C_3}=0.29, \mu_0^{C_4}=0.67, \mu_0^{C_5}=0.07, \mu_0^{C_6}=0.45), \quad (6.16)$$

for which the value of the optimization criterion (6.6) is F=0.0004.

The obtained null solution allows us to arrange for the genetic search for the solution set $S(\mathbf{R},\boldsymbol{\mu}^E)$, which is completely determined by the greatest solution

$$\bar{\boldsymbol{\mu}}^C=(\bar{\mu}^{C_1}=0.78,\ \bar{\mu}^{C_2}=0.12,\ \bar{\mu}^{C_3}=0.29,\ \bar{\mu}^{C_4}=0.67,\ \bar{\mu}^{C_5}=0.12,\ \bar{\mu}^{C_6}=0.78)$$

and the three lower solutions $S^*=\{\underline{\boldsymbol{\mu}}_1^C,\underline{\boldsymbol{\mu}}_2^C,\underline{\boldsymbol{\mu}}_3^C\}$

$$\underline{\boldsymbol{\mu}}_1^C=(\underline{\mu}_1^{C_1}=0.78,\ \underline{\mu}_1^{C_2}=0,\ \underline{\mu}_1^{C_3}=0.29,\ \underline{\mu}_1^{C_4}=0,\ \underline{\mu}_1^{C_5}=0,\ \underline{\mu}_1^{C_6}=0.67);$$

$$\underline{\boldsymbol{\mu}}_2^C=(\underline{\mu}_2^{C_1}=0.78,\ \underline{\mu}_2^{C_2}=0,\ \underline{\mu}_2^{C_3}=0.29,\ \underline{\mu}_2^{C_4}=0.67,\ \underline{\mu}_2^{C_5}=0,\ \underline{\mu}_2^{C_6}=0);$$

$$\underline{\boldsymbol{\mu}}_3^C=(\underline{\mu}_3^{C_1}=0,\ \underline{\mu}_3^{C_2}=0,\ \underline{\mu}_3^{C_3}=0.29,\ \underline{\mu}_3^{C_4}=0,\ \underline{\mu}_3^{C_5}=0,\ \underline{\mu}_3^{C_6}=0.78).$$

Thus, the solution of fuzzy relational equations (6.15) can be represented in the form of intervals:

$$S(\mathbf{R},\boldsymbol{\mu}^E)=\{\mu^{C_1}=0.78; \mu^{C_2}\in[0,0.12]; \mu^{C_3}=0.29; \mu^{C_4}\in[0,0.67]; \mu^{C_5}\in[0,0.12]; \mu^{C_6}\in[0.67,0.78]\}$$
$$\cup\{\mu^{C_1}=0.78;\ \mu^{C_2}\in[0,0.12];\ \mu^{C_3}=0.29;\ \mu^{C_4}=0.67;\ \mu^{C_5}\in[0,0.12];\ \mu^{C_6}\in[0,0.78]\}$$
$$\cup\{\mu^{C_1}\in[0,0.78];\ \mu^{C_2}\in[0,0.12];\ \mu^{C_3}=0.29;\ \mu^{C_4}\in[0,0.67];\ \mu^{C_5}\in[0,0.12];\ \mu^{C_6}=0.78\}. \quad (6.17)$$

The intervals of the values of the input variable for each interval in solution (6.17) can be defined with the help of the membership functions in Fig. 6.12,a:

- x^*=0.060 or $x^*\in$ [0.060, 1.0] for C_1;
- $x^*\in$ [0.418, 1.0] for C_2;
- x^*=0.264 or x^*=0.628 for C_3;
- x^*=0.628, $x^*\in$ [0, 0.628], x^*=0.794 or $x^*\in$ [0.794, 1.0] for C_4;
- $x^*\in$ [0, 0.610] for C_5;
- $x^*\in$ [0.971, 0.978], $x^*\in$ [0, 0.978] or x^*=0.978 for C_6.

The restoration of the input set for y^*=0.23, i.e. points (0.264, 0.230), (0.628, 0.230), (0.794, 0.230) and (0.978, 0.230), is shown by the continuous line in Fig. 6.12, a, in which the values of the causes and effects significances measures are marked. The rest of the found input values correspond to other values of the output variable with the same measures of effects significances. The restoration of these points is shown by the dotted line in Fig. 6.12,a.

Assume the value of the output variable has changed from y^*=0.23 to y^*=0.24 (Fig. 6.12,b). For the new value, the fuzzy effects vector is

$$\boldsymbol{\mu}^E(y^*)=(\mu^{E_1}=0.23;\ \mu^{E_2}=0.62;\ \mu^{E_3}=0.82;\ \mu^{E_4}=0.11).$$

A neural adjustment of the null solution (6.16) has yielded a fuzzy causes vector

$$\boldsymbol{\mu}_0^C=(\mu_0^{C_1}=0.17, \mu_0^{C_2}=0.04, \mu_0^{C_3}=0.23, \mu_0^{C_4}=0.82, \mu_0^{C_5}=0.09, \mu_0^{C_6}=0.62),$$

for which the value of the optimization criterion (6.6) has constituted F=0.0001.

The resulting null solution has allowed adjustment of the bounds in the solution (6.17) and generation of the set of solutions $S(\mathbf{R},\boldsymbol{\mu}^E)$ determined by the greatest solution

$$\bar{\boldsymbol{\mu}}^C=(\bar{\mu}^{C_1}=0.23,\ \bar{\mu}^{C_2}=0.12,\ \bar{\mu}^{C_3}=0.23,\ \bar{\mu}^{C_4}=0.82,\ \bar{\mu}^{C_5}=0.12,\ \bar{\mu}^{C_6}=0.62)$$

and the two lower solutions $S^*=\{\underline{\boldsymbol{\mu}}_1^C,\underline{\boldsymbol{\mu}}_2^C\}$

$$\underline{\boldsymbol{\mu}}_1^C=(\underline{\mu}_1^{C_1}=0.23,\ \underline{\mu}_1^{C_2}=0,\ \underline{\mu}_1^{C_3}=0,\ \underline{\mu}_1^{C_4}=0.82,\ \underline{\mu}_1^{C_5}=0,\ \underline{\mu}_1^{C_6}=0.62);$$
$$\underline{\boldsymbol{\mu}}_2^C=(\underline{\mu}_2^{C_1}=0,\ \underline{\mu}_2^{C_2}=0,\ \underline{\mu}_2^{C_3}=0.23,\ \underline{\mu}_2^{C_4}=0.82,\ \underline{\mu}_2^{C_5}=0,\ \underline{\mu}_2^{C_6}=0.62).$$

Thus, the solution of fuzzy relational equations (6.15) for the new value can be represented in the form of intervals:

$$S(\mathbf{R},\boldsymbol{\mu}^E)=\{\mu^{C_1}=0.23;\ \mu^{C_2}\in[0,0.12];\ \mu^{C_3}\in[0,0.23];\ \mu^{C_4}=0.82;\ \mu^{C_5}\in[0,0.12];\ \mu^{C_6}=0.62\}$$
$$\cup\{\mu^{C_1}\in[0,0.23];\ \mu^{C_2}\in[0,0.12];\ \mu^{C_3}=0.23;\ \mu^{C_4}=0.82;\ \mu^{C_5}\in[0,0.12];\ \mu^{C_6}=0.62\}. \quad (6.18)$$

Solution (6.18) differs from (6.17) in the significance measures of the causes C_1, C_3, C_4 and C_6, for which the ranges of the input variable have been determined using the membership functions in Fig. 6.12,b:

- x^*=0.208 or $x^*\in$ [0.208, 1.0] for C_1;
- x^*=0.236, $x^*\in$ [0, 0.236], x^*=0.656 or $x^*\in$ [0.656, 1.0] for C_3;

- x^* =0.656 or x^* =0.766 for C_4;
- x^* =0.968 for C_6.

The restoration of the input set for y^* =0.24, i.e., points (0.236, 0.240), (0.656, 0.240), (0.766, 0.240), is shown in Fig. 6.12,b.

6.10 Example 5: Oil Pump Diagnosis

Let us consider the algorithm's performance having the recourse to the example of the fuel pump faults causes diagnosis.

Input parameters are (variation ranges are indicated in parentheses):

x_1 – engine speed (2600 – 3200 rpm);

x_2 – filter clear area (30 – 45 cm^2/kw);

x_3 – throat ring side clearance (0.1 – 0.3 mm);

x_4 – suction conduit leakage (0.5 – 2.0 cm^3/h);

x_5 – force main resistance (1.2–3.4 kg/cm^2).

The fault causes to be identified (input term-assessments) are: c_{11} – engine speed x_1 drop; c_{21} – decrease of clear area x_2, i.e. filter clogging; c_{31} (c_{32}) – decrease (increase) of side clearance x_3, i.e. assembling defect (throat ring wear-out); c_{41} – increase of leakage x_4, i.e. fuel escape; c_{51} – high resistance of the force main x_5.

Output parameters are (variation ranges are indicated in parentheses):

y_1 – productivity (20–45 m^3/h);

y_2 – force main pressure (3.7–5.5 kg/cm^2);

y_3 – consumed power (15–30 kw);

y_4 – suction conduit pressure (0.5–1.0 kg/cm^2).

The observed effects (output term-assessments) are: e_{11} – productivity y_1 fall; e_{21} (e_{22}) – force main pressure y_2 drop (rise); e_{31} (e_{32}) – consumed power y_3 drop (rise); e_{41} – pressure in suction conduit y_4 rise.

We shall define the set of causes and effects in the following way:

$$\{C_1, C_2, C_3, C_4, C_5, C_6\} = \{c_{11}, c_{21}, c_{31}, c_{32}, c_{41}, c_{51}\};$$

$$\{E_1, E_2, E_3, E_4, E_5, E_6\} = \{e_{11}, e_{21}, e_{22}, e_{31}, e_{32}, e_{41}\}.$$

"Causes-effects" relations were formed on the basis of expert assessments. For example, the procedure of fuzzy relations construction for effect E_1 consists of the following. Cause C_3 is the least important for effect E_1. Therefore, we start forming the matrix of paired comparisons $\mathbf{A}_1$ (6.11) from the 3rd row. This row is formed by an expert and contains the assessments, which define the degree of advantage of the rest of the causes over C_3. Not a single cause has an absolute advantage over C_3. Therefore, matrix $\mathbf{A}_1$ contains a fictitious cause C_7, where C_7 has *absolute advantage* over C_3. Matrix $\mathbf{A}_1$ (6.11) is defined by the known 3rd row as follows:

$\mathbf{A}_1 =$

	C_1	C_2	C_3	C_4	C_5	C_6	C_7
C_1	1	7/2	1/2	4	3	1	9/2
C_2	2/7	1	1/7	8/7	6/7	2/7	9/7
C_3	2	7	1	8	6	2	9
C_4	1/4	7/8	1/8	1	3/4	1/4	9/8
C_5	1/3	7/6	1/6	4/3	1	1/3	3/2
C_6	1	7/2	1/2	4	3	1	9/2
C_7	2/9	7/9	1/9	8/9	2/3	2/9	1

Matrix $\mathbf{A}_1$ allows us to construct fuzzy set E_1 (6.10) using formula (6.12). The degrees of membership r_{I1} of causes C_I to fuzzy set E_1 are defined as follows:

$r_{11} = (1+7/2+1/2+4+3+1+9/2)^{-1} = 0.06;$

$r_{21} = (2/7+1+1/7+8/7+6/7+2/7+9/7)^{-1} = 0.20;$

$r_{31} = (2+7+1+8+6+2+9)^{-1} = 0.03;$

$r_{41} = (1/4+7/8+1/8+1+3/4+1/4+9/8)^{-1} = 0.23;$

$r_{51} = (1/3+7/6+1/6+4/3+1+1/3+3/2)^{-1} = 0.17;$

$r_{61} = (1+7/2+1/2+4+3+1+9/2)^{-1} = 0.06;$

$r_{71} = (2/9+7/9+1/9+8/9+2/3+2/9+1)^{-1} = 0.26.$

The obtained membership degrees should be normalized, i.e. r_{11} =0.06/0.26 $\approx$ 0.23; r_{21} =0.20/0.26 $\approx$ 0.77; r_{31} =0.03/0.26=0.11; r_{41} =0.23/0.26 $\approx$ 0.88; r_{51} =0.17/0.26 $\approx$ 0.65; r_{61} =0.06/0.26=0.23.

Thus, fuzzy set E_1, whose elements correspond to the 1st column of the fuzzy relational matrix, takes the form:

$$E_1 = \left\{ \frac{0.23}{C_1}, \frac{0.77}{C_2}, \frac{0.11}{C_3}, \frac{0.88}{C_4}, \frac{0.65}{C_5}, \frac{0.23}{C_6} \right\}.$$

The resulting expert fuzzy relational matrix takes the form:

R =

	E_1	E_2	E_3	E_4	E_5	E_6
C_1	0.23	0.90	0.44	0.88	0.11	0.76
C_2	0.77	0.21	0.89	0.23	0.22	0.32
C_3	0.11	0.45	0.22	0.69	0.89	0.24
C_4	0.88	0.21	0.67	0.12	0.11	0.68
C_5	0.65	0.10	0.33	0.12	0.11	0.88
C_6	0.23	0.55	0.11	0.81	0.40	0.12

For the fuzzy model tuning we used the results of diagnosis for 340 pumps. The results of the fuzzy model tuning are given in Tables 6.3, 6.4 and in Fig. 6.13.

Table 6.3. Parameters of the membership functions for the causes and effects fuzzy terms after genetic tuning

Parameter	Fuzzy terms					
	C_1	C_2	C_3	C_4	C_5	C_6
β -	2700	34.75	0.11	0.26	1.84	3.15
σ -	107.12	3.18	0.04	0.05	0.33	0.65
Parameter	Fuzzy terms					
	E_1	E_2	E_3	E_4	E_5	E_6
β -	22.79	3.84	5.32	15.94	28.84	0.89
σ -	5.02	0.92	0.35	3.76	1.85	0.16

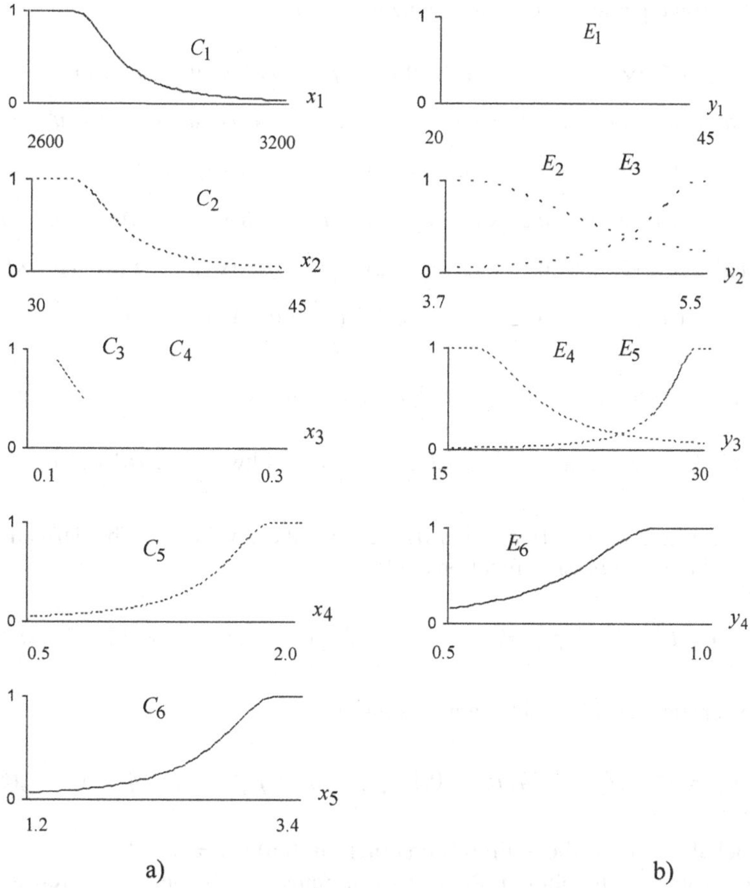

Fig. 6.13. Membership functions of the causes (a) and effects (b) fuzzy terms after tuning

Table 6.4. Parameters of the membership functions for the causes and effects fuzzy terms after neural tuning

Parameter	Fuzzy terms					
	C_1	C_2	C_3	C_4	C_5	C_6
β -	2700	32.27	0.11	0.28	1.82	3.19
σ -	104.57	2.94	0.03	0,06	0.31	0.54
Parameter	Fuzzy terms					
	E_1	E_2	E_3	E_4	E_5	E_6
β -	22.98	3.86	5.37	16.45	28.92	0.89
σ -	4.93	0.87	0.38	3.54	1.82	0.17

Diagnostic equations after tuning take the form:

$$\mu^{E_1} = (\mu^{C_1} \wedge 0.21) \vee (\mu^{C_2} \wedge 0.78) \vee (\mu^{C_3} \wedge 0.15) \vee (\mu^{C_4} \wedge 0.84) \vee (\mu^{C_5} \wedge 0.73) \vee (\mu^{C_6} \wedge 0.18)$$

$$\mu^{E_2} = (\mu^{C_1} \wedge 0.97) \vee (\mu^{C_2} \wedge 0.20) \vee (\mu^{C_3} \wedge 0.43) \vee (\mu^{C_4} \wedge 0.18) \vee (\mu^{C_5} \wedge 0.14) \vee (\mu^{C_6} \wedge 0.58)$$

$$\mu^{E_3} = (\mu^{C_1} \wedge 0.48) \vee (\mu^{C_2} \wedge 0.59) \vee (\mu^{C_3} \wedge 0.85) \vee (\mu^{C_4} \wedge 0.63) \vee (\mu^{C_5} \wedge 0.34) \vee (\mu^{C_6} \wedge 0.12)$$

$$\mu^{E_4} = (\mu^{C_1} \wedge 0.94) \vee (\mu^{C_2} \wedge 0.21) \vee (\mu^{C_3} \wedge 0.64) \vee (\mu^{C_4} \wedge 0.18) \vee (\mu^{C_5} \wedge 0.16) \vee (\mu^{C_6} \wedge 0.74)$$

$$\mu^{E_5} = (\mu^{C_1} \wedge 0.16) \vee (\mu^{C_2} \wedge 0.14) \vee (\mu^{C_3} \wedge 0.92) \vee (\mu^{C_4} \wedge 0.08) \vee (\mu^{C_5} \wedge 0.10) \vee (\mu^{C_6} \wedge 0.41)$$

$$\mu^{E_6} = (\mu^{C_1} \wedge 0.64) \vee (\mu^{C_2} \wedge 0.82) \vee (\mu^{C_3} \wedge 0.21) \vee (\mu^{C_4} \wedge 0.72) \vee (\mu^{C_5} \wedge 0.99) \vee (\mu^{C_6} \wedge 0.09)$$

(6.19)

Let us represent the vector of the observed parameters for a specific pump:

$$\mathbf{Y}^* = (y_1^* = 26.12 \text{ m}^3\text{/h};\ y_2^* = 5.08 \text{ kg/cm}^2;\ y_3^* = 24 \text{ kw};\ y_4^* = 0.781 \text{ kg/cm}^2).$$

The measures of the effects significances for these values can be defined with the help of the membership functions in Fig. 6.13,b:

$$\boldsymbol{\mu}^E(\mathbf{Y}^*) = (\mu^{E_1} = 0.71;\ \mu^{E_2} = 0.34;\ \mu^{E_3} = 0.63;\ \mu^{E_4} = 0.18;\ \mu^{E_5} = 0.12;\ \mu^{E_6} = 0.71).$$

The genetic algorithm yields a null solution

$$\boldsymbol{\mu}_0^C = (\mu_0^{C_1} = 0.26,\ \mu_0^{C_2} = 0.54,\ \mu_0^{C_3} = 0.14,\ \mu_0^{C_4} = 0.69,\ \mu_0^{C_5} = 0.71,\ \mu_0^{C_6} = 0.08)\ ,(6.20)$$

for which the value of the optimization criterion (6.6) is F=0.0144.

The obtained null solution allows us to arrange for the genetic search for the solution set $S(\mathbf{R}, \boldsymbol{\mu}^E)$, which is completely determined by the greatest solution

$$\bar{\boldsymbol{\mu}}^C = (\bar{\mu}^{C_1} = 0.26,\ \bar{\mu}^{C_2} = 0.71,\ \bar{\mu}^{C_3} = 0.16,\ \bar{\mu}^{C_4} = 0.71,\ \bar{\mu}^{C_5} = 0.71,\ \bar{\mu}^{C_6} = 0.16)$$

and the three lower solutions $S^* = \{\underline{\boldsymbol{\mu}}_1^C, \underline{\boldsymbol{\mu}}_2^C, \underline{\boldsymbol{\mu}}_3^C\}$

$$\underline{\boldsymbol{\mu}}_1^C = (\underline{\mu}_1^{C_1} = 0.26,\ \underline{\mu}_1^{C_2} = 0.71,\ \underline{\mu}_1^{C_3} = 0,\ \underline{\mu}_1^{C_4} = 0.63,\ \underline{\mu}_1^{C_5} = 0,\ \underline{\mu}_1^{C_6} = 0);$$
$$\underline{\boldsymbol{\mu}}_2^C = (\underline{\mu}_2^{C_1} = 0.26,\ \underline{\mu}_2^{C_2} = 0,\ \underline{\mu}_2^{C_3} = 0,\ \underline{\mu}_2^{C_4} = 0.71,\ \underline{\mu}_2^{C_5} = 0,\ \underline{\mu}_2^{C_6} = 0);$$
$$\underline{\boldsymbol{\mu}}_3^C = (\underline{\mu}_3^{C_1} = 0.26,\ \underline{\mu}_3^{C_2} = 0,\ \underline{\mu}_3^{C_3} = 0,\ \underline{\mu}_3^{C_4} = 0.63,\ \underline{\mu}_3^{C_5} = 0.71,\ \underline{\mu}_3^{C_6} = 0).$$

Thus, the solution of fuzzy relational equations (6.19) can be represented in the form of intervals:

$$S(\mathbf{R},\boldsymbol{\mu}^E)=\{\mu^{C_1}=0.26;\ \mu^{C_2}=0.71;\ \mu^{C_3}\in[0,0.16];\ \mu^{C_4}\in[0.63,0.71];\ \mu^{C_5}\in[0,0.71];\ \mu^{C_6}\in[0,0.16]\}$$
$$\bigcup\{\mu^{C_1}=0.26;\ \mu^{C_2}\in[0,0.71];\ \mu^{C_3}\in[0,0.16];\ \mu^{C_4}=0.71;\ \mu^{C_5}\in[0,0.71];\ \mu^{C_6}\in[0,0.16]\}$$
$$\bigcup\{\mu^{C_1}=0.26;\ \mu^{C_2}\in[0,0.71];\ \mu^{C_3}\in[0,0.16];\ \mu^{C_4}\in[0.63,0.71];\ \mu^{C_5}=0.71;\ \mu^{C_6}\in[0,0.16]\}. \quad (6.21)$$

The intervals of the values of the input variables for each interval in solution (6.21) can be defined with the help of the membership functions in Fig. 6.13,b:

- x_1^*=2877 rpm for C_1;
- x_2^*=34.15 or $x_2^* \in$ [34.15, 45] cm^2/kw for C_2;
- $x_3^* \in$ [0.178, 0.300] mm for C_3;
- x_3^*=0.242 or $x_3^* \in$ [0.234, 0.242] mm for C_4;
- x_4^*=1.62 or $x_4^* \in$ [0.5, 1.62] cm^3/h for C_5;
- $x_5^* \in$ [1.2, 1.95] kg/cm^2 for C_6.

The obtained solution allows the analyst to make the preliminary conclusions. The cause of the observed pump state should be located and identified as the filter clogging, the throat ring wear-out or fuel escape in the suction conduit (clear area decreased up to 34.15-45 cm^2/kw, side clearance increased up to 0.234-0.242 mm, and leakage increased up to 0.5-1.62 cm^3/h), since the significance measures of the causes C_2, C_4 and C_5 are sufficiently high. An assembly defect of the throat ring for the side clearance within 0.178-0.300 mm should be excluded since the significance measure of the cause C_3 is small. The engine speed reduced to 2877 rpm can also tell on the pump's proper functioning, the significance measure of which is indicative of the cause C_1. Resistance of the force main increased up to 1.2-1.95 kg/cm^2 practically has no influence on the pump fault, so that the significance measure of cause C_6 is small.

Assume a repeated measurement has revealed a decrease in the pump delivery up to y_1^*=24.97 m^3/h and an increase in the suction pressure up to y_4^*=0.792 kg/cm^2, the values of μ^{E_1} increasing up to 0.86, μ^{E_6} up to 0.75, and the values of other parameters remaining unchanged.

A neural adjustment of the null solution (6.20) has yielded a fuzzy causes vector

$$\boldsymbol{\mu}_0^C=(\mu_0^{C_1}=0.26,\ \mu_0^{C_2}=0.17,\ \mu_0^{C_3}=0.10,\ \mu_0^{C_4}=0.93,\ \mu_0^{C_5}=0.75,\ \mu_0^{C_6}=0.05),$$

for which the value of the optimization criterion (6.6) has constituted F=0.0148.

The resulting null solution has allowed adjustment of the bounds in the solution (6.21) and generation of the set of solutions $S(\mathbf{R},\boldsymbol{\mu}^E)$ determined by the greatest solution

$$\bar{\boldsymbol{\mu}}^C=(\bar{\mu}^{C_1}=0.26,\ \bar{\mu}^{C_2}=0.75,\ \bar{\mu}^{C_3}=0.16,\ \bar{\mu}^{C_4}=1.00,\ \bar{\mu}^{C_5}=0.75,\ \bar{\mu}^{C_6}=0.16)$$

and the two lower solutions $S^*=\{\underline{\boldsymbol{\mu}}_1^C,\underline{\boldsymbol{\mu}}_2^C\}$

$$\underline{\boldsymbol{\mu}}_1^C=(\underline{\mu}_1^{C_1}=0.26,\ \underline{\mu}_1^{C_2}=0.75,\ \underline{\mu}_1^{C_3}=0,\ \underline{\mu}_1^{C_4}=0.84,\ \underline{\mu}_1^{C_5}=0,\ \underline{\mu}_1^{C_6}=0);$$

$$\underline{\boldsymbol{\mu}}_2^C=(\underline{\mu}_2^{C_1}=0.26,\ \underline{\mu}_2^{C_2}=0,\ \underline{\mu}_2^{C_3}=0,\ \underline{\mu}_2^{C_4}=0.84,\ \underline{\mu}_2^{C_5}=0.75,\ \underline{\mu}_2^{C_6}=0).$$

Thus, the solution of fuzzy relational equations (6.19) can be represented in the form of intervals:

$$S(\mathbf{R},\boldsymbol{\mu}^E)=\{\mu^{C_1}=0.26;\ \mu^{C_2}=0.75;\ \mu^{C_3}\in[0,0.16];\ \mu^{C_4}\in[0.84,1.00];\ \mu^{C_5}\in[0,0.75];\ \mu^{C_6}\in[0,0.16]\}$$
$$\cup\{\mu^{C_1}=0.26;\ \mu^{C_2}\in[0,0.75];\ \mu^{C_3}\in[0,0.16];\ \mu^{C_4}\in[0.84,1.00];\ \mu^{C_5}=0.75;\ \mu^{C_6}\in[0,0.16]\}. \qquad (6.22)$$

Solution (6.22) differs from (6.21) in the significance measures of the causes C_2, C_4 and C_5. The ranges of input variables have been determined for these causes using the membership functions in Fig. 6.13,a:

- $x_2^*=33.97$ or $x_2^*\in[33.97, 45]$ cm^2/kw for C_2;
- $x_3^*\in[0.254, 0.300]$ mm for C_4;
- $x_4^*=1.64$ or $x_4^*\in[0.5, 1.64]$ cm^3/h for C_5.

The solution obtained allows for the final conclusions. The state of the pump being observed is due to the throat ring wear-out (the side clearance increased to 0.254-0.300 mm), since the significance measure of the cause C_4 is maximal. The causes of the pump failure are still the filter clogging and fuel escape in the suction conduit (the flow area decreased to 33.97-45 cm^2/kw and the leakage increased to 0.5-1.64 cm^3/h), since the significance measures of the causes C_2 and C_5 are reasonably high. The values of other parameters have not changed.

To test the fuzzy model we used the results of diagnosis for 250 pumps with different kinds of faults. The tuning algorithm efficiency characteristics for the testing data are given in Table 6.5. Attaining an average accuracy rate of 95% required 30 min of the operation of a genetic algorithm and 4 min of the operation of a neural network (Intel Core 2 Duo P7350 2.0 GHz).

Table 6.5. Tuning algorithm efficiency characteristics

Causes (diagnoses)	Number of cases in the data sample	Probability of the correct diagnosis		
		Before tuning	After tuning	
			Null solution (genetic algorithm)	Refined diagnoses (neural network)
C_1	105	83 / 105 = 0.79	99 / 105 = 0.94	103 / 105 = 0.98
C_2	203	164 / 203 = 0.81	186 / 203 = 0.92	197 / 203 = 0.97
C_3	59	52 / 59 = 0.88	54 / 59 = 0.91	57 / 59 = 0.97
C_4	187	154 / 187 = 0.82	174 / 187 = 0.93	178 / 187 = 0.95
C_5	94	85 / 94 = 0.90	90 / 94 = 0.96	93 / 94 = 0.99
C_6	75	64 / 75 = 0.85	69 / 75 = 0.92	73 / 75 = 0.97

References

1. Sanchez, E.: Medical diagnosis and composite fuzzy relations. In: Gupta, M.M., Yager, R.R. (eds.) Advances in Fuzzy Set Theory and Applications, pp. 437–444. North-Holland, Amsterdam (1979)
2. Sanchez, E.: Inverses of fuzzy relations. Applications to possibility distributions and medical diagnosis. Fuzzy Sets and Systems 2(1), 75–86 (1979)
3. Zadeh, L.: The Concept of Linguistic Variable and It's Application to Approximate Decision Making, p. 176. Mir, Moscow (1976)
4. Peeva, K., Kyosev, Y.: Fuzzy Relational Calculus Theory, Applications and Software, CD-ROM, p. 292. World Scientific Publishing Company (2004), http://mathworks.net
5. Sanchez, E.: Resolution of composite fuzzy relation equations. Information and Control 30(1), 38–48 (1976)
6. Miyakoshi, M., Shimbo, M.: Lower solutions of systems of fuzzy equations. Fuzzy Sets and Systems 19, 37–46 (1986)
7. Di Nola, A., Sessa, S., Pedrycz, W., Sanchez, E.: Fuzzy Relation Equations and Their Applications to Knowledge Engineering. Kluwer Academic Press, Dordrecht (1989)
8. De Baets, B.: Analytical solution methods for fuzzy relational equations. In: Dubois, D., Prade, H. (eds.) Fundamentals of Fuzzy Sets. The Handbooks of Fuzzy Sets Series, vol. 1, pp. 291–340. Kluwer Academic Publishers (2000)
9. Pedrycz, W.: Processing in relational structures: fuzzy relational equations. Fuzzy Sets and Systems 40(1), 77–106 (1991)
10. Gottwald, S., Pedrycz, W.: Solvability of fuzzy relational equations and manipulation of fuzzy data. Fuzzy Sets and Systems 18(1), 45–65 (1986)
11. Neundorf, D., Bohm, R.: Solvability criteria for systems of fuzzy relation equations. Fuzzy Sets and Systems 80(3), 345–352 (1996)
12. Gottwald, S., Perfilieva, I.: Solvability and approximate solvability of fuzzy relation equations. Intern. J. General Systems 32(4), 361–372 (2003)

13. Pedrycz, W.: Inverse problem in fuzzy relational equations. Fuzzy Sets and Systems 36(2), 277–291 (1990)
14. Pedrycz, W.: Approximate solutions of fuzzy relational equations. Fuzzy Sets and Systems 28(2), 183–202 (1988)
15. Pedrycz, W.: Neurocomputations in relational systems. IEEE Trans. Pattern Analysis and Machine Intelligence 13, 289–297 (1991)
16. Pedrycz, W.: Optimization schemes for decomposition of fuzzy relations. Fuzzy Sets and Systems 100(1-3), 301–325 (1998)
17. Sanchez, E.: Fuzzy genetic algorithms in soft computing environment. In: Proc. Fifth IFSA World Congress, XLIV-L (1993)
18. Pedrycz, W.: Fuzzy models and relational equations. Mathematical Modelling 9(6), 427–434 (1987)
19. Pedrycz, W.: Fuzzy modelling: fundamentals, construction and evaluation. Fuzzy Sets and Systems 41, 1–15 (1991)
20. Pedrycz, W.: Genetic algorithms for learning in fuzzy relational structures. Fuzzy Sets and Systems 69(1), 37–52 (1995)
21. Blanco, A., Delgado, M., Requena, J.: Identification of fuzzy relational equations by fuzzy neural networks. Fuzzy Sets and Systems 71(2), 215–226 (1995)
22. Ciaramella, A., Tagliaferri, R., Pedrycz, W., Di Nola, A.: Fuzzy relational neural network. International Journal of Approximate Reasoning 41(2), 146–163 (2006)
23. Rotshtein, A.: Design and tuning of fuzzy rule-based systems for medical diagnosis. In: Teodorescu, N.-H., Kandel, A., Gain, L. (eds.) Fuzzy and Neuro-Fuzzy Systems in Medicine, pp. 243–289. CRC Press (1998)
24. Rotshtein, A., Posner, M., Rakytyanska, H.: Cause and effect analysis by fuzzy relational equations and a genetic algorithm. Reliability Engineering and System Safety 91(9), 1095–1101 (2006)
25. Rotshtein, A., Rakytyanska, H.: Genetic Algorithm for Fuzzy Logical Equations Solving in Diagnostic Expert Systems. In: Monostori, L., Váncza, J., Ali, M. (eds.) IEA/AIE 2001. LNCS (LNAI), vol. 2070, pp. 349–358. Springer, Heidelberg (2001)
26. Rotshtein, A.P., Rakytyanska, H.B.: Fuzzy Relation-based Diagnosis. Automation and Remote Control 68(12), 2198–2213 (2007)
27. Rotshtein, A.P., Rakytyanska, H.B.: Diagnosis Problem Solving using Fuzzy Relations. IEEE Transactions on Fuzzy Systems 16(3), 664–675 (2008)
28. Rotshtein, A.P., Rakytyanska, H.B.: Adaptive Diagnostic System based on Fuzzy Relations. Cybernetics and Systems Analysis 45(4), 623–637 (2009)
29. Rotshtein, A., Katel'nikov, D.: Identification of non-linear objects by fuzzy knowledge bases. Cybernetics and Systems Analysis 34(5), 676–683 (1998)
30. Rotshtein, A., Mityushkin, Y.: Neurolinguistic identification of nonlinear dependencies. Cybernetics and Systems Analysis 36(2), 179–187 (2000)
31. Gen, M., Cheng, R.: Genetic Algorithms and Engineering Design. John Wiley & Sons (1997)
32. Tsypkin, Y.Z.: Information Theory of Identification, p. 320. Nauka, Moscow (1984) (in Russian)
33. Rotshtein, A.: Modification of Saaty method for the construction of fuzzy set membership functions. In: Proc. FUZZY 1997, Intern. Conf. on Fuzzy Logic and Its Applications, Zichron Yaakov, Israel, pp. 125–130 (1997)
34. Saaty, T.L.: Mathematical Models of Arms Control and Disarmament. John Wiley & Sons (1968)

Chapter 7
Inverse Inference Based on Fuzzy Rules

The wide class of the problems, arising from engineering, medicine, economics and other domains, belongs to the class of inverse problems [1]. The typical representative of the inverse problem is the problem of medical and technical diagnosis, which amounts to the restoration and the identification of the unknown causes of the disease or the failure through the observed effects, i.e. the symptoms or the external signs of the failure. The diagnosis problem, which is based on a cause and effect analysis and abductive reasoning can be formally described by neural networks [2] or Bayesian networks [3, 4]. In the cases, when domain experts are involved in developing cause-effect connections, the dependency between unobserved and observed parameters can be modelled using the means of fuzzy sets theory [5, 6]: fuzzy relations and fuzzy IF-THEN rules. Fuzzy relational calculus plays the central role as a uniform platform for inverse problem resolution on various fuzzy approximation operators [7, 8]. In the case of a multiple variable linguistic model, the cause-effect dependency is extended to the multidimensional fuzzy relational structure [6], and the problem of inputs restoration and identification amounts to solving a system of multidimensional fuzzy relational equations [9, 10]. Fuzzy IF-THEN rules enable us to consider complex combinations in cause-effect connections as being simpler and more natural, which are difficult to model with fuzzy relations. In rule-based models, an inputs-outputs connection is described by a hierarchical system of simplified fuzzy relational equations with max-min and dual min-max laws of composition [11 – 13].

In works [14 – 16], an expert system using a genetic algorithm [17] as a tool to solve the diagnosis problem was proposed. The diagnosis problem based on a cause and effect analysis was formally described by the single input single output fuzzy relation approximator [18 – 20]. This chapter proposes an approach for inverse problem solution based on the description of the interconnection between unobserved and observed parameters of an object (causes and effects) with the help of fuzzy IF-THEN rules. The problem consists of not only solving a system of fuzzy logical equations, which correspond to IF-THEN rules, but also in selection of such forms of the fuzzy terms membership functions and such weights of the fuzzy IF–THEN rules, which provide maximal proximity between model and real results of diagnosis [21].

The essence of the proposed approach consists of formulating and solving the optimization problems, which, on the one hand, find the roots of fuzzy logical equations, corresponding to IF-THEN rules, and, on the other hand, tune the fuzzy

A.P. Rotshtein et al.: Fuzzy Evidence in Identif., Forecast. and Diagn., STUDFUZZ 275, pp. 193 – 233.
springerlink.com

model using the readily available experimental data. The hybrid genetic and neuro approach is proposed for solving the formulated optimization problems.

This chapter is written on the basis of [11 – 13].

7.1 Diagnostic Approximator Based on Fuzzy Rules

The diagnosis object is treated as a black box with n inputs and m outputs (Fig. 7.1). Outputs of the object are associated with the observed effects (symptoms). Inputs correspond to the causes of the observed effects (diagnoses). The problem of diagnosis consists of restoration and identification of the causes (inputs) through the observed effects (outputs). Inputs and outputs can be considered as linguistic variables given on the corresponding universal sets. Fuzzy terms are used for these linguistic variables evaluation.

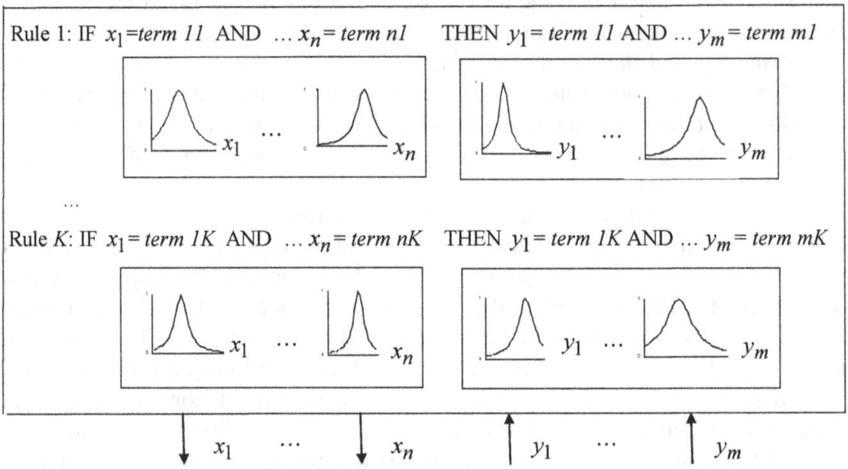

Fig. 7.1. Object of diagnosis

We shall denote the following:

$\{x_1, x_2, ..., x_n\}$ is the set of input parameters, $x_i \in [\underline{x}_i, \overline{x}_i]$, $i = \overline{1, n}$;

$\{y_1, y_2, ..., y_m\}$ is the set of output parameters, $y_j \in [\underline{y}_j, \overline{y}_j]$, $j = \overline{1, m}$;

$\{c_{i1}, c_{i2}, ..., c_{ik_i}\}$ is the set of linguistic terms for parameter x_i evaluation, $i = \overline{1, n}$;

$\{e_{j1}, e_{j2}, ..., e_{jq_j}\}$ is the set of linguistic terms for parameter y_j evaluation, $j = \overline{1, m}$.

Each term-assessment is described with the help of a fuzzy set:

$$c_{il} = \{(x_i, \mu^{c_{il}}(x_i))\}, \quad i = \overline{1,n}, \; l = \overline{1,k_i};$$

$$e_{jp} = \{(y_j, \mu^{e_{jp}}(y_j))\}, \quad j = \overline{1,m}, \; p = \overline{1,q_j}.$$

where $\mu^{c_{il}}(x_i)$ is a membership function of variable x_i to the term-assessment c_{il}, $i = \overline{1,n}$, $l = \overline{1,k_i}$;

$\mu^{e_{jp}}(y_j)$ is a membership function of variable y_j to the term-assessment e_{jp}, $j = \overline{1,m}$, $p = \overline{1,q_j}$.

We shall redenote the set of input and output terms-assessments in the following way:

$\{C_1, C_2, ..., C_N\} = \{c_{11}, c_{12}, ..., c_{1k_1}, ..., c_{n1}, c_{n2}, ..., c_{nk_n}\}$ is the set of terms for input parameters evaluation, where $N = k_1 + k_2 + ... + k_n$;

$\{E_1, E_2, ..., E_M\} = \{e_{11}, e_{12}, ..., e_{1q_1}, ..., e_{m1}, e_{m2}, ..., e_{mq_m}\}$ is the set of terms for output parameters evaluation, where $M = q_1 + q_2 + ... + q_m$.

Set $\{C_I, I = \overline{1,N}\}$ is called fuzzy causes (diagnoses), and set $\{E_J, J = \overline{1,M}\}$ is called fuzzy effects (symptoms).

Causes - effects interconnection can be represented using the expert matrix of knowledge (Table 7.1).

Table 7.1. Fuzzy knowledge base

№ rule	Inputs				Outputs						
	x_1	x_2	...	x_n	y_1	Weight	y_2	Weight		y_m	Weight
1	a_{11}	a_{21}	...	a_{n1}	b_{11}	w_{11}	b_{21}	w_{21}	...	b_{m1}	w_{m1}
2	a_{12}	a_{22}	...	a_{n2}	b_{12}	w_{12}	b_{22}	w_{22}	...	b_{m2}	w_{m2}
...	...	...	...	...	...	...	...	...	...	...	...
K	a_{1K}	a_{2K}	...	a_{nK}	b_{1K}	w_{1K}	b_{2K}	w_{2K}	...	b_{mK}	w_{mK}

The fuzzy knowledge base below corresponds to this matrix:

Rule l: IF $x_1 = a_{1l}$ AND $x_2 = a_{2l}$ ... AND $x_n = a_{nl}$

THEN $y_1 = b_{1l}$ (with weight w_{1l})

AND $y_2 = b_{2l}$ (with weight w_{2l}) ...

AND $y_m = b_{ml}$ (with weight w_{ml}), $l = \overline{1,K}$; (7.1) where a_{il} is a fuzzy term for variable x_i evaluation in the rule with number l;

b_{jl} is a fuzzy term for variable y_j evaluation in the rule with number l;

w_{jl} is a rule weight, i.e. a number in the range [0, 1], characterizing the measure of confidence of an expert relative to the statement with number jl;

K is the number of fuzzy rules.

The problem of diagnosis is set in the following way: it is necessary to restore and identify the values of the input parameters $(x_1^*, x_2^*, ..., x_n^*)$ through the values of the observed output parameters $(y_1^*, y_2^*, ..., y_m^*)$.

7.2 Interconnection of Fuzzy Rules and Relations

This fuzzy rules base is modelled by the fuzzy relational matrices presented in Table 7.2. These relational matrices can be translated as a set of fuzzy IF-THEN rules

IF $\mathbf{X} = A_L$

(i.e., $x_1 = C_1$ (with weight v_{1L})

AND … $x_i = C_I$ (with weight v_{IL}) …

AND $x_n = C_N$ (with weight v_{NL}))

THEN $y_j = E_J$ (with weight r_{LJ}) (7.2)

Here A_L is the combination of input terms in rule L, $L = \overline{1, K}$.

Table 7.2. Fuzzy relational matrices

	IF inputs					THEN outputs				
	x_1	…	x_i	…	x_n	y_1	…	y_j	…	y_m
	c_{11} … c_{1k_1}	…	c_{i1} … c_{ik_i}	…	c_{n1} … c_{nk_n}	e_{11} … e_{1q_1}	…	e_{j1} … e_{jq_j}	…	e_{m1} … e_{mq_m}
	C_1	…	C_I	…	C_N	E_1	…	E_J	…	E_M
A_1	v_{11}	…	v_{I1}	…	v_{N1}	r_{11}	…	r_{1J}	…	r_{1M}
…	…	…	…	…	…	…	…	…	…	…
A_L	v_{1L}	…	v_{IL}	…	v_{NL}	r_{L1}	…	r_{LJ}	…	r_{LM}
…	…	…	…	…	…	…	…	…	…	…
A_K	v_{1K}	…	v_{IK}	…	v_{NK}	r_{K1}	…	r_{KJ}	…	r_{KM}

"Causes – rules – effects" interconnection is given by the hierarchical system of relational matrices $\mathbf{V} \subseteq C_I \times A_L = [v_{IL}, I = \overline{1,N}, L = \overline{1,K}]$ and $\mathbf{R} \subseteq A_L \times E_J = [r_{LJ}, L = \overline{1,K}, J = \overline{1,M}]$. An element of binary matrix $\mathbf{V}$ is the weight of term $v_{IL} \in \{0,1\}$, where $v_{IL} = 1(0)$ if term C_I is present (absent) in the causes combination A_L. An element of fuzzy relational matrix $\mathbf{R}$ is the weight of rule $r_{LJ} \in [0, 1]$, characterizing the degree to which causes combination A_L influences upon the rise of effect E_J.

Given the matrices **R** and **V**, the "causes-effects" dependency can be described with the help of Zadeh's compositional rule of inference [5]

$$\boldsymbol{\mu}^E = \boldsymbol{\mu}^A \circ \mathbf{R}, \tag{7.3}$$

where

$$\boldsymbol{\mu}^A = \boldsymbol{\mu}^C \bullet \overline{\mathbf{V}}. \tag{7.4}$$

Here $\overline{\mathbf{V}}$ is the complement of the matrix of terms weights $\mathbf{V}$;

$\boldsymbol{\mu}^C = (\mu^{C_1}, \mu^{C_2}, ..., \mu^{C_N})$ is the fuzzy causes vector with elements $\mu^{C_I} \in [0, 1]$, interpreted as some significance measures of C_I causes;

$\boldsymbol{\mu}^E = (\mu^{E_1}, \mu^{E_2}, ..., \mu^{E_M})$ is the fuzzy effects vector with elements $\mu^{E_J} \in [0, 1]$, interpreted as some significance measures of E_J effects;

$\boldsymbol{\mu}^A = (\mu^{A_1}, \mu^{A_2}, ..., \mu^{A_K})$ is the fuzzy causes combinations vector with elements $\mu^{A_L} \in [0, 1]$, interpreted as some significance measures of A_L causes combinations;

- $(\circ)$ is the operation of *min-max* (*max-min*) composition [5].

Finding vector $\boldsymbol{\mu}^C$ amounts to the solution of the hierarchical system of simplified fuzzy relational equations with *max-min* and dual *min-max* laws of composition

$$\mu^{E_1} = (\mu^{A_1} \wedge r_{11}) \vee (\mu^{A_2} \wedge r_{21}) \vee ... \vee (\mu^{A_K} \wedge r_{K1})$$

$$\mu^{E_2} = (\mu^{A_1} \wedge r_{12}) \vee (\mu^{A_2} \wedge r_{22}) \vee ... \vee (\mu^{A_K} \wedge r_{K2})$$

$$...$$

$$\mu^{E_M} = (\mu^{A_1} \wedge r_{1M}) \vee (\mu^{A_2} \wedge r_{2M}) \vee ... \vee (\mu^{A_K} \wedge r_{KM}), \tag{7.5}$$

where

$$\mu^{A_1} = (\mu^{C_1} \vee \bar{v}_{11}) \wedge (\mu^{C_2} \vee \bar{v}_{21}) \wedge ... \wedge (\mu^{C_N} \vee \bar{v}_{N1})$$

$$\mu^{A_2} = (\mu^{C_1} \vee \bar{v}_{12}) \wedge (\mu^{C_2} \vee \bar{v}_{22}) \wedge ... \wedge (\mu^{C_N} \vee \bar{v}_{N2})$$

$$...$$

$$\mu^{A_K} = (\mu^{C_1} \vee \bar{v}_{1K}) \wedge (\mu^{C_2} \vee \bar{v}_{2K}) \wedge ... \wedge (\mu^{C_N} \vee \bar{v}_{NK}), \tag{7.6}$$

which is derived from relations (7.3) and (7.4).

Since the operations $\vee$ and $\wedge$ are replaced by *max* and *min* in fuzzy set theory [5], systems (7.5) and (7.6) can be rearranged as:

$$\mu^{E_J} = \max_{L=\overline{1,K}}(\min(\mu^{A_L}, r_{LJ})),\ J = \overline{1,M}\ , \tag{7.7}$$

where

$$\mu^{A_L} = \min_{I=\overline{1,N}}(\max(\mu^{C_I}, \bar{v}_{IL})),\ L = \overline{1,K} \tag{7.8}$$

or

$$\mu^{E_J} = \max_{L=\overline{1,K}}\left(\min\left(\min_{I=\overline{1,N}}(\max(\mu^{C_I}, \bar{v}_{IL})), r_{LJ} \right)\right),\ J = \overline{1,M}\ . \tag{7.9}$$

To translate the specific values of the input and output variables into the measures of the causes and effects significances it is necessary to define a membership function of linguistic terms C_I and E_J, $I = \overline{1,N}$, $J = \overline{1,M}$, used in the fuzzy rules (7.1). We use a bell-shaped membership function model of variable u to arbitrary term T in the form:

$$\mu^T(u) = \frac{1}{1+\left(\frac{u-\beta}{\sigma}\right)^2}, \tag{7.10}$$

where β is a coordinate of function maximum, $\mu^T(\beta)=1$; σ is a parameter of concentration-extension.

Correlations (7.9), (7.10) define the generalized fuzzy model of diagnosis as follows:

$$\mathbf{\mu}^E(\mathbf{Y}, \mathbf{B}_E, \mathbf{\Omega}_E) = F_Y(\mathbf{X}, \mathbf{R}, \mathbf{B}_C, \mathbf{\Omega}_C), \tag{7.11}$$

where $\mathbf{R} = (r_{11}, r_{12}, ..., r_{1K}, ..., r_{m1}, r_{m2}, ..., r_{mK})$ is the vector of rules weights;

$\mathbf{B}_C = (\beta^{C_1}, \beta^{C_2}, ..., \beta^{C_N})$ and $\mathbf{\Omega}_C = (\sigma^{C_1}, \sigma^{C_2}, ..., \sigma^{C_N})$ are the vectors of β - and σ - parameters for fuzzy causes C_1, C_2,..., C_N membership functions;

$\mathbf{B}_E = (\beta^{E_1}, \beta^{E_2}, ..., \beta^{E_M})$ and $\mathbf{\Omega}_E = (\sigma^{E_1}, \sigma^{E_2}, ..., \sigma^{E_M})$ are the vectors of β - and σ - parameters for fuzzy effects E_1, E_2,..., E_M membership functions;

F_Y is the operator of inputs-outputs connection, corresponding to formulae (7.9), (7.10).

7.3 Optimization Problem for Fuzzy Rules Based Inverse Inference

Following the approach, proposed in [14 – 16], the problem of solving fuzzy logic equations (7.9) is formulated as follows. Fuzzy causes vector $\mathbf{\mu}^C = (\mu^{C_1}, \mu^{C_2}, ..., \mu^{C_N})$

should be found which satisfies the constraints $\mu^{C_I} \in [0, 1]$, $I = \overline{1,N}$, and also provides the least distance between observed and model fuzzy effects vectors:

$$F = \sum_{J=1}^{M} \left[\mu^{E_J} - \max_{L=\overline{1,K}} \left(\min \left(\min_{I=\overline{1,N}} (\max(\mu^{C_I}, \bar{v}_{IL})), r_{LJ} \right) \right) \right]^2 = \min_{\mathbf{\mu}^C} . \quad (7.12)$$

Solving hierarchical system of fuzzy relational equations (7.9) is accomplished by way of consequent solving system (7.7) with *max-min* law of composition and system (7.8) with *min-max* law of composition.

The problem of solving fuzzy logic equations (7.7) is formulated as follows. Fuzzy causes combinations vector $\mathbf{\mu}^A = (\mu^{A_1}, \mu^{A_2}, ..., \mu^{A_K})$ should be found which satisfies the constraints $\mu^{A_L} \in [0, 1]$, $L = \overline{1,K}$, and also provides the least distance between observed and model fuzzy effects vectors:

$$F_1 = \sum_{J=1}^{M} \left[\mu^{E_J} - \max_{L=\overline{1,K}} (\min(\mu^{A_L}, r_{LJ})) \right]^2 = \min_{\mathbf{\mu}^A} \quad (7.13)$$

The problem of solving fuzzy logic equations (7.8) is formulated as follows. Fuzzy causes vector $\mathbf{\mu}^C = (\mu^{C_1}, \mu^{C_2}, ..., \mu^{C_N})$, should be found which satisfies the constraints $\mu^{C_I} \in [0, 1]$, $I = \overline{1,N}$, and also provides the least distance between observed and model fuzzy causes combinations vectors:

$$F_2 = \sum_{L=1}^{K} \left[\mu^{A_L} - \min_{I=\overline{1,N}} (\max(\mu^{C_I}, \bar{v}_{IL})) \right]^2 = \min_{\mathbf{\mu}^C} \quad (7.14)$$

Following [8], in the general case, system (7.7) has a solution set $S(\mathbf{R}, \mathbf{\mu}^E)$, which is completely characterized by the unique greatest solution $\bar{\mathbf{\mu}}^A$ and the set of lower solutions $S^*(\mathbf{R}, \mathbf{\mu}^E) = \{ \underline{\mathbf{\mu}}_k^A, k = \overline{1,T} \}$:

$$S(\mathbf{R}, \mathbf{\mu}^E) = \bigcup_{\underline{\mathbf{\mu}}_k^A \in S^*} \left[\underline{\mathbf{\mu}}_k^A, \bar{\mathbf{\mu}}^A \right]. \quad (7.15)$$

Here $\bar{\mathbf{\mu}}^A = (\bar{\mu}^{A_1}, \bar{\mu}^{A_2}, ..., \bar{\mu}^{A_K})$ and $\underline{\mathbf{\mu}}_k^A = (\underline{\mu}_k^{A_1}, \underline{\mu}_k^{A_2}, ..., \underline{\mu}_k^{A_K})$ are the vectors of the upper and lower bounds of causes combinations A_L significance measures, where the union is taken over all $\underline{\mathbf{\mu}}_k^A \in S^*(\mathbf{R}, \mathbf{\mu}^E)$.

For the greatest solution $\bar{\boldsymbol{\mu}}^{A}$, system (7.8) has a solution set $\bar{D}(\bar{\boldsymbol{\mu}}^{A})$, which is completely characterized by the unique least solution $\underline{\boldsymbol{\mu}}^{C}$ and the set of upper solutions $\bar{D}^{*}(\bar{\boldsymbol{\mu}}^{A})=\{\bar{\boldsymbol{\mu}}_{l}^{C}, l=\overline{1,H}\}$:

$$\bar{D}\left(\bar{\boldsymbol{\mu}}^{A}\right) = \bigcup_{\bar{\boldsymbol{\mu}}_{l}^{C} \in \bar{D}^{*}} \left[\underline{\boldsymbol{\mu}}^{C}, \bar{\boldsymbol{\mu}}_{l}^{C}\right]. \tag{7.16}$$

Here $\underline{\boldsymbol{\mu}}^{C}=(\underline{\mu}^{C_1}, \underline{\mu}^{C_2}, ..., \underline{\mu}^{C_N})$ and $\bar{\boldsymbol{\mu}}_{l}^{C}=(\bar{\mu}_{l}^{C_1}, \bar{\mu}_{l}^{C_2}, ..., \bar{\mu}_{l}^{C_N})$ are the vectors of the lower and upper bounds of causes C_I significance measures, where the union is taken over all $\bar{\boldsymbol{\mu}}_{l}^{C} \in \bar{D}^{*}\left(\bar{\boldsymbol{\mu}}^{A}\right)$.

For each lower solution $\underline{\boldsymbol{\mu}}_{k}^{A}$, $k=\overline{1,T}$, system (7.8) has a solution set $\underline{D}_{k}(\underline{\boldsymbol{\mu}}_{k}^{A})$, which is completely characterized by the unique least solution $\underline{\boldsymbol{\mu}}_{k}^{C}$ and the set of upper solutions $\underline{D}_{k}^{*}(\underline{\boldsymbol{\mu}}_{k}^{A})=\left\{\bar{\boldsymbol{\mu}}_{kl}^{C}, l=\overline{1,H_k}\right\}$:

$$\underline{D}_{k}(\underline{\boldsymbol{\mu}}_{k}^{A}) = \bigcup_{\bar{\boldsymbol{\mu}}_{kl}^{C} \in \underline{D}_{k}^{*}} \left[\underline{\boldsymbol{\mu}}_{k}^{C}, \bar{\boldsymbol{\mu}}_{kl}^{C}\right]. \tag{7.17}$$

Here $\underline{\boldsymbol{\mu}}_{k}^{C}=(\underline{\mu}_{k}^{C_1}, \underline{\mu}_{k}^{C_2}, ..., \underline{\mu}_{k}^{C_N})$ and $\bar{\boldsymbol{\mu}}_{kl}^{C}=(\bar{\mu}_{kl}^{C_1}, \bar{\mu}_{kl}^{C_2}, ..., \bar{\mu}_{kl}^{C_N})$ are the vectors of the lower and upper bounds of causes C_I significance measures, where the union is taken over all $\bar{\boldsymbol{\mu}}_{kl}^{C} \in \underline{D}_{k}^{*}(\underline{\boldsymbol{\mu}}_{k}^{A})$.

Following [14 – 16], formation of diagnostic results begins with the search for the null solution $\boldsymbol{\mu}_{0}^{C}=(\mu_{0}^{C_1}, \mu_{0}^{C_2}, ..., \mu_{0}^{C_N})$ of optimization problem (7.12). Formation of intervals (7.15) begins with the search for the null vector of the causes combinations significances measures $\boldsymbol{\mu}_{0}^{A}(\boldsymbol{\mu}_{0}^{C})=(\mu_{0}^{A_1}, \mu_{0}^{A_2}, ..., \mu_{0}^{A_K})$, which corresponds to the obtained null solution $\boldsymbol{\mu}_{0}^{C}$. The upper bound ($\bar{\mu}^{A_L}$) is found in the range $[\mu_{0}^{A_L}, 1]$. The lower bound ($\underline{\mu}_{k}^{A_L}$) for $k=1$ is found in the range $[0, \mu_{0}^{A_L}]$, and for $k>1$ – in the range $[0, \bar{\mu}^{A_L}]$, where the minimal solutions $\underline{\boldsymbol{\mu}}_{p}^{A}$, $p<k$, are excluded from the search space.

Let $\boldsymbol{\mu}^{A}(t)=(\mu^{A_1}(t), \mu^{A_2}(t), ..., \mu^{A_K}(t))$ be some t-th solution of optimization problem (7.13). While searching for upper bounds ($\bar{\mu}^{A_L}$) it is suggested that

$\mu^{A_L}(t) \geq \mu^{A_L}(t-1)$, and while searching for lower bounds ($\underline{\mu}_k^{A_L}$) it is suggested that $\mu^{A_L}(t) \leq \mu^{A_L}(t-1)$ (Fig. 7.2a). The definition of the upper (lower) bounds follows the rule: if $\boldsymbol{\mu}^A(t) \neq \boldsymbol{\mu}^A(t-1)$, then $\overline{\mu}^{A_L}$ ($\underline{\mu}_k^{A_L}$)= $\mu^{A_L}(t)$, $L = \overline{1,K}$. If $\boldsymbol{\mu}^A(t) = \boldsymbol{\mu}^A(t-1)$, then the search for the interval solution $[\underline{\boldsymbol{\mu}}_k^A, \overline{\boldsymbol{\mu}}^A]$ is stopped. Formation of intervals (7.15) will go on until the condition $\underline{\boldsymbol{\mu}}_k^A \neq \underline{\boldsymbol{\mu}}_p^A$, $p < k$, has been satisfied.

Formation of intervals (7.16) begins with the search for the null solution $\overline{\boldsymbol{\mu}}_0^C = (\overline{\mu}_0^{C_1}, \overline{\mu}_0^{C_2}, ..., \overline{\mu}_0^{C_N})$ for the greatest solution $\overline{\boldsymbol{\mu}}^A$. The lower bound ($\underline{\mu}^{C_l}$) of solution set (7.16) is found in the range $[0, \overline{\mu}_0^{C_l}]$. The upper bound ($\overline{\mu}_l^{C_l}$) for $l = 1$ is found in the range $[\overline{\mu}_0^{C_l}, 1]$, and for $l > 1$ – in the range $[\underline{\mu}^{C_l}, 1]$, where the maximal solutions $\overline{\mu}_p^{C_l}$, $p < l$, are excluded from the search space.

Formation of intervals (7.17) begins with the search for the null solutions $\underline{\boldsymbol{\mu}}_{0k}^C = (\underline{\mu}_{0k}^{C_1}, \underline{\mu}_{0k}^{C_2}, ..., \underline{\mu}_{0k}^{C_N})$ for each of the lower solutions $\underline{\boldsymbol{\mu}}_k^A$, $k = \overline{1,T}$. The lower bound ($\underline{\mu}_k^{C_l}$) of solution set (7.17) is found in the range $[0, \underline{\mu}_{0k}^{C_l}]$. The upper bound ($\overline{\mu}_{kl}^{C_l}$) for $l = 1$ is found in the range $[\underline{\mu}_{0k}^{C_l}, 1]$, and for $l > 1$ – in the range $[\underline{\mu}_k^{C_l}, 1]$, where the maximal solutions $\overline{\mu}_{kp}^{C_l}$, $p < l$, are excluded from the search space.

Let $\boldsymbol{\mu}^C(t) = (\mu^{C_1}(t), \mu^{C_2}(t), ..., \mu^{C_N}(t))$ be some t-th solution of optimization problem (7.14). While searching for upper bounds ($\overline{\mu}_l^{C_l}$ or $\overline{\mu}_{kl}^{C_l}$) it is suggested that $\mu^{C_l}(t) \geq \mu^{C_l}(t-1)$, and while searching for lower bounds ($\underline{\mu}^{C_l}$ or $\underline{\mu}_k^{C_l}$) it is suggested that $\mu^{C_l}(t) \leq \mu^{C_l}(t-1)$ (Fig. 7.2b,c). The definition of the upper (lower) bounds follows the rule: if $\boldsymbol{\mu}^C(t) \neq \boldsymbol{\mu}^C(t-1)$, then $\overline{\mu}_l^{C_l}$ ($\underline{\mu}^{C_l}$)= $\mu^{C_l}(t)$ or $\overline{\mu}_{kl}^{C_l}$ ($\underline{\mu}_k^{C_l}$)= $\mu^{C_l}(t)$, $I = \overline{1,N}$. If $\boldsymbol{\mu}^C(t) = \boldsymbol{\mu}^C(t-1)$, then the search for the interval solution $[\underline{\boldsymbol{\mu}}_l^C, \overline{\boldsymbol{\mu}}^C]$ or $[\underline{\boldsymbol{\mu}}_{kl}^C, \overline{\boldsymbol{\mu}}_k^C]$ is stopped. Formation of intervals (7.16) and (7.17) will go on until the conditions $\overline{\boldsymbol{\mu}}_l^C \neq \overline{\boldsymbol{\mu}}_p^C$ and $\overline{\boldsymbol{\mu}}_{kl}^C \neq \overline{\boldsymbol{\mu}}_{kp}^C$, $p < l$, have been satisfied.

The hybrid genetic and neuro approach is proposed for solving optimization problems (7.12) – (7.14).

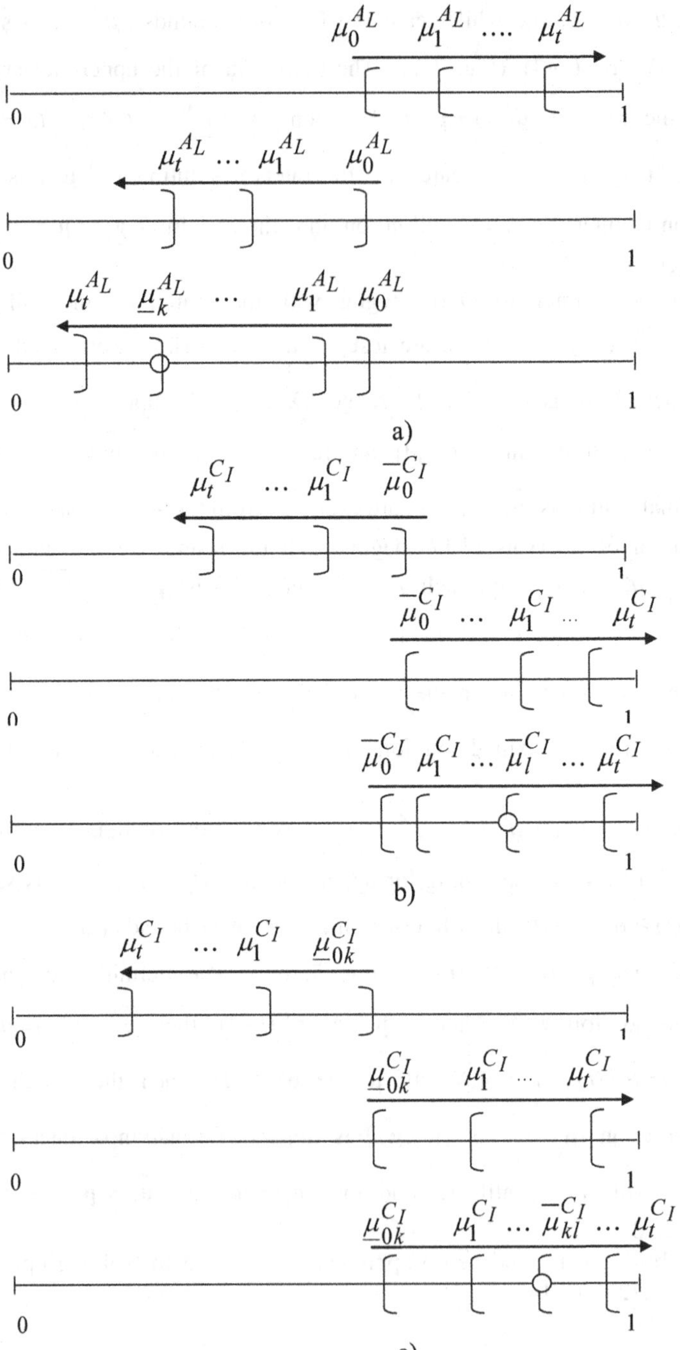

Fig. 7.2. Search for the solution sets (7.15) (a), (7.16) (b), (7.17) (c)

7.4 Genetic Algorithm for Fuzzy Rules Based Inverse Inference

The chromosome needed in the genetic algorithm [14 – 16] for solving optimization problems (7.12) – (7.14) includes the binary codes of parameters μ^{C_I} , $I = \overline{1,N}$, and μ^{A_L} , $L = \overline{1,K}$ (Fig. 7.3).

μ^{C_1}	μ^{C_2}	...	μ^{C_N}	μ^{A_1}	μ^{A_2}	...	μ^{A_K}

Fig. 7.3. Structure of the chromosome

The crossover operation is defined in Fig. 7.4, and is carried out by way of exchanging genes inside each of the solutions μ^{C_I} and μ^{A_L} . The points of crossover shown in dotted lines are selected randomly. Upper symbols (1 and 2) in the vectors of parameters correspond to the first and second chromosomes-parents.

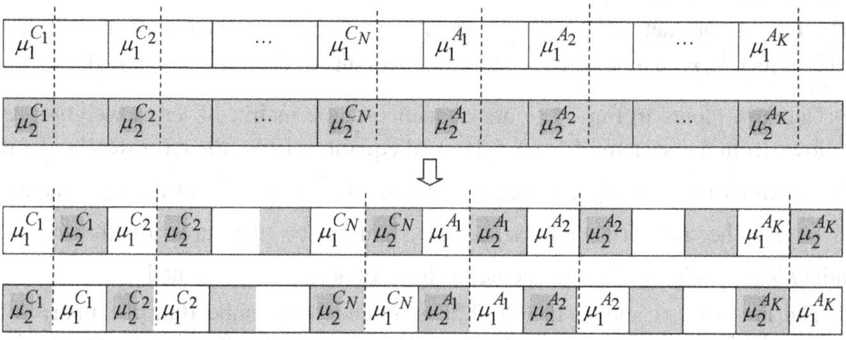

Fig. 7.4. Structure of the crossover operation

A mutation operation implies random change (with some probability) of chromosome elements

$$Mu\ (\mu^{C_I}) = RANDOM\left(\left[\ \underline{\mu}^{C_I}, \overline{\mu}^{C_I}\ \right]\right);$$

$$Mu\ (\mu^{A_L}) = RANDOM\left(\left[\ \underline{\mu}^{A_L}, \overline{\mu}^{A_L}\ \right]\right).$$

where $RANDOM([\underline{x}, \overline{x}])$ denotes a random number within the interval $[\underline{x}, \overline{x}]$.

Fitness function is built on the basis of criteria (7.12) – (7.14).

7.5 Neuro-fuzzy Network for Fuzzy Rules Based Inverse Inference

The neuro-fuzzy networks isomorphic to the systems of fuzzy logical equations (7.7) – (7.9), are presented in Fig. 7.5, a-c, respectively, and the elements of the neuro-fuzzy networks are shown in Table 3.1 [16].

The network in Fig. 7.5,a is designed so that the adjusted weights of arcs are the unknown significance measures of causes combinations μ^{A_L}, $L = \overline{1,K}$. The network in Fig. 7.5b is designed so that the adjusted weights of arcs are the unknown significance measures of causes μ^{C_I}, $I = \overline{1,N}$.

Network inputs in Fig. 7.5a are elements of the matrix of rules weights. As follows from the system of fuzzy relational equations (7.7), the rule weight r_{LJ} is the significance measure of the effect μ^{E_J} provided that the significance measure of the causes combination μ^{A_L} is equal to unity, and the significance measures of other combinations are equal to zero, i. e., $r_{LJ} = \mu^{E_J}$ ($\mu^{A_L}=1$, $\mu^{A_P}=0$), $P = \overline{1,K}$, $P \neq L$. At the network outputs, actual significance measures of the effects $\max\limits_{L=\overline{1,K}}[\min(\mu^{A_L}, r_{LJ})]$ obtained for the actual weights of arcs μ^{A_L} are united.

Network inputs in Fig. 7.5,b are elements of the matrix of terms weights. As follows from the system of fuzzy relational equations (7.8), the term weight v_{IL} is the maximal possible significance measure of the cause μ^{C_I} in the combination μ^{A_L}. At the network outputs, actual significance measures of the causes $\min(\mu^{C_I}, v_{IL})$ obtained for the actual weights of arcs μ^{C_I} are united.

The neuro-fuzzy model in Fig. 7.5c is obtained by embedding the matrix of fuzzy relations into the neural network so that the adjusted weights of arcs are the unknown significance measures of the causes μ^{C_I}, $I = \overline{1,N}$. Network inputs in Fig. 7.5,c are elements of the matrix of terms weights. At the network outputs, actual significance measures of the effects $\max\limits_{L=\overline{1,K}}\left(\min\left(\min\limits_{I=\overline{1,N}}(\mu^{C_I}, v_{IL}), r_{LJ}\right)\right)$ obtained for the actual weights of arcs μ^{C_I} and r_{LJ} are united.

Thus, the problem of solving the system of fuzzy logic equations (7.9) is reduced to the problem of training of a neuro fuzzy network (see Fig. 7.5c) with the use of points

$$(v_{1L}, v_{2L}, \ldots, v_{NL}, \mu^{E_J}),\ L = \overline{1,K},\ J = \overline{1,M}.$$

The problem of solving the system of fuzzy logic equations (7.7) is reduced to the problem of training of a neuro fuzzy network (see Fig. 7.5a) with the use of points

$$(r_{1J}, r_{2J}, \ldots, r_{KJ}, \mu^{E_J}),\ J = \overline{1,M}.$$

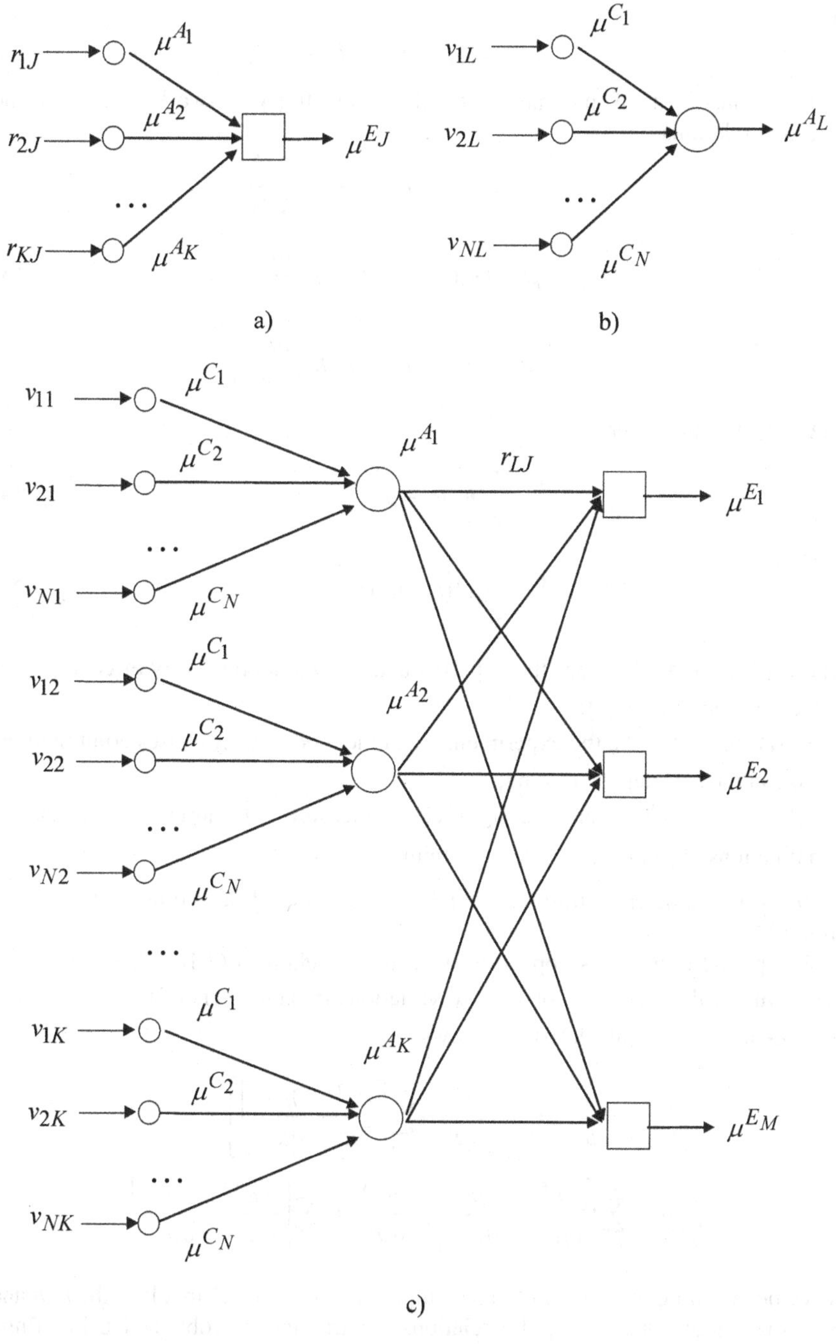

Fig. 7.5. Neuro-fuzzy models of diagnostic equations

The problem of solving the system of fuzzy logic equations (7.8) is reduced to the problem of training of a neuro fuzzy network (see Fig. 7.5b) with the use of points

$$(v_{1L}, v_{2L}, ..., v_{NL}, \mu^{A_L}), \; L = \overline{1,K} .$$

The adjustment of parameters of the neuro-fuzzy networks employs the recurrent relations

$$\mu^{C_I}(t+1) = \mu^{C_I}(t) - \eta \frac{\partial \varepsilon_t^E}{\partial \mu^{C_I}(t)} ,$$

$$\mu^{A_L}(t+1) = \mu^{A_L}(t) - \eta \frac{\partial \varepsilon_t^E}{\partial \mu^{A_L}(t)} , \tag{7.18}$$

$$\mu^{C_I}(t+1) = \mu^{C_I}(t) - \eta \frac{\partial \varepsilon_t^A}{\partial \mu^{C_I}(t)} ,$$

that minimize the criteria

$$\varepsilon_t^E = \frac{1}{2}(\hat{\boldsymbol{\mu}}^E(t) - \boldsymbol{\mu}^E(t))^2 , \tag{7.19}$$

$$\varepsilon_t^A = \frac{1}{2}(\hat{\boldsymbol{\mu}}^A(t) - \boldsymbol{\mu}^A(t))^2 , \tag{7.20}$$

where $\hat{\boldsymbol{\mu}}^E(t)$ and $\boldsymbol{\mu}^E(t)$ are the experimental and the model fuzzy effects vectors at the t-th step of training;

$\hat{\boldsymbol{\mu}}^A(t)$ and $\boldsymbol{\mu}^A(t)$ are the experimental and the model fuzzy causes combinations vectors at the t-th step of training;

$\mu^{C_I}(t)$ and $\mu^{A_L}(t)$ are the significance measures of causes C_I and causes combinations A_L at the t-th step of training;

η is a parameter of training, which can be selected according to the results from [22].

The partial derivatives appearing in recurrent relations (7.18) characterize the sensitivity of the error (ε_t^E or ε_t^A) to variations in parameters of the neuro-fuzzy network and can be calculated as follows:

$$\frac{\partial \varepsilon_t^E}{\partial \mu^{C_I}} = \sum_{J=1}^{M} \left[\frac{\partial \varepsilon_t^E}{\partial \mu^{E_J}} \cdot \sum_{L=1}^{K} \left[\frac{\partial \mu^{E_J}}{\partial \mu^{A_L}} \cdot \frac{\partial \mu^{A_L}}{\partial \mu^{C_I}} \right] \right];$$

$$\frac{\partial \varepsilon_t^E}{\partial \mu^{A_L}} = \sum_{J=1}^{M} \left[\frac{\partial \varepsilon_t^E}{\partial \mu^{E_J}} \cdot \frac{\partial \mu^{E_J}}{\partial \mu^{A_L}} \right]; \quad \frac{\partial \varepsilon_t^A}{\partial \mu^{C_I}} = \sum_{L=1}^{K} \left[\frac{\partial \varepsilon_t^A}{\partial \mu^{A_L}} \cdot \frac{\partial \mu^{A_L}}{\partial \mu^{C_I}} \right].$$

Since determining the element "fuzzy output" from Table 3.1 involves the *min* and *max* fuzzy-logic operations, the relations for training are obtained using finite differences.

7.6 Problem of Fuzzy Rules Tuning

It is assumed that the training data which is given in the form of L pairs of experimental data is known: $\langle \hat{\mathbf{X}}_p, \hat{\mathbf{Y}}_p \rangle$, $p = \overline{1, L}$, where $\hat{\mathbf{X}}_p = (\hat{x}_1^p, \hat{x}_2^p, ..., \hat{x}_n^p)$ and $\hat{\mathbf{Y}}_p = (\hat{y}_1^p, \hat{y}_2^p, ..., \hat{y}_m^p)$ are the vectors of the values of the input and output variables in the experiment number p.

The essence of tuning of the fuzzy model (7.11) consists of finding such null solutions $\boldsymbol{\mu}_0^C(\hat{x}_1^p, \hat{x}_2^p, ..., \hat{x}_n^p)$ of the inverse problem, which minimize criterion (7.12) for all the points of the training data:

$$\sum_{p=1}^{L} [F_Y(\boldsymbol{\mu}_0^C(\hat{x}_1^p, \hat{x}_2^p, ..., \hat{x}_n^p)) - \hat{\boldsymbol{\mu}}^E(\hat{y}_1^p, \hat{y}_2^p, ..., \hat{y}_m^p)]^2 = \min.$$

In other words, the essence of tuning of the fuzzy model (7.11) consists of finding such a vector of rules weights $\mathbf{R}$ and such vectors of membership functions parameters $\mathbf{B}_C$, $\boldsymbol{\Omega}_C$, $\mathbf{B}_E$, $\boldsymbol{\Omega}_E$, which provide the least distance between model and experimental fuzzy effects vectors:

$$\sum_{p=1}^{L} [F_Y(\hat{\mathbf{X}}_p, \mathbf{R}, \mathbf{B}_C, \boldsymbol{\Omega}_C) - \hat{\boldsymbol{\mu}}^E(\hat{\mathbf{Y}}_p, \mathbf{B}_E, \boldsymbol{\Omega}_E)]^2 = \min_{\mathbf{R}, \mathbf{B}_C, \boldsymbol{\Omega}_C, \mathbf{B}_E, \boldsymbol{\Omega}_E} . \quad (7.21)$$

7.7 Genetic Algorithm for Fuzzy Rules Tuning

The chromosome needed in the genetic algorithm [23, 24] for solving the optimization problem (7.21) is defined as the vector-line of binary codes of parameters $\mathbf{R}$, $\mathbf{B}_C$, $\boldsymbol{\Omega}_C$, $\mathbf{B}_E$, $\boldsymbol{\Omega}_E$ (Fig. 7.6).

$\mathbf{R}$	$\mathbf{B}_C$	$\boldsymbol{\Omega}_C$	$\mathbf{B}_E$	$\boldsymbol{\Omega}_E$

Fig. 7.6. Structure of the chromosome

The crossover operation is defined in Fig. 7.7, and is carried out by way of exchanging genes inside the vector of rules weights ($\mathbf{R}$) and each of the vectors of membership functions parameters $\mathbf{B}_C$, $\boldsymbol{\Omega}_C$, $\mathbf{B}_E$, $\boldsymbol{\Omega}_E$. The points of cross-over shown in dotted lines are selected randomly. Upper symbols (1 and 2) in the vectors of parameters correspond to the first and second chromosomes-parents.

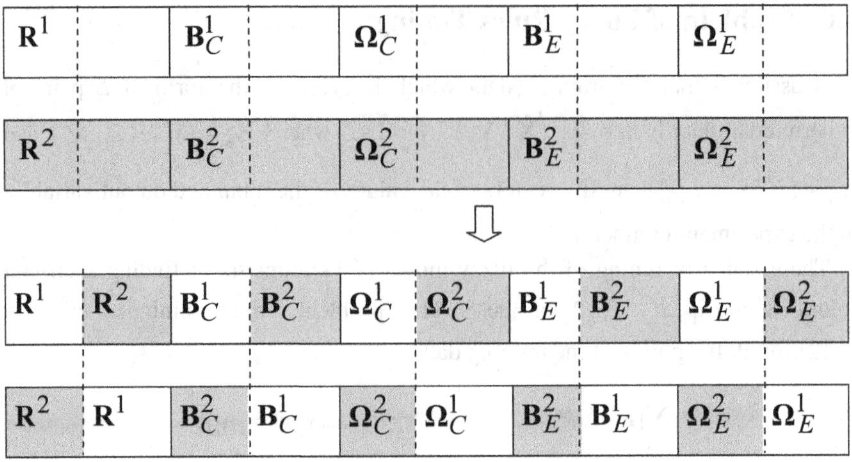

Fig. 7.7. Structure of the crossover operation

A mutation operation implies random change (with some probability) of chromosome elements:

$$Mu\,(\beta^{C_I}) = RANDOM\left(\left[\,\underline{\beta}^{C_I}, \overline{\beta}^{C_I}\,\right]\right);\;\; Mu\,(\sigma^{C_I}) = RANDOM\left(\left[\,\underline{\sigma}^{C_I}, \overline{\sigma}^{C_I}\,\right]\right);$$

$$Mu\,(\beta^{E_J}) = RANDOM\left(\left[\,\underline{\beta}^{E_J}, \overline{\beta}^{E_J}\,\right]\right);\;\; Mu\,(\sigma^{E_J}) = RANDOM\left(\left[\,\underline{\sigma}^{E_J}, \overline{\sigma}^{E_J}\,\right]\right);$$

$$Mu\,(r_{LJ}) = RANDOM\,([\underline{r}_{LJ}, \overline{r}_{LJ}]),$$

where $RANDOM\,([\underline{x}, \overline{x}])$ denotes a random number within the interval $[\underline{x}, \overline{x}]$.

The fitness function is built on the basis of criterion (7.21).

7.8 Adaptive Tuning of Fuzzy Rules

The neuro-fuzzy model of the object of diagnostics is shown in Fig. 7.8, and the nodes are represented in Table 3.1. The neuro-fuzzy model in Fig. 7.8 is obtained by embedding the matrices of fuzzy relations into the neural network so that the weights of arcs subject to tuning are rules weights (fuzzy relations) and the membership functions for causes and effects fuzzy terms [16, 25].

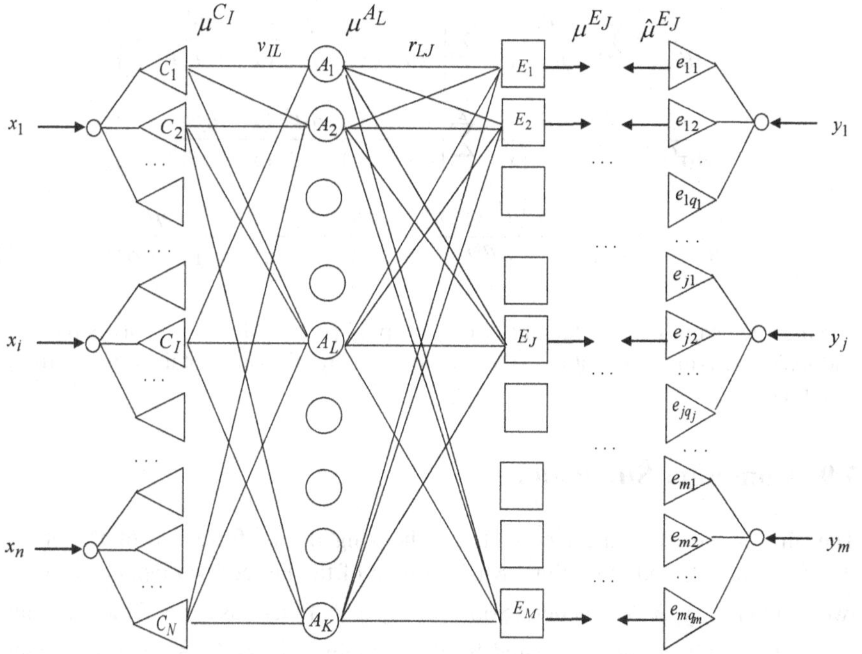

Fig. 7.8. Neuro-fuzzy model of the object of diagnostics

To train the parameters of the neuro-fuzzy network, the recurrent relations:

$$r_{LJ}(t+1) = r_{LJ}(t) - \eta \frac{\partial \varepsilon_t}{\partial r_{LJ}(t)};$$

$$\beta^{C_I}(t+1) = \beta^{C_I}(t) - \eta \frac{\partial \varepsilon_t}{\partial \beta^{C_I}(t)}; \quad \sigma^{C_I}(t+1) = \sigma^{C_I}(t) - \eta \frac{\partial \varepsilon_t}{\partial \sigma^{C_I}(t)};$$

$$\beta^{E_J}(t+1) = \beta^{E_J}(t) - \eta \frac{\partial \varepsilon_t}{\partial \beta^{E_J}(t)}; \quad \sigma^{E_J}(t+1) = \sigma^{E_J}(t) - \eta \frac{\partial \varepsilon_t}{\partial \sigma^{E_J}(t)}, \qquad (7.22)$$

minimizing criterion (7.19) are used, where

$r_{LJ}(t)$ are fuzzy relations (rules weights) at the t-th step of training;

$\beta^{C_I}(t)$, $\sigma^{C_I}(t)$, $\beta^{E_J}(t)$, $\sigma^{E_J}(t)$ are parameters of the membership functions for causes and effects fuzzy terms at the t-th step of training.

The partial derivatives appearing in recurrent relations (7.22) characterize the sensitivity of the error (ε_t) to variations in parameters of the neuro-fuzzy network and can be calculated as follows:

$$\frac{\partial \varepsilon_t}{\partial r_{LJ}} = \frac{\partial \varepsilon_t}{\partial \mu^{E_J}(X)} \cdot \frac{\partial \mu^{E_J}(X)}{\partial r_{LJ}};$$

$$\frac{\partial \varepsilon_t}{\partial \beta^{C_I}} = \sum_{J=1}^{M}\left[\frac{\partial \varepsilon_t}{\partial \mu^{E_J}(x_i)} \cdot \sum_{L=1}^{K}\left[\frac{\partial \mu^{E_J}(x_i)}{\partial \mu^{A_L}(x_i)} \cdot \frac{\partial \mu^{A_L}(x_i)}{\partial \mu^{C_I}(x_i)} \cdot \frac{\partial \mu^{C_I}(x_i)}{\partial \beta^{C_I}}\right]\right];$$

$$\frac{\partial \varepsilon_t}{\partial \sigma^{C_I}} = \sum_{J=1}^{M}\left[\frac{\partial \varepsilon_t}{\partial \mu^{E_J}(x_i)} \cdot \sum_{L=1}^{K}\left[\frac{\partial \mu^{E_J}(x_i)}{\partial \mu^{A_L}(x_i)} \cdot \frac{\partial \mu^{A_L}(x_i)}{\partial \mu^{C_I}(x_i)} \cdot \frac{\partial \mu^{C_I}(x_i)}{\partial \sigma^{C_I}}\right]\right];$$

$$\frac{\partial \varepsilon_t}{\partial \beta^{E_J}} = \frac{\partial \varepsilon_t}{\partial \mu^{E_J}(y_j)} \cdot \frac{\partial \mu^{E_J}(y_j)}{\partial \beta^{E_J}}; \qquad \frac{\partial \varepsilon_t}{\partial \sigma^{E_J}} = \frac{\partial \varepsilon_t}{\partial \mu^{E_J}(y_j)} \cdot \frac{\partial \mu^{E_J}(y_j)}{\partial \sigma^{E_J}}.$$

Since determining the element "fuzzy output" (see Table 3.1) involves the *min* and *max* fuzzy-logic operations, the relations for training are obtained using finite differences.

7.9 Computer Simulations

The aim of the experiment consists of checking the performance of the above proposed models and algorithms with the help of the target "two inputs (x_1, x_2) – two outputs (y_1, y_2)" model. Some analytical functions $y_1 = f_1(x_1, x_2)$ and $y_2 = f_2(x_1, x_2)$ were approximated by the combined fuzzy knowledge base, and served simultaneously as training and testing data generator. The input values (x_1, x_2) restored for each output combination (y_1, y_2) were compared with the target level lines.

The target model is given by the formulae:

$$y_1 = f_1(x_1, x_2) = \frac{1}{10}(2z-0.9)\ (7z-1)\ (17z-19)\ (15z-2)\,, \tag{7.23}$$

$$y_2 = f_2(x_1, x_2) = -\frac{1}{2}y_1 + 1\,,$$

where $z = \dfrac{(x_1-3.0)^2 + (x_2-2.5)^2}{40}$.

The target model is represented in Fig. 7.9.

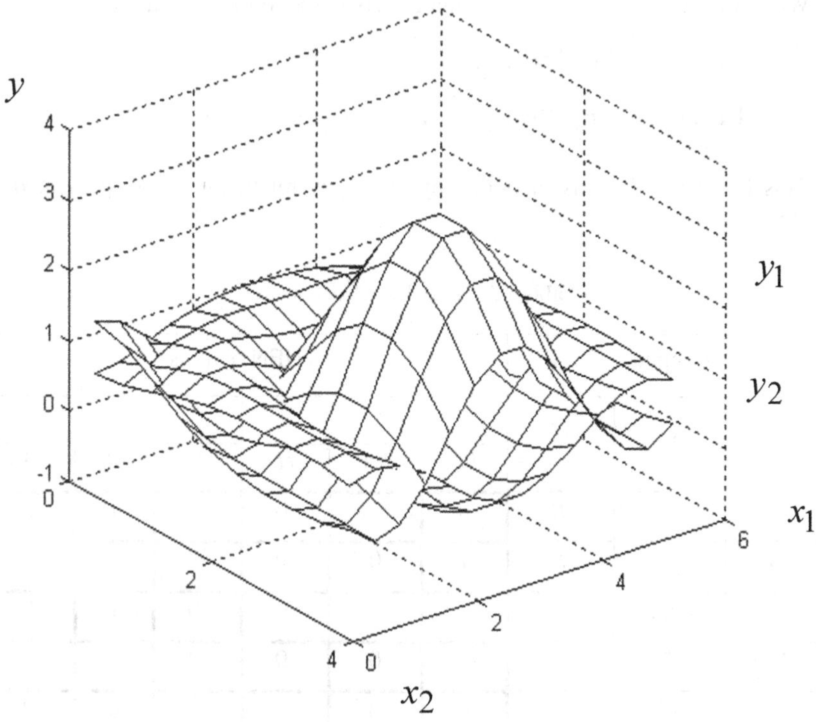

Fig. 7.9. "Inputs-outputs" model-generator

The fuzzy IF-THEN rules correspond to this model:

Rule 1: IF x_1 =*L* AND x_2 =*L* THEN y_1 =*hA* AND y_2 =*lA*;

Rule 2: IF x_1 =*A* AND x_2 =*L* THEN y_1 =*hL* AND y_2 =*A*;

Rule 3: IF x_1 =*H* AND x_2 =*L* THEN y_1 =*hA* AND y_2 =*lA*;

Rule 4: IF x_1 =*L* AND x_2 =*H* THEN y_1 =*hL* AND y_2 =*A*;

Rule 5: IF x_1 =*A* AND x_2 =*H* THEN y_1 =*H* AND y_2 =*L*;

Rule 6: IF x_1 =*H* AND x_2 =*H* THEN y_1 =*hL* AND y_2 =*A*.

where the total number of the input and output terms-assessments consists of: c_{11} *Low (L)*, c_{12} *Average (A)*, c_{13} *High (H)* for x_1, c_{21} *(Low)*, c_{22} *(High)* for x_2; e_{11} *higher than Low (hL)*, e_{12} *higher than Average (hA)*, e_{13} *High (H)* for y_1; e_{21} *Low (L)*, e_{22} *lower than Average (lA)*, e_{23} *Average (A)* for y_2.

We shall define the set of causes and effects in the following way:

$$\{ C_1, C_2, \ldots, C_5 \} = \{ c_{11}, c_{12}, c_{13}, c_{21}, c_{22} \};$$

$$\{ E_1, E_2, \ldots, E_6 \} = \{ e_{11}, e_{12}, e_{13}, e_{21}, e_{22}, e_{23} \}.$$

This fuzzy rule base is modelled by the fuzzy relational matrix presented in Table 7.3.

Table 7.3. Fuzzy knowledge matrix

IF inputs			THEN outputs					
	x_1	x_2	y_1			y_2		
			hL	*hA*	*H*	*L*	*lA*	*A*
A_1	*L*	*L*	0	1	0	0	1	0
A_2	*A*	*L*	1	0	0	0	0	1
A_3	*H*	*L*	0	1	0	0	1	0
A_4	*L*	*H*	1	0	0	0	0	1
A_5	*A*	*H*	0	0	1	1	0	0
A_6	*H*	*H*	1	0	0	0	0	1

The results of the fuzzy model tuning are given in Tables 7.4, 7.5.

Table 7.4. Parameters of the membership functions for the causes fuzzy terms before (after) tuning

Parameter	Fuzzy terms				
	C_1	C_2	C_3	C_4	C_5
β -	0 (0.03)	3.0 (3.03)	6.0 (5.98)	0 (0.02)	3.0 (3.05)
σ -	1.0 (0.71)	2.0 (0.62)	1.0 (0.69)	1.0 (0.73)	2.0 (0.60)

Table 7.5. Parameters of the membership functions for the effects fuzzy terms before (after) tuning

Parameter	Fuzzy terms					
	E_1	E_2	E_3	E_4	E_5	E_6
β -	0 (0.02)	1.0 (1.10)	3.5 (3.36)	-0.7 (-0.67)	0.5 (0.44)	0.8 (0.89)
σ -	0.5 (0.27)	0.5 (0.29)	2.0 (1.91)	2.0 (1.70)	0.5 (0.31)	0.5 (0.25)

Fuzzy logic equations after tuning take the form:

$$\mu^{E_1} = (\mu^{A_2} \wedge 0.75) \vee (\mu^{A_4} \wedge 0.78) \vee (\mu^{A_6} \wedge 0.86)$$

$$\mu^{E_2} = (\mu^{A_1} \wedge 0.80) \vee (\mu^{A_3} \wedge 0.92)$$

$$\mu^{E_3} = (\mu^{A_5} \wedge 0.97)$$

$$\mu^{E_4} = (\mu^{A_1} \wedge 0.50) \vee (\mu^{A_3} \wedge 0.48) \vee (\mu^{A_5} \wedge 0.77)$$

$$\mu^{E_5} = (\mu^{A_1} \wedge 0.76) \vee (\mu^{A_3} \wedge 0.72)$$

$$\mu^{E_6} = (\mu^{A_2} \wedge 0.96) \vee (\mu^{A_4} \wedge 0.82) \vee (\mu^{A_6} \wedge 0.87) \ , \tag{7.24}$$

where

$$\mu^{A_1} = \mu^{C_1} \wedge \mu^{C_4}$$

$$\mu^{A_2} = \mu^{C_2} \wedge \mu^{C_4}$$

$$\mu^{A_3} = \mu^{C_3} \wedge \mu^{C_4}$$

$$\mu^{A_4} = \mu^{C_1} \wedge \mu^{C_5}$$

$$\mu^{A_5} = \mu^{C_2} \wedge \mu^{C_5}$$

$$\mu^{A_6} = \mu^{C_3} \wedge \mu^{C_5} \, . \tag{7.25}$$

The results of solving the problem of inverse inference before and after tuning are shown in Fig. 7.10, 7.11. The same figure depicts the membership functions of the fuzzy terms for the causes and effects before and after tuning.

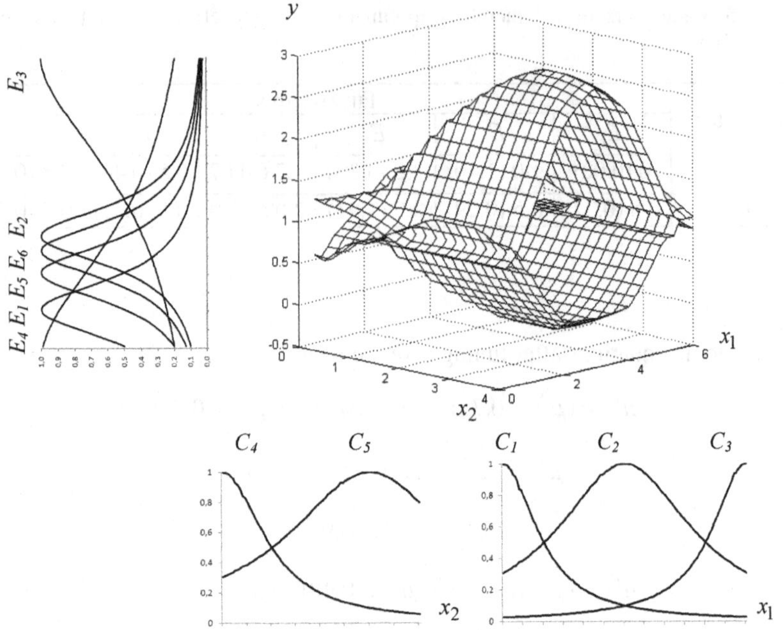

Fig. 7.10. Solution to the problem of inverse fuzzy inference before tuning

Let the specific values of the output variables consist of y_1^*=0.20 and y_2^*=0.80. The fuzzy effects vector for these values can be defined with the help of the membership functions in Fig. 7.11:

$$\mathbf{\mu}^E = (\mu^{E_1}(y_1^*) = 0.69;\ \mu^{E_2}(y_1^*) = 0.09;\ \mu^{E_3}(y_1^*) = 0.27;$$
$$\mu^{E_4}(y_2^*) = 0.57;\ \mu^{E_5}(y_2^*) = 0.43;\ \mu^{E_6}(y_2^*) = 0.89).$$

The genetic algorithm yields a null solution of the optimization problem (7.12)

$$\mathbf{\mu}_0^C = (\mu_0^{C_1} = 0.26,\ \mu_0^{C_2} = 0.93,\ \mu_0^{C_3} = 0.20,\ \mu_0^{C_4} = 0.89,\ \mu_0^{C_5} = 0.42)\,, \qquad (7.26)$$

for which the value of the optimization criterion (7.12) is F=0.1064.

The null vector of the causes combinations significances measures

$$\mathbf{\mu}_0^A = (\mu_0^{A_1} = 0.26,\ \mu_0^{A_2} = 0.89,\ \mu_0^{A_3} = 0.20,\ \mu_0^{A_4} = 0.26,\ \mu_0^{A_5} = 0.42,\ \mu_0^{A_6} = 0.20)$$

corresponds to the obtained null solution.

The obtained null solution allows us to arrange for the genetic search for the solution set $S(\mathbf{R}, \mathbf{\mu}^E)$, which is completely determined by the greatest solution

$$\bar{\mathbf{\mu}}^A = (\ \bar{\mu}^{A_1} = 0.26,\ \bar{\mu}^{A_2} = 0.89,\ \bar{\mu}^{A_3} = 0.26,\ \bar{\mu}^{A_4} = 0.75,\ \bar{\mu}^{A_5} = 0.42,\ \bar{\mu}^{A_6} = 0.75)$$

and the two lower solutions $S^* = \{\underline{\mathbf{\mu}}_1^A, \underline{\mathbf{\mu}}_2^A\}$

$$\underline{\mathbf{\mu}}_1^A = (\underline{\mu}_1^{A_1} = 0.26,\ \underline{\mu}_1^{A_2} = 0.89,\ \underline{\mu}_1^{A_3} = 0,\ \underline{\mu}_1^{A_4} = 0,\ \underline{\mu}_1^{A_5} = 0.42,\ \underline{\mu}_1^{A_6} = 0);$$

$$\underline{\mathbf{\mu}}_2^A = (\underline{\mu}_2^{A_1} = 0,\ \underline{\mu}_2^{A_2} = 0.89,\ \underline{\mu}_2^{A_3} = 0.26,\ \underline{\mu}_2^{A_4} = 0,\ \underline{\mu}_2^{A_5} = 0.42,\ \underline{\mu}_2^{A_6} = 0).$$

Thus, the solution of fuzzy relational equations (7.24) can be represented in the form of intervals:

$$S(\mathbf{R}, \mathbf{\mu}^E) = \{\mu^{A_1} = 0.26,\ \mu^{A_2} = 0.89,\ \mu^{A_3} \in [0, 0.26],\ \mu^{A_4} \in [0, 0.75],\ \mu^{A_5} = 0.42,\ \mu^{A_6} \in [0, 0.75]\}$$
$$\cup \{\mu^{A_1} \in [0, 0.26],\ \mu^{A_2} = 0.89,\ \mu^{A_3} = 0.26,\ \mu^{A_4} \in [0, 0.75],\ \mu^{A_5} = 0.42,\ \mu^{A_6} \in [0, 0.75]\}. \tag{7.27}$$

We next apply the genetic algorithm for solving the optimization problem (7.14) for the greatest solution $\overline{\mathbf{\mu}}^A$ and the two lower solutions $\underline{\mathbf{\mu}}_1^A$ and $\underline{\mathbf{\mu}}_2^A$.

For the greatest solution $\overline{\mathbf{\mu}}^A$, the genetic algorithm yields a null solution of the optimization problem (7.14)

$$\overline{\mathbf{\mu}}_0^C = (\overline{\mu}_0^{C_1} = 0.49,\ \overline{\mu}_0^{C_2} = 0.96,\ \overline{\mu}_0^{C_3} = 0.49,\ \overline{\mu}_0^{C_4} = 0.90,\ \overline{\mu}_0^{C_5} = 0.49), \tag{7.28}$$

for which the value of the optimization criterion (7.14) is F=0.2459.

The obtained null solution allows us to arrange for the genetic search for the solution set $\overline{D}(\overline{\mathbf{\mu}}^A)$, which is completely determined by the least solution

$$\underline{\mathbf{\mu}}^C = (\underline{\mu}^{C_1} = 0.49,\ \underline{\mu}^{C_2} = 0.89,\ \underline{\mu}^{C_3} = 0.49,\ \underline{\mu}^{C_4} = 0.89,\ \underline{\mu}^{C_5} = 0.49)$$

and the two upper solutions $\overline{D}^* = \{\overline{\mathbf{\mu}}_1^C, \overline{\mathbf{\mu}}_2^C\}$

$$\overline{\mathbf{\mu}}_1^C = (\overline{\mu}_1^{C_1} = 0.49,\ \overline{\mu}_1^{C_2} = 0.89,\ \overline{\mu}_1^{C_3} = 0.49,\ \overline{\mu}_1^{C_4} = 1.0,\ \overline{\mu}_1^{C_5} = 0.49);$$

$$\overline{\mathbf{\mu}}_2^C = (\overline{\mu}_2^{C_1} = 0.49,\ \overline{\mu}_2^{C_2} = 1.0,\ \overline{\mu}_2^{C_3} = 0.49,\ \overline{\mu}_2^{C_4} = 0.89,\ \overline{\mu}_2^{C_5} = 0.49).$$

Thus, the solution of fuzzy relational equations (7.25) for the greatest solution $\overline{\mathbf{\mu}}^A$ can be represented in the form of intervals:

$$\overline{D}(\overline{\mathbf{\mu}}^A) = \{\mu^{C_1} = 0.49,\ \mu^{C_2} = 0.89,\ \mu^{C_3} = 0.49,\ \mu^{C_4} \in [0.89, 1.0],\ \mu^{C_5} = 0.49\}$$
$$\cup \{\mu^{C_1} = 0.49,\ \mu^{C_2} \in [0.89, 1.0],\ \mu^{C_3} = 0.49,\ \mu^{C_4} = 0.89,\ \mu^{C_5} = 0.49\}. \tag{7.29}$$

For the first lower solution $\underline{\mathbf{\mu}}_1^A$, the genetic algorithm yields a null solution of the optimization problem (7.14)

$$\underline{\mathbf{\mu}}_{01}^C = (\underline{\mu}_{01}^{C_1} = 0.13,\ \underline{\mu}_{01}^{C_2} = 0.89,\ \underline{\mu}_{01}^{C_3} = 0,\ \underline{\mu}_{01}^{C_4} = 0.94,\ \underline{\mu}_{01}^{C_5} = 0.42), \tag{7.30}$$

for which the value of the optimization criterion (7.14) is F=0.0338.

The obtained null solution allows us to arrange for the genetic search for the solution set $\underline{D}_1(\underline{\mathbf{\mu}}_1^A)$, which is completely determined by the least solution

$$\underline{\mathbf{\mu}}^C = (\underline{\mu}^{C_1}=0.13,\ \underline{\mu}^{C_2}=0.89,\ \underline{\mu}^{C_3}=0,\ \underline{\mu}^{C_4}=0.89,\ \underline{\mu}^{C_5}=0.42)$$

and the two upper solutions $\underline{D}_1^* = \{\overline{\mathbf{\mu}}_1^C, \overline{\mathbf{\mu}}_2^C\}$

$$\overline{\mathbf{\mu}}_1^C = (\overline{\mu}_1^{C_1}=0.13,\ \overline{\mu}_1^{C_2}=0.89,\ \overline{\mu}_1^{C_3}=0,\ \overline{\mu}_1^{C_4}=1.0,\ \overline{\mu}_1^{C_5}=0.42);$$

$$\overline{\mathbf{\mu}}_2^C = (\overline{\mu}_2^{C_1}=0.13,\ \overline{\mu}_2^{C_2}=1.0,\ \overline{\mu}_2^{C_3}=0,\ \overline{\mu}_2^{C_4}=0.89,\ \overline{\mu}_2^{C_5}=0.42).$$

Thus, the solution of fuzzy relational equations (7.25) for the first lower solution $\underline{\mathbf{\mu}}_1^A$ can be represented in the form of intervals:

$$\underline{D}_1(\underline{\mathbf{\mu}}_1^A) = \{\mu^{C_1}=0.13,\ \mu^{C_2}=0.89,\ \mu^{C_3}=0,\ \mu^{C_4}\in[0.89, 1.0],\ \mu^{C_5}=0.42\}$$
$$\cup\{\mu^{C_1}=0.13,\ \mu^{C_2}\in[0.89, 1.0],\ \mu^{C_3}=0,\ \mu^{C_4}=0.89,\ \mu^{C_5}=0.42\}. \quad (7.31)$$

For the second lower solution $\underline{\mathbf{\mu}}_2^A$, the genetic algorithm yields a null solution of the optimization problem (7.14)

$$\underline{\mathbf{\mu}}_{02}^C = (\underline{\mu}_{02}^{C_1}=0,\ \underline{\mu}_{02}^{C_2}=0.97,\ \underline{\mu}_{02}^{C_3}=0.13,\ \underline{\mu}_{02}^{C_4}=0.89,\ \underline{\mu}_{02}^{C_5}=0.42), \quad (7.32)$$

for which the value of the optimization criterion (7.14) is F=0.0338.

The obtained null solution allows us to arrange for the genetic search for the solution set $\underline{D}_2(\underline{\mathbf{\mu}}_2^A)$, which is completely determined by the least solution

$$\underline{\mathbf{\mu}}^C = (\underline{\mu}^{C_1}=0,\ \underline{\mu}^{C_2}=0.89,\ \underline{\mu}^{C_3}=0.13,\ \underline{\mu}^{C_4}=0.89,\ \underline{\mu}^{C_5}=0.42)$$

and the two upper solutions $\underline{D}_2^* = \{\overline{\mathbf{\mu}}_1^C, \overline{\mathbf{\mu}}_2^C\}$

$$\overline{\mathbf{\mu}}_1^C = (\overline{\mu}_1^{C_1}=0,\ \overline{\mu}_1^{C_2}=0.89,\ \overline{\mu}_1^{C_3}=0.13,\ \overline{\mu}_1^{C_4}=1.0,\ \overline{\mu}_1^{C_5}=0.42);$$

$$\overline{\mathbf{\mu}}_2^C = (\overline{\mu}_2^{C_1}=0,\ \overline{\mu}_2^{C_2}=1.0,\ \overline{\mu}_2^{C_3}=0.13,\ \overline{\mu}_2^{C_4}=0.89,\ \overline{\mu}_2^{C_5}=0.42).$$

Thus, the solution of fuzzy relational equations (7.25) for the second lower solution $\underline{\mathbf{\mu}}_2^A$ can be represented in the form of intervals:

$$\underline{D}_2(\underline{\mathbf{\mu}}_2^A) = \{\mu^{C_1}=0,\ \mu^{C_2}=0.89,\ \mu^{C_3}=0.13,\ \mu^{C_4}\in[0.89, 1.0],\ \mu^{C_5}=0.42\}$$
$$\cup\{\mu^{C_1}=0,\ \mu^{C_2}\in[0.89, 1.0],\ \mu^{C_3}=0.13,\ \mu^{C_4}=0.89,\ \mu^{C_5}=0.42\}. \quad (7.33)$$

The intervals of the values of the input variable for each interval in solutions (7.29), (7.31), (7.33) can be defined with the help of the membership functions in Fig. 7.11:

- x_1^* =0.75 or x_1^* =1.85 or x_1^* =6.00 for C_1;
- $x_1^* \in [2.81, 3.25]$ for C_2;
- x_1^* =5.27 or x_1^* =4.20 or x_1^* =0 for C_3;
- $x_2^* \in [0, 0.27]$ for C_4;
- x_2^* =2.44 and x_2^* =3.66 or x_2^* =2.35 and x_2^* =3.75 for C_5.

The restoration of the input set for y_1^* =0.20 and y_2^* =0.80 is shown in Fig. 7.11, in which the values of the causes $C_1 \div C_5$ and effects $E_1 \div E_6$ significances measures are marked. The comparison of the target and restored level lines for y_1^* =0.20 and y_2^* =0.80 is shown in Fig. 7.12.

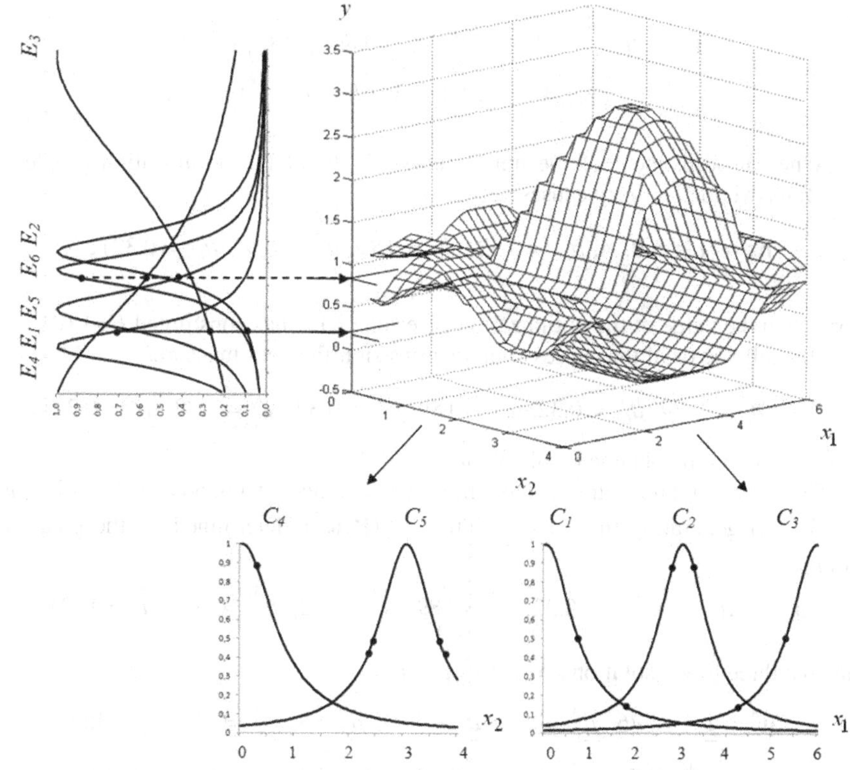

Fig. 7.11. Solution to the problem of inverse fuzzy inference for y_1^* =0.20 and y_2^* =0.80

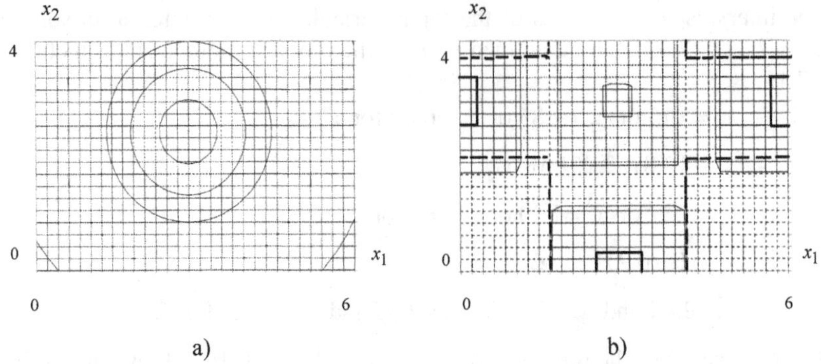

Fig. 7.12. Comparison of the target (a) and restored (b) level linesfor y_1^*=0.20 and y_2^*=0.80

Let the values of the output variables have changed with y_1^*=0.20 and y_2^*=0.80 to y_1^*=1.00 and y_2^*=0.60 (Fig. 7.13). For the new values, the fuzzy effects vector is

$$\mathbf{\mu}^E = (\mu^{E_1}(y_1^*)=0.07;\ \mu^{E_2}(y_1^*)=0.89;\ \mu^{E_3}(y_1^*)=0.40;$$
$$\mu^{E_4}(y_2^*)=0.64;\ \mu^{E_5}(y_2^*)=0.79;\ \mu^{E_6}(y_2^*)=0.43).$$

A neural adjustment of the null solution (7.26) of the optimization problem (7.12) has yielded a fuzzy causes vector

$$\mathbf{\mu}_0^C = (\mu_0^{C_1}=0.84,\ \mu_0^{C_2}=0.32,\ \mu_0^{C_3}=0.89,\ \mu_0^{C_4}=0.95,\ \mu_0^{C_5}=0.32)\ ,$$

for which the value of the optimization criterion (7.12) has constituted F=0.1015.

The null vector of the causes combinations significances measures

$$\mathbf{\mu}_0^A = (\mu_0^{A_1}=0.84,\ \mu_0^{A_2}=0.32,\ \mu_0^{A_3}=0.89,\ \mu_0^{A_4}=0.32,\ \mu_0^{A_5}=0.32,\ \mu_0^{A_6}=0.32)\ ,$$

corresponds to the obtained null solution.

The resultant null solution has allowed adjusting the bounds in the solution (7.27) and generating the set of solutions $S(\mathbf{R},\mathbf{\mu}^E)$ determined by the greatest solution

$$\bar{\mathbf{\mu}}^A = (\bar{\mu}^{A_1}=1.0,\ \bar{\mu}^{A_2}=0.32,\ \bar{\mu}^{A_3}=0.89,\ \bar{\mu}^{A_4}=0.32,\ \bar{\mu}^{A_5}=0.32,\ \bar{\mu}^{A_6}=0.32)$$

and the three lower solutions $S^* = \{\underline{\mathbf{\mu}}_1^A, \underline{\mathbf{\mu}}_2^A, \underline{\mathbf{\mu}}_3^A\}$

$$\underline{\mathbf{\mu}}_1^A = (\underline{\mu}_1^{A_1}=0.76,\ \underline{\mu}_1^{A_2}=0.32,\ \underline{\mu}_1^{A_3}=0.89,\ \underline{\mu}_1^{A_4}=0,\ \underline{\mu}_1^{A_5}=0.32,\ \underline{\mu}_1^{A_6}=0);$$
$$\underline{\mathbf{\mu}}_2^A = (\underline{\mu}_2^{A_1}=0.76,\ \underline{\mu}_2^{A_2}=0,\ \underline{\mu}_2^{A_3}=0.89,\ \underline{\mu}_2^{A_4}=0.32,\ \underline{\mu}_2^{A_5}=0.32,\ \underline{\mu}_2^{A_6}=0);$$
$$\underline{\mathbf{\mu}}_3^A = (\underline{\mu}_3^{A_1}=0.76,\ \underline{\mu}_3^{A_2}=0,\ \underline{\mu}_3^{A_3}=0.89,\ \underline{\mu}_3^{A_4}=0,\ \underline{\mu}_3^{A_5}=0.32,\ \underline{\mu}_3^{A_6}=0.32).$$

Thus, the solution of fuzzy relational equations (7.24) for the new values can be represented in the form of intervals:

$$S(\mathbf{R},\boldsymbol{\mu}^E)=\{\mu^{A_1}\in[0.76,1.0],\ \mu^{A_2}=0.32,\ \mu^{A_3}=0.89,\ \mu^{A_4}\in[0,0.32],\ \mu^{A_5}=0.32,\ \mu^{A_6}\in[0,0.32]\}$$

$$\cup\{\mu^{A_1}\in[0.76,1.0],\ \mu^{A_2}\in[0,0.32],\ \mu^{A_3}=0.89,\ \mu^{A_4}=0.32,\ \mu^{A_5}=0.32,\ \mu^{A_6}\in[0,0.32]\}$$

$$\cup\{\mu^{A_1}\in[0.76,1.0],\ \mu^{A_2}\in[0,0.32],\ \mu^{A_3}=0.89,\ \mu^{A_4}\in[0,0.32],\ \mu^{A_5}=0.32,\ \mu^{A_6}=0.32]\}.$$

For the greatest solution $\bar{\boldsymbol{\mu}}^A$, a neural adjustment of the null solution (7.28) has yielded a fuzzy causes vector

$$\bar{\boldsymbol{\mu}}_0^C=(\bar{\mu}_0^{C_1}=1.0,\ \bar{\mu}_0^{C_2}=0.32,\ \bar{\mu}_0^{C_3}=0.89,\ \bar{\mu}_0^{C_4}=1.0,\ \bar{\mu}_0^{C_5}=0.32),$$

for which the value of the optimization criterion (7.14) has constituted F=0.0.

The resultant null solution has allowed adjusting the bounds in the solution (7.29) and generating the set of solutions $\overline{D}(\bar{\boldsymbol{\mu}}^A)$ determined by the unique (null) solution

$$\overline{D}(\bar{\boldsymbol{\mu}}^A)=\{\mu^{C_1}=1.0,\ \mu^{C_2}=0.32,\ \mu^{C_3}=0.89,\ \mu^{C_4}=1.0,\ \mu^{C_5}=0.32\}. \quad (7.34)$$

For the first lower solution $\underline{\boldsymbol{\mu}}_1^A$, a neural adjustment of the null solution (7.30) has yielded a fuzzy causes vector

$$\underline{\boldsymbol{\mu}}_{01}^C=(\underline{\mu}_{01}^{C_1}=0.76,\ \underline{\mu}_{01}^{C_2}=0.32,\ \underline{\mu}_{01}^{C_3}=0.89,\ \underline{\mu}_{01}^{C_4}=0.92,\ \underline{\mu}_{01}^{C_5}=0.11),$$

for which the value of the optimization criterion (7.14) has constituted F=0.0683.

The resulting null solution has allowed adjusting the bounds in the solution (7.31) and generating the set of solutions $\underline{D}_1(\underline{\boldsymbol{\mu}}_1^A)$, which is completely determined by the least solution

$$\underline{\boldsymbol{\mu}}^C=(\underline{\mu}^{C_1}=0.76,\ \underline{\mu}^{C_2}=0.32,\ \underline{\mu}^{C_3}=0.89,\ \underline{\mu}^{C_4}=0.89,\ \underline{\mu}^{C_5}=0.11)$$

and the two upper solutions $\underline{D}_1^*=\{\bar{\boldsymbol{\mu}}_1^C,\bar{\boldsymbol{\mu}}_2^C\}$

$$\bar{\boldsymbol{\mu}}_1^C=(\bar{\mu}_1^{C_1}=0.76,\ \bar{\mu}_1^{C_2}=0.32,\ \bar{\mu}_1^{C_3}=0.89,\ \bar{\mu}_1^{C_4}=1.0,\ \bar{\mu}_1^{C_5}=0.11);$$

$$\bar{\boldsymbol{\mu}}_2^C=(\bar{\mu}_2^{C_1}=0.76,\ \bar{\mu}_2^{C_2}=0.32,\ \bar{\mu}_2^{C_3}=1.0,\ \bar{\mu}_2^{C_4}=0.89,\ \bar{\mu}_2^{C_5}=0.11).$$

Thus, the solution of fuzzy relational equations (7.24) for the first lower solution $\underline{\boldsymbol{\mu}}_1^A$ can be represented in the form of intervals:

$$\underline{D}_1(\underline{\boldsymbol{\mu}}_1^A)=\{\mu^{C_1}=0.76,\ \mu^{C_2}=0.32,\ \mu^{C_3}=0.89,\ \mu^{C_4}\in[0.89,1.0],\ \mu^{C_5}=0.11\}$$

$$\cup\{\mu^{C_1}=0.76,\ \mu^{C_2}=0.32,\ \mu^{C_3}\in[0.89,1.0],\ \mu^{C_4}=0.89,\ \mu^{C_5}=0.11\}. \quad (7.35)$$

For the second lower solution $\underline{\mathbf{\mu}}_2^A$, a neural adjustment of the null solution (7.32) has yielded a fuzzy causes vector

$$\underline{\mathbf{\mu}}_{02}^C = (\underline{\mu}_{02}^{C_1} = 0.76, \underline{\mu}_{02}^{C_2} = 0.16, \underline{\mu}_{02}^{C_3} = 0.89, \underline{\mu}_{02}^{C_4} = 1.0, \underline{\mu}_{02}^{C_5} = 0.16),$$

and for the third lower solution $\underline{\mathbf{\mu}}_3^A$, a neural adjustment of the null solution (7.32) has yielded a fuzzy causes vector

$$\underline{\mathbf{\mu}}_{03}^C = (\underline{\mu}_{03}^{C_1} = 0.76, \underline{\mu}_{03}^{C_2} = 0.16, \underline{\mu}_{03}^{C_3} = 0.96, \underline{\mu}_{03}^{C_4} = 0.89, \underline{\mu}_{03}^{C_5} = 0.16),$$

for which the value of the optimization criterion (7.14) has constituted F=0.1024.

The resulting null solutions have allowed adjusting the bounds in the solution (7.33) and generating the sets of solutions $D_2(\underline{\mathbf{\mu}}_2^A)$ and $D_3(\underline{\mathbf{\mu}}_3^A)$, which are completely determined by the least solution

$$\underline{\mathbf{\mu}}^C = (\underline{\mu}^{C_1} = 0.76, \underline{\mu}^{C_2} = 0.16, \underline{\mu}^{C_3} = 0.89, \underline{\mu}^{C_4} = 0.89, \underline{\mu}^{C_5} = 0.16)$$

and the two upper solutions $\underline{D}_2^* = \underline{D}_3^* = \{\bar{\mathbf{\mu}}_1^C, \bar{\mathbf{\mu}}_2^C\}$

$$\bar{\mathbf{\mu}}_1^C = (\bar{\mu}_1^{C_1} = 0.76, \bar{\mu}_1^{C_2} = 0.16, \bar{\mu}_1^{C_3} = 0.89, \bar{\mu}_1^{C_4} = 1.0, \bar{\mu}_1^{C_5} = 0.16);$$

$$\bar{\mathbf{\mu}}_2^C = (\bar{\mu}_2^{C_1} = 0.76, \bar{\mu}_2^{C_2} = 0.16, \bar{\mu}_2^{C_3} = 1.0, \bar{\mu}_2^{C_4} = 0.89, \bar{\mu}_2^{C_5} = 0.16).$$

Thus, the solution of fuzzy relational equations (7.24) for the second and third lower solutions $\underline{\mathbf{\mu}}_2^A$ and $\underline{\mathbf{\mu}}_3^A$ can be represented in the form of intervals:

$$\underline{D}_2(\underline{\mathbf{\mu}}_2^A) = \underline{D}_3(\underline{\mathbf{\mu}}_3^A) = \{\mu^{C_1} = 0.76, \mu^{C_2} = 0.16, \mu^{C_3} = 0.89, \mu^{C_4} \in [0.89, 1.0], \mu^{C_5} = 0.16\}$$
$$\cup \{\mu^{C_1} = 0.76, \mu^{C_2} = 0.16, \mu^{C_3} \in [0.89, 1.0], \mu^{C_4} = 0.89, \mu^{C_5} = 0.16\}. \quad (7.36)$$

The intervals of the values of the input variable for each interval in solutions (7.34), (7.35), (7.36) can be defined with the help of the membership functions in Fig. 7.13:

- x_1^* =0.43 or x_1^* =0 for C_1;
- x_1^* =2.12 and x_1^* =3.93 or x_1^* =1.60 and x_1^* =4.45 for C_2;
- $x_1^* \in [5.74, 6.0]$ for C_3;
- $x_2^* \in [0, 0.27]$ for C_4;
- x_2^* =2.17 and x_2^* =3.92 or x_2^* =1.68 or x_2^* =1.35 for C_5.

The restoration of the input set for y_1^*=1.00 and y_2^*=0.60 is shown in Fig. 7.13, in which the values of the causes $C_1 \div C_5$ and effects $E_1 \div E_6$ significances measures are marked. The comparison of the target and restored level lines for y_1^*=1.00 and y_2^*=0.60 is shown in Fig. 7.14.

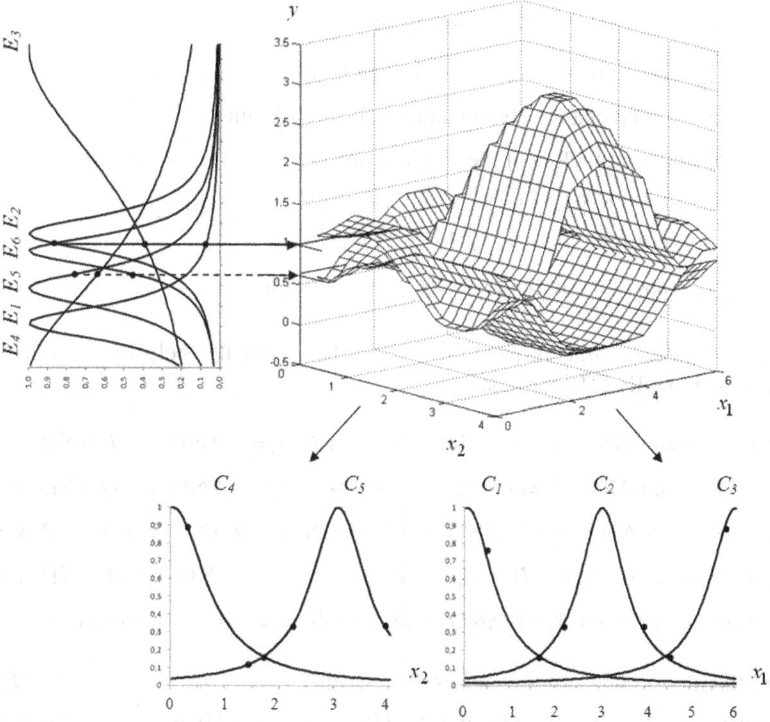

Fig. 7.13. Solution to the problem of inverse fuzzy inference for y_1^*=1.00 and y_2^*=0.60

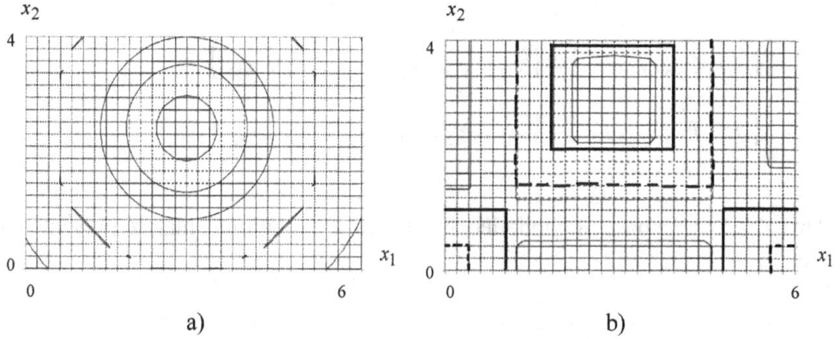

Fig. 7.14. Comparison of the target (a) and restored (b) level lines for y_1^*=1.00 (____) and y_2^*=0.60 (_ _ _)

7.10 Example 6: Hydro Elevator Diagnosis

Let us consider the algorithm's performance having the recourse to the example of the hydraulic elevator faults causes diagnosis. Input parameters of the hydro elevator are (variation ranges are indicated in parentheses):

x_1 – engine speed (30 – 50 r.p.s);

x_2 – inlet pressure (0.02 – 0.15 kg/cm^2);

x_3 – feed change gear clearance (0.1 – 0.3 mm).

Output parameters of the elevator are:

y_1 – productivity (13 – 24 l/min);

y_2 – consumed power (2.1 – 3.0 kw);

y_3 – suction conduit pressure (0.5 – 1 kg/cm^2).

"Causes-effects" interconnection is described with the help of the following system of fuzzy IF-THEN rules:

Rule 1: IF x_1 =*I* AND x_2 =*I* AND x_3=*I* THEN y_1=*D* AND y_2=*I* AND y_3=*D*;

Rule 2: IF x_1 =*D* AND x_2 =*D* AND x_3=*D* THEN y_1=*D* AND y_2=*D* AND y_3=*I*;

Rule 3: IF x_1 =*I* AND x_2 =*I* AND x_3=*D* THEN y_1=*D* AND y_2=*D* AND y_3=*D*;

Rule 4: IF x_1 =*I* AND x_2 =*D* AND x_3=*D* THEN y_1=*I* AND y_2=*D* AND y_3=*D*;

Rule 5: IF x_1 =*D* AND x_2 =*I* AND x_3=*D* THEN y_1=*I* AND y_2=*D* AND y_3=*I*.

where the total number of the causes and effects consists of: c_{11} *Decrease* (*D*) and c_{12} *Increase* (*I*) for x_1; c_{21} (*D*) and c_{21} (*I*) for x_2; c_{31} (*D*) and c_{32} (*I*) for x_3; e_{11} (*D*) and e_{12} (*I*) for y_1; e_{21} (*D*) and e_{22} (*I*) for y_2; e_{31} (*D*) and e_{32} (*I*) for y_3.

We shall define the set of causes and effects in the following way:

$$\{ C_1, C_2, \ldots, C_6 \} = \{ c_{11},\ c_{12},\ c_{21},\ c_{22},\ c_{31},\ c_{32} \};$$

$$\{ E_1, E_2, \ldots, E_6 \} = \{ e_{11},\ e_{12},\ e_{21},\ e_{22},\ e_{31},\ e_{32} \}.$$

This fuzzy rule base is modelled by the fuzzy relational matrix presented in Table 7.6.

Table 7.6. Fuzzy knowledge matrix

IF inputs				THEN outputs					
	x_1	x_2	x_3	y_1		y_2		y_3	
				D	I	D	I	D	I
A_1	I	I	I	1	0	0	1	1	0
A_2	D	D	D	1	0	1	0	0	1
A_3	I	I	D	1	0	1	0	1	0
A_4	I	D	D	0	1	1	0	1	0
A_5	D	I	D	0	1	1	0	0	1

For the fuzzy model tuning we used the results of diagnosis for 200 hydraulic elevators. The results of the fuzzy model tuning are given in Tables 7.7, 7.8 and in Fig. 7.15.

Table 7.7. Parameters of the membership functions for the causes fuzzy terms after tuning

Parameter	Fuzzy terms					
	C_1	C_2	C_3	C_4	C_5	C_6
β -	32.15	48.65	0.021	0.144	0.11	0.27
σ -	7.75	6.27	0.054	0.048	0.06	0.08

Table 7.8. Parameters of the membership functions for the effects fuzzy terms after tuning

Parameter	Fuzzy terms					
	E_1	E_2	E_3	E_4	E_5	E_6
β -	13.58	21.43	2.24	2.85	0.53	0.98
σ -	4.76	4.58	0.35	0.17	0.31	0.22

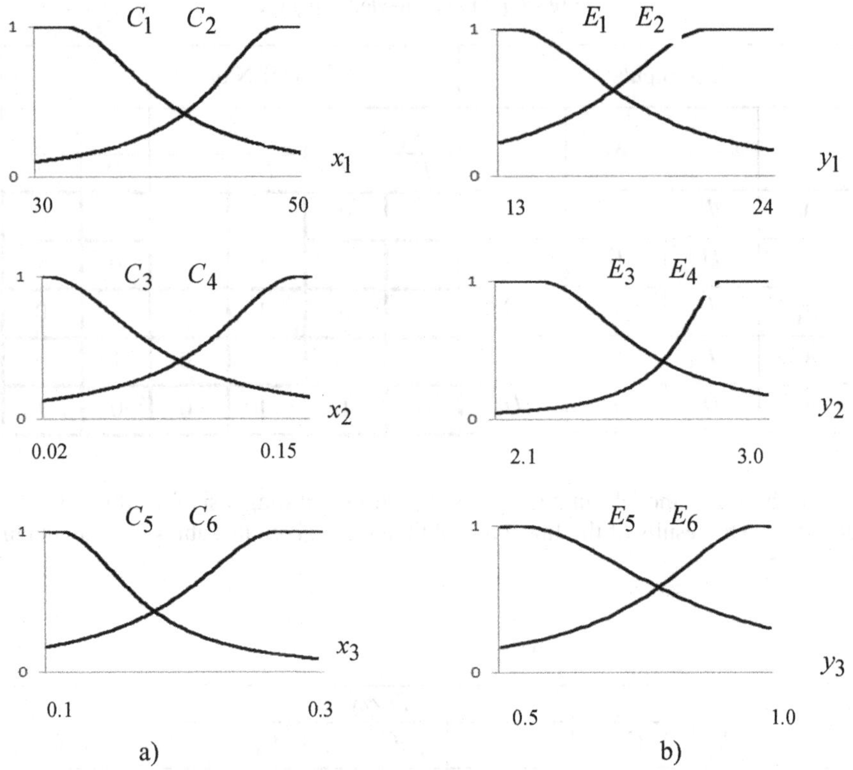

Fig. 7.15. Membership functions of the causes (a) and effects (b) fuzzy terms after tuning

Diagnostic equations after tuning take the form:

$$\mu^{E_1} = (\mu^{A_1} \wedge 0.97) \vee (\mu^{A_2} \wedge 0.65) \vee (\mu^{A_3} \wedge 0.77)$$

$$\mu^{E_2} = (\mu^{A_4} \wedge 1.00) \vee (\mu^{A_5} \wedge 0.46)$$

$$\mu^{E_3} = (\mu^{A_2} \wedge 0.99) \vee (\mu^{A_3} \wedge 0.80) \vee (\mu^{A_4} \wedge 0.69) \vee (\mu^{A_5} \wedge 0.93)$$

$$\mu^{E_4} = (\mu^{A_1} \wedge 0.96)$$

$$\mu^{E_5} = (\mu^{A_1} \wedge 0.72) \vee (\mu^{A_3} \wedge 0.47) \vee (\mu^{A_4} \wedge 0.76)$$

$$\mu^{E_6} = (\mu^{A_2} \wedge 0.92) \vee (\mu^{A_5} \wedge 0.87), \tag{7.37}$$

where

$$\mu^{A_1} = \mu^{C_2} \wedge \mu^{C_4} \wedge \mu^{C_6}$$

$$\mu^{A_2} = \mu^{C_1} \wedge \mu^{C_3} \wedge \mu^{C_5}$$

$$\mu^{A_3} = \mu^{C_2} \wedge \mu^{C_4} \wedge \mu^{C_5}$$

$$\mu^{A_4} = \mu^{C_2} \wedge \mu^{C_3} \wedge \mu^{C_5}$$

$$\mu^{A_5} = \mu^{C_1} \wedge \mu^{C_4} \wedge \mu^{C_5}. \tag{7.38}$$

Let us represent the vector of the observed parameters for a specific elevator:

$$\mathbf{Y}^* = (y_1^* = 17.10 \text{ l/min};\ y_2^* = 2.45 \text{ kw};\ y_3^* = 0.87 \text{ kg/cm}^2).$$

The measures of the effects significances for these values can be defined with the help of the membership functions in Fig. 7.15,b:

$$\boldsymbol{\mu}^E = (\mu^{E_1}(y_1^*) = 0.65;\ \mu^{E_2}(y_1^*) = 0.53;$$

$$\mu^{E_3}(y_2^*) = 0.74;\ \mu^{E_4}(y_2^*) = 0.15;$$

$$\mu^{E_5}(y_3^*) = 0.45;\ \mu^{E_6}(y_3^*) = 0.80).$$

The genetic algorithm yields a null solution of the optimization problem (7.12)

$$\boldsymbol{\mu}_0^C = (\mu_0^{C_1} = 0.77, \mu_0^{C_2} = 0.49, \mu_0^{C_3} = 0.77, \mu_0^{C_4} = 0.62, \mu_0^{C_5} = 0.77, \mu_0^{C_6} = 0.15)\,, \tag{7.39}$$

for which the value of the optimization criterion (7.12) takes the value of F=0.0050.

The null vector of the causes combinations significances measures

$$\boldsymbol{\mu}_0^A = (\mu_0^{A_1} = 0.15, \mu_0^{A_2} = 0.77, \mu_0^{A_3} = 0.49, \mu_0^{A_4} = 0.49, \mu_0^{A_5} = 0.62)$$

corresponds to the obtained null solution.

The obtained null solution allows us to arrange for the genetic search for the solution set $S(\mathbf{R}, \boldsymbol{\mu}^E)$, which is completely determined by the greatest solution

$$\bar{\boldsymbol{\mu}}^A = (\bar{\mu}^{A_1} = 0.15,\ \bar{\mu}^{A_2} = 0.77,\ \bar{\mu}^{A_3} = 0.65,\ \bar{\mu}^{A_4} = 0.49,\ \bar{\mu}^{A_5} = 0.77)$$

and the two lower solutions $S^* = \{\underline{\boldsymbol{\mu}}_1^A, \underline{\boldsymbol{\mu}}_2^A\}$

$$\underline{\boldsymbol{\mu}}_1^A = (\underline{\mu}_1^{A_1} = 0.15,\ \underline{\mu}_1^{A_2} = 0.77,\ \underline{\mu}_1^{A_3} = 0,\ \underline{\mu}_1^{A_4} = 0.49,\ \underline{\mu}_1^{A_5} = 0);$$

$$\underline{\boldsymbol{\mu}}_2^A = (\underline{\mu}_2^{A_1} = 0.15,\ \underline{\mu}_2^{A_2} = 0,\ \underline{\mu}_2^{A_3} = 0.65,\ \underline{\mu}_2^{A_4} = 0.49,\ \underline{\mu}_2^{A_5} = 0.77).$$

Thus, the solution of fuzzy relational equations (7.37) can be represented in the form of intervals:

$$S(\mathbf{R},\boldsymbol{\mu}^E)=\{\mu^{A_1}=0.15,\ \mu^{A_2}=0.77,\ \mu^{A_3}\in[0,0.65],\ \mu^{A_4}=0.49,\ \mu^{A_5}\in[0,0.77]\}$$
$$\cup\{\mu^{A_1}=0.15,\ \mu^{A_2}\in[0,0.77],\ \mu^{A_3}=0.65,\ \mu^{A_4}=0.49,\ \mu^{A_5}=0.77\}. \qquad (7.40)$$

We next apply the genetic algorithm for solving optimization problem (7.14) for the greatest solution $\bar{\boldsymbol{\mu}}^A$ and the two lower solutions $\underline{\boldsymbol{\mu}}_1^A$ and $\underline{\boldsymbol{\mu}}_2^A$.

For the greatest solution $\bar{\boldsymbol{\mu}}^A$, the genetic algorithm yields a null solution of the optimization problem (7.14)

$$\bar{\boldsymbol{\mu}}_0^C=(\bar{\mu}_0^{C_1}=0.77,\bar{\mu}_0^{C_2}=0.57,\bar{\mu}_0^{C_3}=0.94,\bar{\mu}_0^{C_4}=0.80,\bar{\mu}_0^{C_5}=0.79,\bar{\mu}_0^{C_6}=0.15), \qquad (7.41)$$

for which the value of the optimization criterion (7.14) is F=0.0128.

The obtained null solution allows us to arrange for the genetic search for the solution set $\bar{D}(\bar{\boldsymbol{\mu}}^A)$, which is completely determined by the least solution

$$\underline{\boldsymbol{\mu}}^C=(\underline{\mu}^{C_1}=0.77,\ \underline{\mu}^{C_2}=0.57,\ \underline{\mu}^{C_3}=0.77,\ \underline{\mu}^{C_4}=0.77,\ \underline{\mu}^{C_5}=0.77,\ \underline{\mu}^{C_6}=0.15)$$

and the four upper solutions $\bar{D}^*=\{\bar{\boldsymbol{\mu}}_1^C,\bar{\boldsymbol{\mu}}_2^C,\bar{\boldsymbol{\mu}}_3^C,\bar{\boldsymbol{\mu}}_4^C\}$

$$\bar{\boldsymbol{\mu}}_1^C=(\bar{\mu}_1^{C_1}=0.77,\ \bar{\mu}_1^{C_2}=0.57,\ \bar{\mu}_1^{C_3}=1.0,\ \bar{\mu}_1^{C_4}=1.0,\ \bar{\mu}_1^{C_5}=1.0,\ \bar{\mu}_1^{C_6}=0.15);$$
$$\bar{\boldsymbol{\mu}}_2^C=(\bar{\mu}_2^{C_1}=1.0,\ \bar{\mu}_2^{C_2}=0.57,\ \bar{\mu}_2^{C_3}=0.77,\ \bar{\mu}_2^{C_4}=1.0,\ \bar{\mu}_2^{C_5}=1.0,\ \bar{\mu}_2^{C_6}=0.15);$$
$$\bar{\boldsymbol{\mu}}_3^C=(\bar{\mu}_3^{C_1}=1.0,\ \bar{\mu}_3^{C_2}=0.57,\ \bar{\mu}_3^{C_3}=1.0,\ \bar{\mu}_3^{C_4}=0.77,\ \bar{\mu}_3^{C_5}=1.0,\ \bar{\mu}_3^{C_6}=0.15);$$
$$\bar{\boldsymbol{\mu}}_4^C=(\bar{\mu}_4^{C_1}=1.0,\ \bar{\mu}_4^{C_2}=0.57,\ \bar{\mu}_4^{C_3}=1.0,\ \bar{\mu}_4^{C_4}=1.0,\ \bar{\mu}_4^{C_5}=0.77,\ \bar{\mu}_4^{C_6}=0.15).$$

Thus, the solution of fuzzy relational equations (7.38) for the greatest solution $\bar{\boldsymbol{\mu}}^A$ can be represented in the form of intervals:

$$\bar{D}(\bar{\boldsymbol{\mu}}^A)=\{\mu^{C_1}=0.77,\mu^{C_2}=0.57,\mu^{C_3}\in[0.77,1.0],\mu^{C_4}\in[0.77,1.0],\mu^{C_5}\in[0.77,1.0],\mu^{C_6}=0.15\}$$
$$\cup\{\mu^{C_1}\in[0.77,1.0],\mu^{C_2}=0.57,\mu^{C_3}=0.77,\mu^{C_4}\in[0.77,1.0],\mu^{C_5}\in[0.77,1.0],\mu^{C_6}=0.15\}$$
$$\cup\{\mu^{C_1}\in[0.77,1.0],\mu^{C_2}=0.57,\mu^{C_3}\in[0.77,1.0],\mu^{C_4}=0.77,\mu^{C_5}\in[0.77,1.0],\mu^{C_6}=0.15\}$$
$$\cup\{\mu^{C_1}\in[0.77,1.0],\mu^{C_2}=0.57,\mu^{C_3}\in[0.77,1.0],\mu^{C_4}\in[0.77,1.0],\mu^{C_5}=0.77,\mu^{C_6}=0.15\}. \qquad (7.42)$$

For the first lower solution $\underline{\boldsymbol{\mu}}_1^A$, the genetic algorithm yields a null solution of the optimization problem (7.14)

$$\underline{\boldsymbol{\mu}}_{01}^C=(\underline{\mu}_{01}^{C_1}=0.77,\underline{\mu}_{01}^{C_2}=0.49,\underline{\mu}_{01}^{C_3}=0.84,\underline{\mu}_{01}^{C_4}=0,\underline{\mu}_{01}^{C_5}=0.92,\underline{\mu}_{01}^{C_6}=0), \qquad (7.43)$$

for which the value of the optimization criterion (7.14) is F=0.0225.

The obtained null solution allows us to arrange for the genetic search for the solution set $\underline{D}_1(\underline{\boldsymbol{\mu}}_1^A)$, which is completely determined by the least solution

$$\underline{\boldsymbol{\mu}}^C = (\underline{\mu}^{C_1}=0.77,\ \underline{\mu}^{C_2}=0.49,\ \underline{\mu}^{C_3}=0.77,\ \underline{\mu}^{C_4}=0,\ \underline{\mu}^{C_5}=0.77,\ \underline{\mu}^{C_6}=0)$$

and the three upper solutions $\underline{D}_1^* = \{\overline{\boldsymbol{\mu}}_1^C, \overline{\boldsymbol{\mu}}_2^C, \overline{\boldsymbol{\mu}}_3^C\}$

$$\overline{\boldsymbol{\mu}}_1^C = (\overline{\mu}_1^{C_1}=0.77,\ \overline{\mu}_1^{C_2}=0.49,\ \overline{\mu}_1^{C_3}=1.0,\ \overline{\mu}_1^{C_4}=0,\ \overline{\mu}_1^{C_5}=1.0,\ \overline{\mu}_1^{C_6}=0);$$

$$\overline{\boldsymbol{\mu}}_2^C = (\overline{\mu}_2^{C_1}=1.0,\ \overline{\mu}_2^{C_2}=0.49,\ \overline{\mu}_2^{C_3}=0.77,\ \overline{\mu}_2^{C_4}=0,\ \overline{\mu}_2^{C_5}=1.0,\ \overline{\mu}_2^{C_6}=0);$$

$$\overline{\boldsymbol{\mu}}_3^C = (\overline{\mu}_3^{C_1}=1.0,\ \overline{\mu}_3^{C_2}=0.49,\ \overline{\mu}_3^{C_3}=1.0,\ \overline{\mu}_3^{C_4}=0,\ \overline{\mu}_3^{C_5}=0.77,\ \overline{\mu}_3^{C_6}=0).$$

Thus, the solution of fuzzy relational equations (7.38) for the first lower solution $\underline{\boldsymbol{\mu}}_1^A$ can be represented in the form of intervals:

$$\underline{D}_1(\underline{\boldsymbol{\mu}}_1^A) = \{\mu^{C_1}=0.77,\ \mu^{C_2}=0.49,\ \mu^{C_3}\in[0.77, 1.0],\ \mu^{C_4}=0,\ \mu^{C_5}\in[0.77, 1.0],\ \mu^{C_6}=0\}$$
$$\cup\{\mu^{C_1}\in[0.77, 1.0],\ \mu^{C_2}=0.49,\ \mu^{C_3}=0.77,\ \mu^{C_4}=0,\ \mu^{C_5}\in[0.77, 1.0],\ \mu^{C_6}=0\}$$
$$\cup\{\mu^{C_1}\in[0.77, 1.0],\ \mu^{C_2}=0.49,\ \mu^{C_3}\in[0.77, 1.0],\ \mu^{C_4}=0,\ \mu^{C_5}=0.77,\ \mu^{C_6}=0\}. \quad (7.44)$$

For the second lower solution $\underline{\boldsymbol{\mu}}_2^A$, the genetic algorithm yields a null solution of the optimization problem (7.14)

$$\underline{\boldsymbol{\mu}}_{02}^C = (\underline{\mu}_{02}^{C_1}=0.77, \underline{\mu}_{02}^{C_2}=0.65, \underline{\mu}_{02}^{C_3}=0.25, \underline{\mu}_{02}^{C_4}=0.97, \underline{\mu}_{02}^{C_5}=0.85, \underline{\mu}_{02}^{C_6}=0.15), \quad (7.45)$$

for which the value of the optimization criterion (7.14) is F=0.1201.

The obtained null solution allows us to arrange for the genetic search for the solution set $\underline{D}_2(\underline{\boldsymbol{\mu}}_2^A)$, which is completely determined by the least solution

$$\underline{\boldsymbol{\mu}}^C = (\underline{\mu}^{C_1}=0.77,\ \underline{\mu}^{C_2}=0.65,\ \underline{\mu}^{C_3}=0.25,\ \underline{\mu}^{C_4}=0.77,\ \underline{\mu}^{C_5}=0.77,\ \underline{\mu}^{C_6}=0.15)$$

and the three upper solutions $\underline{D}_2^* = \{\overline{\boldsymbol{\mu}}_1^C, \overline{\boldsymbol{\mu}}_2^C, \overline{\boldsymbol{\mu}}_3^C\}$

$$\overline{\boldsymbol{\mu}}_1^C = (\overline{\mu}_1^{C_1}=0.77,\ \overline{\mu}_1^{C_2}=0.65,\ \overline{\mu}_1^{C_3}=0.25,\ \overline{\mu}_1^{C_4}=1.0,\ \overline{\mu}_1^{C_5}=1.0,\ \overline{\mu}_1^{C_6}=0.15);$$

$$\overline{\boldsymbol{\mu}}_2^C = (\overline{\mu}_2^{C_1}=1.0,\ \overline{\mu}_2^{C_2}=0.65,\ \overline{\mu}_2^{C_3}=0.25,\ \overline{\mu}_2^{C_4}=0.77,\ \overline{\mu}_2^{C_5}=1.0,\ \overline{\mu}_2^{C_6}=0.15);$$

$$\overline{\boldsymbol{\mu}}_3^C = (\overline{\mu}_3^{C_1}=1.0,\ \overline{\mu}_3^{C_2}=0.65,\ \overline{\mu}_3^{C_3}=0.25,\ \overline{\mu}_3^{C_4}=1.0,\ \overline{\mu}_3^{C_5}=0.77,\ \overline{\mu}_3^{C_6}=0.15).$$

Thus, the solution of fuzzy relational equations (7.38) for the second lower solution $\underline{\boldsymbol{\mu}}_2^A$ can be represented in the form of intervals:

$$\underline{D}_2(\underline{\boldsymbol{\mu}}_2^A)=\{\mu^{C_1}=0.77,\ \mu^{C_2}=0.65,\ \mu^{C_3}=0.25,\ \mu^{C_4}\in[0.77,1.0],\ \mu^{C_5}\in[0.77,1.0],\ \mu^{C_6}=0.15\}$$

$$\cup\{\mu^{C_1}\in[0.77,1.0],\ \mu^{C_2}=0.65,\ \mu^{C_3}=0.25,\ \mu^{C_4}=0.77,\ \mu^{C_5}\in[0.77,1.0],\ \mu^{C_6}=0.15\}$$

$$\cup\{\mu^{C_1}\in[0.77,1.0],\ \mu^{C_2}=0.65,\ \mu^{C_3}=0.25,\ \mu^{C_4}\in[0.77,1.0],\ \mu^{C_5}=0.77,\ \mu^{C_6}=0.15\}. \quad (7.46)$$

Following the solutions (7.42), (7.44), (7.46), the causes C_1, C_3, C_4 and C_5 are the causes of the observed elevator state, so that $\mu^{C_1}>\mu^{C_2}$, $\mu^{C_3}=\mu^{C_4}$, $\mu^{C_5}>\mu^{C_6}$. The intervals of the values of the input variables for these causes can be defined with the help of the membership functions in Fig. 7.15,a:

- $x_1^* \in [30.0, 36.4]$ r.p.s for C_1;
- $x_2^* \in [0.020, 0.050]$ kg/cm^2 for C_3 and $x_2^* \in [0.118, 0.150]$ kg/cm^2 for C_4;
- $x_3^* \in [0.100, 0.143]$ mm for C_5.

The obtained solution allows the analyst to make the preliminary conclusions. The elevator failure may be because of the engine speed reduced to 30 – 36 r.p.s, the inlet pressure decreased to 0.020 – 0.050 kg/cm^2 or increased to 0.118 – 0.150 kg/cm^2, and the feed change gear clearance decreased to 100 – 143 mkm.

Assume a repeated measurement has revealed an increase in the elevator productivity up to y_1^*=18.80 l/min, an increase of the consumed power up to y_2^*=2.51 kw, and a decrease in the suction pressure up to y_3^*=0.75 kg/cm^2.

For the new values, the fuzzy effects vector is

$$\boldsymbol{\mu}^E=(\mu^{E_1}(y_1^*)=0.45;\ \mu^{E_2}(y_1^*)=0.75;$$
$$\mu^{E_3}(y_2^*)=0.63;\ \mu^{E_4}(y_2^*)=0.20;$$
$$\mu^{E_5}(y_3^*)=0.67;\ \mu^{E_6}(y_3^*)=0.48).$$

A neural adjustment of the null solution (7.39) of the optimization problem (7.12) has yielded a fuzzy causes vector

$$\boldsymbol{\mu}_0^C=(\mu_0^{C_1}=0.46,\ \mu_0^{C_2}=0.69,\ \mu_0^{C_3}=0.75,\ \mu_0^{C_4}=0.25,\ \mu_0^{C_5}=0.92,\ \mu_0^{C_6}=0.20),$$

for which the value of the optimization criterion (7.12) has constituted F=0.0094.

The null vector of the causes combinations significances measures

$$\boldsymbol{\mu}_0^A=(\mu_0^{A_1}=0.20,\ \mu_0^{A_2}=0.46,\ \mu_0^{A_3}=0.25,\ \mu_0^{A_4}=0.69,\ \mu_0^{A_5}=0.25),$$

corresponds to the obtained null solution.

The resulting null solution has allowed adjusting the bounds in the solution (7.40) and generating the set of solutions $S(\mathbf{R},\boldsymbol{\mu}^E)$ determined by the greatest solution

$$\bar{\boldsymbol{\mu}}^A = (\bar{\mu}^{A_1} = 0.20,\ \bar{\mu}^{A_2} = 0.46,\ \bar{\mu}^{A_3} = 0.46,\ \bar{\mu}^{A_4} = 0.69,\ \bar{\mu}^{A_5} = 0.46)$$

and the two lower solutions $S^* = \{\underline{\boldsymbol{\mu}}_1^A, \underline{\boldsymbol{\mu}}_2^A\}$

$$\underline{\boldsymbol{\mu}}_1^A = (\underline{\mu}_1^{A_1} = 0.20,\ \underline{\mu}_1^{A_2} = 0.46,\ \underline{\mu}_1^{A_3} = 0,\ \underline{\mu}_1^{A_4} = 0.69,\ \underline{\mu}_1^{A_5} = 0);$$

$$\underline{\boldsymbol{\mu}}_2^A = (\underline{\mu}_2^{A_1} = 0.20,\ \underline{\mu}_2^{A_2} = 0,\ \underline{\mu}_2^{A_3} = 0.46,\ \underline{\mu}_2^{A_4} = 0.69,\ \underline{\mu}_2^{A_5} = 0.46).$$

Thus, the solution of fuzzy relational equations (7.37) for the new values can be represented in the form of intervals:

$$S(\mathbf{R},\boldsymbol{\mu}^E) = \{\mu^{A_1} = 0.20,\ \mu^{A_2} = 0.46,\ \mu^{A_3} \in [0, 0.46],\ \mu^{A_4} = 0.69,\ \mu^{A_5} \in [0, 0.46]\}$$
$$\cup \{\mu^{A_1} = 0.20,\ \mu^{A_2} \in [0, 0.46],\ \mu^{A_3} = 0.46,\ \mu^{A_4} = 0.69,\ \mu^{A_5} = 0.46\}.$$

For the greatest solution $\bar{\boldsymbol{\mu}}^A$, a neural adjustment of the null solution (7.41) has yielded a fuzzy causes vector

$$\bar{\boldsymbol{\mu}}_0^C = (\bar{\mu}_0^{C_1} = 0.46, \bar{\mu}_0^{C_2} = 0.78, \bar{\mu}_0^{C_3} = 0.69, \bar{\mu}_0^{C_4} = 0.46, \bar{\mu}_0^{C_5} = 0.91, \bar{\mu}_0^{C_6} = 0.20),$$

for which the value of the optimization criterion (7.14) has constituted F=0.

The resultant null solution has allowed adjusting the bounds in the solution (7.42) and generating the set of solutions $\bar{D}(\bar{\boldsymbol{\mu}}^A)$, which is completely determined by the least solution

$$\underline{\boldsymbol{\mu}}^C = (\underline{\mu}^{C_1} = 0.46,\ \underline{\mu}^{C_2} = 0.69,\ \underline{\mu}^{C_3} = 0.69,\ \underline{\mu}^{C_4} = 0.46,\ \underline{\mu}^{C_5} = 0.69,\ \underline{\mu}^{C_6} = 0.20)$$

and the three upper solutions $\bar{D}^* = \{\bar{\boldsymbol{\mu}}_1^C, \bar{\boldsymbol{\mu}}_2^C, \bar{\boldsymbol{\mu}}_3^C\}$

$$\bar{\boldsymbol{\mu}}_1^C = (\bar{\mu}_1^{C_1} = 0.46,\ \bar{\mu}_1^{C_2} = 0.69,\ \bar{\mu}_1^{C_3} = 1.0,\ \bar{\mu}_1^{C_4} = 0.46,\ \bar{\mu}_1^{C_5} = 1.0,\ \bar{\mu}_1^{C_6} = 0.20);$$

$$\bar{\boldsymbol{\mu}}_2^C = (\bar{\mu}_2^{C_1} = 0.46,\ \bar{\mu}_2^{C_2} = 1.0,\ \bar{\mu}_2^{C_3} = 0.69,\ \bar{\mu}_2^{C_4} = 0.46,\ \bar{\mu}_2^{C_5} = 1.0,\ \bar{\mu}_2^{C_6} = 0.20);$$

$$\bar{\boldsymbol{\mu}}_3^C = (\bar{\mu}_3^{C_1} = 0.46,\ \bar{\mu}_3^{C_2} = 1.0,\ \bar{\mu}_3^{C_3} = 1.0,\ \bar{\mu}_3^{C_4} = 0.46,\ \bar{\mu}_3^{C_5} = 0.69,\ \bar{\mu}_3^{C_6} = 0.20).$$

Thus, the solution of fuzzy relational equations (7.38) for the greatest solution $\bar{\boldsymbol{\mu}}^A$ can be represented in the form of intervals:

$$\begin{aligned}\bar{D}(\bar{\boldsymbol{\mu}}^A)=&\{\mu^{C_1}=0.46,\ \mu^{C_2}=0.69,\ \mu^{C_3}\in[0.69,1.0],\ \mu^{C_4}=0.46,\ \mu^{C_5}\in[0.69,1.0],\ \mu^{C_6}=0.20\}\\ &\cup\{\mu^{C_1}=0.46,\ \mu^{C_2}\in[0.69,1.0],\ \mu^{C_3}=0.69,\ \mu^{C_4}=0.46,\ \mu^{C_5}\in[0.69,1.0],\ \mu^{C_6}=0.20\}\\ &\cup\{\mu^{C_1}=0.46,\ \mu^{C_2}\in[0.69,1.0],\ \mu^{C_3}\in[0.69,1.0],\ \mu^{C_4}=0.46,\ \mu^{C_5}=0.69,\ \mu^{C_6}=0.20\}\end{aligned} \tag{7.47}$$

For the first lower solution $\underline{\boldsymbol{\mu}}_1^A$, a neural adjustment of the null solution (7.43) has yielded a fuzzy causes vector

$$\underline{\boldsymbol{\mu}}_{01}^C=(\underline{\mu}_{01}^{C_1}=0.46,\ \underline{\mu}_{01}^{C_2}=0.92,\ \underline{\mu}_{01}^{C_3}=0.86,\ \underline{\mu}_{01}^{C_4}=0.10,\ \underline{\mu}_{01}^{C_5}=0.69,\ \underline{\mu}_{01}^{C_6}=0.10),$$

for which the value of the optimization criterion (7.14) has constituted F=0.0300.

The resulting null solution has allowed adjusting the bounds in the solution (7.44) and generating the set of solutions $\underline{D}_1(\underline{\boldsymbol{\mu}}_1^A)$, which is completely determined by the least solution

$$\underline{\boldsymbol{\mu}}^C=(\underline{\mu}^{C_1}=0.46,\ \underline{\mu}^{C_2}=0.69,\ \underline{\mu}^{C_3}=0.69,\ \underline{\mu}^{C_4}=0.10,\ \underline{\mu}^{C_5}=0.69,\ \underline{\mu}^{C_6}=0.10)$$

and the three upper solutions $\underline{D}_1^*=\{\bar{\boldsymbol{\mu}}_1^C,\bar{\boldsymbol{\mu}}_2^C,\bar{\boldsymbol{\mu}}_3^C\}$

$$\begin{aligned}\bar{\boldsymbol{\mu}}_1^C&=(\bar{\mu}_1^{C_1}=0.46,\ \bar{\mu}_1^{C_2}=0.69,\ \bar{\mu}_1^{C_3}=1.0,\ \bar{\mu}_1^{C_4}=0.10,\ \bar{\mu}_1^{C_5}=1.0,\ \bar{\mu}_1^{C_6}=0.10);\\ \bar{\boldsymbol{\mu}}_2^C&=(\bar{\mu}_2^{C_1}=0.46,\ \bar{\mu}_2^{C_2}=1.0,\ \bar{\mu}_2^{C_3}=0.69,\ \bar{\mu}_2^{C_4}=0.10,\ \bar{\mu}_2^{C_5}=1.0,\ \bar{\mu}_2^{C_6}=0.10);\\ \bar{\boldsymbol{\mu}}_3^C&=(\bar{\mu}_3^{C_1}=0.46,\ \bar{\mu}_3^{C_2}=1.0,\ \bar{\mu}_3^{C_3}=1.0,\ \bar{\mu}_3^{C_4}=0.10,\ \bar{\mu}_3^{C_5}=0.69,\ \bar{\mu}_3^{C_6}=0.10).\end{aligned}$$

Thus, the solution of fuzzy relational equations (7.38) for the first lower solution $\underline{\boldsymbol{\mu}}_1^A$ can be represented in the form of intervals:

$$\begin{aligned}\underline{D}_1(\underline{\boldsymbol{\mu}}_1^A)=&\{\mu^{C_1}=0.46,\ \mu^{C_2}=0.69,\ \mu^{C_3}\in[0.69,1.0],\ \mu^{C_4}=0.10,\ \mu^{C_5}\in[0.69,1.0],\ \mu^{C_6}=0.10\}\\ &\cup\{\mu^{C_1}=0.46,\ \mu^{C_2}\in[0.69,1.0],\ \mu^{C_3}=0.69,\ \mu^{C_4}=0.10,\ \mu^{C_5}\in[0.69,1.0],\ \mu^{C_6}=0.10\}\\ &\cup\{\mu^{C_1}=0.46,\ \mu^{C_2}\in[0.69,1.0],\ \mu^{C_3}\in[0.69,1.0],\ \mu^{C_4}=0.10,\ \mu^{C_5}=0.69,\ \mu^{C_6}=0.10\}.\end{aligned} \tag{7.48}$$

For the second lower solution $\underline{\mathbf{\mu}}_2^A$, a neural adjustment of the null solution (7.45) has yielded a fuzzy causes vector

$$\underline{\mathbf{\mu}}_{02}^{C} = (\underline{\mu}_{02}^{C_1} = 0.23, \underline{\mu}_{02}^{C_2} = 0.69, \underline{\mu}_{02}^{C_3} = 0.83, \underline{\mu}_{02}^{C_4} = 0.46, \underline{\mu}_{02}^{C_5} = 0.97, \underline{\mu}_{02}^{C_6} = 0.20),$$

for which the value of the optimization criterion (7.14) has constituted F=0.1058.

The resulting null solution has allowed adjusting the bounds in the solution (7.46) and generating the sets of solutions $\underline{D}_2(\underline{\mathbf{\mu}}_2^A)$, which is completely determined by the least solution

$$\underline{\mathbf{\mu}}^C = (\underline{\mu}^{C_1} = 0.23, \underline{\mu}^{C_2} = 0.69, \underline{\mu}^{C_3} = 0.69, \underline{\mu}^{C_4} = 0.46, \underline{\mu}^{C_5} = 0.69, \underline{\mu}^{C_6} = 0.20)$$

and the three upper solutions $\underline{D}_2^* = \{\overline{\mathbf{\mu}}_1^C, \overline{\mathbf{\mu}}_2^C, \overline{\mathbf{\mu}}_3^C\}$

$$\overline{\mathbf{\mu}}_1^C = (\overline{\mu}_1^{C_1} = 0.23, \overline{\mu}_1^{C_2} = 0.69, \overline{\mu}_1^{C_3} = 1.0, \overline{\mu}_1^{C_4} = 0.46, \overline{\mu}_1^{C_5} = 1.0, \overline{\mu}_1^{C_6} = 0.20);$$

$$\overline{\mathbf{\mu}}_2^C = (\overline{\mu}_2^{C_1} = 0.23, \overline{\mu}_2^{C_2} = 1.0, \overline{\mu}_2^{C_3} = 0.69, \overline{\mu}_2^{C_4} = 0.46, \overline{\mu}_2^{C_5} = 1.0, \overline{\mu}_2^{C_6} = 0.20);$$

$$\overline{\mathbf{\mu}}_3^C = (\overline{\mu}_3^{C_1} = 0.23, \overline{\mu}_3^{C_2} = 1.0, \overline{\mu}_3^{C_3} = 1.0, \overline{\mu}_3^{C_4} = 0.46, \overline{\mu}_3^{C_5} = 0.69, \overline{\mu}_3^{C_6} = 0.20).$$

Thus, the solution of fuzzy relational equations (7.38) for the second lower solution $\underline{\mathbf{\mu}}_2^A$ can be represented in the form of intervals:

$$\begin{aligned}\underline{D}_2(\underline{\mathbf{\mu}}_2^A) = &\{\mu^{C_1} = 0.23, \mu^{C_2} = 0.69, \mu^{C_3} \in [0.69, 1.0], \mu^{C_4} = 0.46, \mu^{C_5} \in [0.69, 1.0], \mu^{C_6} = 0.20\}\\ &\bigcup\{\mu^{C_1} = 0.23, \mu^{C_2} \in [0.69, 1.0], \mu^{C_3} = 0.69, \mu^{C_4} = 0.46, \mu^{C_5} \in [0.69, 1.0], \mu^{C_6} = 0.20\}\\ &\bigcup\{\mu^{C_1} = 0.23, \mu^{C_2} \in [0.69, 1.0], \mu^{C_3} \in [0.69, 1.0], \mu^{C_4} = 0.46, \mu^{C_5} = 0.69, \mu^{C_6} = 0.20\}.\end{aligned} \tag{7.49}$$

Following the resultant solutions (7.47), (7.48), (7.49), the causes C_2, C_3 and C_5 are the causes of the observed elevator state, since $\mu^{C_2} > \mu^{C_1}$, $\mu^{C_3} > \mu^{C_4}$, $\mu^{C_5} > \mu^{C_6}$. The intervals of the values of the input variables for these causes can be defined with the help of the membership functions in Fig. 7.15,a:

- $x_1^* \in [44.4, 50.0]$ r.p.s for C_2;
- $x_2^* \in [0.020, 0.057]$ kg/cm^2 for C_3;
- $x_3^* \in [0.100, 0.150]$ mm for C_5.

The solution obtained allows the final conclusions. Thus, the causes of the observed elevator state should be located and identified as the increase of the engine speed to 45-50 r.p.s, the decrease of the inlet pressure to 0.020 – 0.057 kg/cm^2, and the decrease of the feed change gear clearance to 100-150 mk.

To test the fuzzy model we used the results of diagnosis for 192 elevators with different kinds of faults. The tuning algorithm efficiency characteristics for the testing data are given in Table 7.9. Attaining a 96% correctness of the diagnostics required 30 min of operation of the genetic algorithm and 7 min of operation of the neural network (Intel Core 2 Duo P7350 2.0 GHz).

Table 7.9. Tuning algorithm efficiency characteristics

Causes (diagnoses)	Number of cases in the data sample	Probability of the correct diagnosis		
		Before tuning	After tuning	
			Null solution (genetic algorithm)	Refined diagnoses (neural network)
C_1	104	86 / 104=0.82	96 / 104=0.92	101 / 104=0.97
C_2	88	67 / 88=0.76	80 / 88=0.91	84 / 88=0.95
C_3	92	74 / 92=0.80	82 / 92=0.89	88 / 92=0.95
C_4	100	70 / 100=0.70	93 / 100=0.93	97 / 100=0.97
C_5	122	103 / 122=0.84	109 / 122=0.89	117 / 122=0.96
C_6	70	51 / 70=0.73	61 / 70=0.87	68 / 70=0.97

References

1. Tihonov, A.N., Arsenin, V.Y.: Methods of solving incorrect problems. Science (1974) (in Russian)
2. Abdelbar, A.M., Andrews, E., Wunsch, D.: Abductive reasoning with recurrent neural networks. Neural Networks 16, 665–673 (2003)
3. Gelsema, E.S.: Abductive reasoning in Bayesian belief networks using a genetic algorithm. Pattern Recognition Letters 16(8), 865–871 (1995)
4. Li, H.-L., Kao, H.-Y.: Constrained abductive reasoning with fuzzy parameters in Bayesian networks. Computers & Operations Research 32(1), 87–105 (2005)
5. Zadeh, L.: The Concept of Linguistic Variable and It's Application to Approximate Decision Making. Mir, Moscow (1976)
6. Yager, R.R., Filev, D.P.: Essentials of Fuzzy Modelling and Control. John Willey & Sons, New York (1994)
7. Di Nola, A., Sessa, S., Pedrycz, W., Sanchez, E.: Fuzzy relation equations and their applications to knowledge engineering. Kluwer Academic Press, Dordrecht (1989)

8. Peeva, K., Kyosev, Y.: Fuzzy relational calculus theory, applications and software. World Scientific, New York (2004) CD-ROM, http://mathworks.net
9. Pedrycz, W.: Inverse problem in fuzzy relational equations. Fuzzy Sets Systems 36, 277–291 (1990)
10. Pedrycz, W.: Approximate solutions of fuzzy relational equations. Fuzzy Sets Systems 28(2), 183–202 (1988)
11. Rotshtein, A., Rakytyanska, H.: Diagnosis based on fuzzy IF-THEN rules and genetic algorithms. In: Hippe, Z.S., Kulikowski, J.L. (eds.) Human-Computer Systems Interaction. Backgrounds and Applications, Part I. AISC, vol. 60, pp. 541–556. Springer, Heidelberg (2009)
12. Rotshtein, A.P., Rakytyanska, H.B.: Fuzzy genetic object identification: Multiple inputs/Multiple outputs case. In: Hippe, Z.S., Kulikowski, J.L., Mroczek, T. (eds.) Human - Computer Systems Interaction, Part II. AISC, vol. 99, pp. 375–394. Springer, Heidelberg (2012)
13. Rotshtein, A., Rakytyanska, H.: Fuzzy logic and the least squares method in diagnosis problem solving. In: Sarma, R.D. (ed.) Genetic Diagnoses, Nova Science Publishers, N.-Y (2011)
14. Rotshtein, A., Rakytyanska, H.: Fuzzy relation-based diagnosis. Automation and Remote Control 68(12), 2198–2213 (2007)
15. Rotshtein, A., Rakytyanska, H.: Diagnosis problem solving using fuzzy relations. IEEE Transactions on Fuzzy Systems 16(3), 664–675 (2008)
16. Rotshtein, A., Rakytyanska, H.: Adaptive diagnostic system based on fuzzy relations. Cybernetics and Systems Analysis 45(4), 623–637 (2009)
17. Rotshtein, A., Posner, M., Rakytyanska, H.: Cause and effect analysis by fuzzy relational equations and a genetic algorithm. Reliability Engineering and System Safety 91(9), 1095–1101 (2006)
18. Higashi, M., Klir, G.J.: Identification of fuzzy relation systems. IEEE Transactions Systems, Man, and Cybernetics 14, 349–355 (1984)
19. Pedrycz, W.: An identification algorithm in fuzzy relational systems. Fuzzy Sets and Systems 13, 153–167 (1984)
20. Branco, P.J., Dente, J.A.: A fuzzy relational identification algorithm and its application to predict the behaviour of a motor drive system. Fuzzy Sets Systems 109(3), 343–354 (2000)
21. Rotshtein, A.: Design and Tuning of Fuzzy Rule-based Systems for Medical Diagnosis. In: Teodorescu, N.-H., Kandel, A., Gain, L. (eds.) Fuzzy and Neuro-Fuzzy Systems in Medicine, pp. 243–289. CRC Press, Boca-Raton (1998)
22. Tsypkin, Y.Z.: Information Theory of Identification, p. 320. Nauka, Moscow (1984) (in Russian)
23. Gen, M., Cheng, R.: Genetic Algorithms and Engineering Design. John Wiley & Sons (1997)
24. Cordon, O., Herrera, F., Hoffmann, F., Magdalena, L.: Genetic Fuzzy Systems. In: Evolutionary Tuning and Learning of Fuzzy Knowledge Bases, p. 492. World Scientific (2001)
25. Rotshtein, A., Mityushkin, Y.: Neurolinguistic identification of nonlinear dependencies. Cybernetics and Systems Analysis 36(2), 179–187 (2000)

Chapter 8
Fuzzy Relations Extraction from Experimental Data

In this chapter, a problem of fuzzy genetic object identification expressed mathematically in terms of fuzzy relational equations is considered.

Fuzzy relational calculus [1, 2] provides a powerful theoretical background for knowledge extraction from data. Some fuzzy rule base is modelled by a fuzzy relational matrix, discovering the structure of the data set [3 – 5]. Fuzzy relational equations, which connect membership functions of input and output variables, are built on the basis of a fuzzy relational matrix and Zadeh's compositional rule of inference [6, 7]. The identification problem consists of extraction of an unknown relational matrix which can be translated as a set of fuzzy IF-THEN rules. In fuzzy relational calculus this type of problem relates to inverse problem resolution for the composite fuzzy relational equations [2]. Solvability and approximate solvability conditions of the composite fuzzy relational equations are considered in [2, 8, 9]. While the theoretical foundations of fuzzy relational equations are well developed, they call for more efficient use of their potential in system modeling. The non-optimizing approach [10] is widely used for fuzzy relational identification. Such adaptive recursive techniques are of interest for the most of on-line applications [11 – 13]. Under general conditions, an optimization environment is the convenient tool for fuzzy relational identification [14]. An approach for identification of fuzzy relational models by fuzzy neural networks is proposed in [15 – 17].

The genetic algorithm as a tool to solve the fuzzy relational equations was proposed in [18]. The genetic algorithm [19 – 21] allows us to solve the inverse problem which consists of the restoration of the unknown values of the vector of the unobserved parameters through the known values of the vector of the observed parameters and the known fuzzy relational matrix. In this chapter, the genetic algorithm [19 – 21] is adapted to identify the relational matrix for the given inputs-outputs data set. The algorithm for fuzzy relation matrix identification is accomplished in two stages. At the first stage, parameters of membership functions included in the fuzzy knowledge base and rules weights are defined using the genetic algorithm [22]. In this case, proximity of linguistic approximation results and experimental data is the criterion of extracted relations quality. It is shown here that in comparison with [22] the non-unique set of IF-THEN rules can be extracted from the given data. Following [18 – 21], at the

A.P. Rotshtein et al.: Fuzzy Evidence in Identif., Forecast. and Diagn., STUDFUZZ 275, pp. 235–258.
springerlink.com

second stage the obtained null solution allows us to arrange the genetic search for the complete solution set, which is determined by the unique maximum matrix and a set of minimum matrices. After linguistic interpretation the resulting solution can be represented as a set of possible rules collections, discovering the structure of the given data.

The approach proposed is illustrated by the computer experiment and the example of medical diagnosis. This chapter is written on the basis of [23].

8.1 "Multiple Inputs – Multiple Outputs" Object

Let us consider an object

$$\mathbf{Y} = f(\mathbf{X}) \tag{8.1}$$

with n inputs $\mathbf{X} = (x_1, x_2, ..., x_n)$ and m outputs $\mathbf{Y} = (y_1, y_2, ..., y_m)$, for which the following is known:

- intervals of inputs and outputs change

$$x_i \in [\underline{x}_i, \overline{x}_i],\ i = \overline{1,n};\ y_j \in [\underline{y}_j, \overline{y}_j],\ j = \overline{1,m};$$

- classes of decisions e_{jp} for evaluation of output variable y_j, $j = \overline{1,m}$, formed by digitizing the range $[\underline{y}_j, \overline{y}_j]$ into q_j levels

$$[\underline{y}_j, \overline{y}_j] = \underbrace{[\underline{y}_j, \overline{y}_{j1})}_{e_{j1}} \cup ... \cup \underbrace{[\underline{y}_{jp}, \overline{y}_{jp})}_{e_{jp}} \cup ... \cup \underbrace{[\underline{y}_{jq_j}, \overline{y}_j]}_{e_{jq_j}};$$

- training data in the form of L pairs of "inputs-outputs" experimental data

$$\langle \hat{\mathbf{X}}_s, \hat{\mathbf{Y}}_s \rangle,\quad s = \overline{1,L},$$

where $\hat{\mathbf{X}}_s = (\hat{x}_1^s, \hat{x}_2^s, ..., \hat{x}_n^s)$ and $\hat{\mathbf{Y}}_s = (\hat{y}_1^s, \hat{y}_2^s, ..., \hat{y}_m^s)$ are the vectors of the values of the input and output variables in the experiment number s.

It is necessary to transfer the available training data into the following system of IF-THEN rules [7]:

Rule l: IF $x_1 = a_{1l}$ AND ... $x_i = a_{il}$ AND ... $x_n = a_{nl}$

THEN $y_1 = b_{1l}$ AND ... $y_j = b_{jl}$ AND ... $y_m = b_{ml}$, $l = \overline{1,N}$, (8.2)

where a_{il} is the fuzzy term describing a variable x_i in rule l, $i = \overline{1,n}$;

b_{jl} is the fuzzy term describing a variable y_j in rule l, $j = \overline{1,m}$;

N is the number of rules.

8.2 Fuzzy Rules, Relations and Relational Equations

This fuzzy rule base is modelled by the fuzzy relational matrix presented in Table 8.1.

Table 8.1. Fuzzy knowledge base

<table>
<tr><td colspan="6">IF inputs</td><td colspan="11">THEN outputs</td></tr>
<tr><td rowspan="3"></td><td rowspan="3">x_1</td><td rowspan="3">…</td><td rowspan="3">x_i</td><td rowspan="3">…</td><td rowspan="3">x_n</td><td colspan="3">y_1</td><td>…</td><td colspan="3">y_j</td><td>…</td><td colspan="3">y_m</td></tr>
<tr><td>e_{11}</td><td>…</td><td>e_{1q_1}</td><td>…</td><td>e_{j1}</td><td>…</td><td>e_{jq_j}</td><td>…</td><td>e_{m1}</td><td>…</td><td>e_{mq_m}</td></tr>
<tr><td>E_1</td><td colspan="4">…</td><td>E_k</td><td colspan="4">…</td><td>E_M</td></tr>
<tr><td>C_1</td><td>a_{11}</td><td>…</td><td>a_{i1}</td><td>…</td><td>a_{n1}</td><td>r_{11}</td><td colspan="4">…</td><td>r_{1k}</td><td colspan="4">…</td><td>r_{1M}</td></tr>
<tr><td>…</td><td>…</td><td>…</td><td></td><td>…</td><td>…</td><td>…</td><td colspan="4">…</td><td>…</td><td colspan="4">…</td><td>…</td></tr>
<tr><td>C_l</td><td>a_{1l}</td><td>…</td><td>a_{il}</td><td>…</td><td>a_{nl}</td><td>r_{l1}</td><td colspan="4">…</td><td>r_{lk}</td><td colspan="4">…</td><td>r_{lM}</td></tr>
<tr><td>…</td><td>…</td><td>…</td><td></td><td>…</td><td>…</td><td>…</td><td colspan="4">…</td><td>…</td><td colspan="4">…</td><td>…</td></tr>
<tr><td>C_N</td><td>a_{1N}</td><td>…</td><td>a_{iN}</td><td>…</td><td>a_{nN}</td><td>r_{N1}</td><td colspan="4">…</td><td>r_{Nk}</td><td colspan="4">…</td><td>r_{NM}</td></tr>
</table>

This relational matrix can be translated as a set of fuzzy IF-THEN rules

$$\text{Rule } l: \text{IF } \mathbf{X} = C_l \text{ THEN } y_j = e_{jp} \text{ with weight } r_{l,jp}, \tag{8.3}$$

where C_l is the combination of input terms in rule l, $l = \overline{1,N}$;

$r_{l,jp}$ is the relation $C_l \times e_{jp}$, $j = \overline{1,m}$, $p = \overline{1,q_j}$, interpreted as the rule weight.

We shall redenote the set of classes of output variables as $\{E_1, E_2, ..., E_M\} = \{e_{11}, e_{12}, ..., e_{1q_1}, ..., e_{m1}, e_{m2}, ..., e_{mq_m}\}$, where $M = q_1 + q_2 + ... + q_m$.

In the presence of relational matrix

$$\mathbf{R} \subseteq C_l \times E_k = [r_{lk}, \ l = \overline{1,N}, \ k = \overline{1,M}]$$

the "inputs-outputs" dependency can be described with the help of Zadeh's compositional rule of inference [6]

$$\boldsymbol{\mu}^E(\mathbf{Y}) = \boldsymbol{\mu}^C(\mathbf{X}) \circ \mathbf{R}, \tag{8.4}$$

where $\boldsymbol{\mu}^C(\mathbf{X}) = (\mu^{C_1}, \mu^{C_2}, ..., \mu^{C_N})$ is the vector of membership degrees of vector $\mathbf{X}$ to input combinations C_l;

$\boldsymbol{\mu}^E(\mathbf{Y}) = (\mu^{E_1}, \mu^{E_2}, ..., \mu^{E_M})$ is the vector of membership degrees of variables y_j to classes e_{jp};

$\circ$ is the operation of *max-min* composition [6].

The system of fuzzy relational equations is derived from relation (8.4):

$$\mu^{e_{jp}}(y_j) = (\mu^{C_1}(\mathbf{X}) \wedge r_{1,jp}) \vee (\mu^{C_2}(\mathbf{X}) \wedge r_{2,jp}) \vee ... \vee (\mu^{C_N}(\mathbf{X}) \wedge r_{N,jp}),$$

where

$$\mu^{C_l}(\mathbf{X}) = \mu^{a_{1l}}(x_1) \wedge \mu^{a_{2l}}(x_2) \wedge ... \wedge \mu^{a_{nl}}(x_n), \quad l = \overline{1,N};$$

or

$$\mu^{e_{jp}}(y_j) = \underset{l=\overline{1,N}}{\vee} ((\underset{i=\overline{1,n}}{\wedge} \mu^{a_{il}}(x_i)) \wedge r_{l,jp}). \tag{8.5}$$

Here

$\mu^{a_{il}}(x_i)$ is a membership function of a variable x_i to the fuzzy term a_{il};

$\mu^{e_{jp}}(y_j)$ is a membership function of a variable y_j to the class e_{jp}.

Taking into account the fact that operations $\vee$ and $\wedge$ are replaced by *max* and *min* in fuzzy set theory, system (8.5) is rewritten in the form

$$\mu^{e_{jp}}(y_j) = \max_{l=\overline{1,N}} \left(\min \left(\min_{i=\overline{1,n}} [\mu^{a_{il}}(x_i)], r_{l,jp} \right) \right). \tag{8.6}$$

We use a bell-shaped membership function model of variable u to arbitrary term T in the form [22]:

$$\mu^T(u) = \frac{1}{1 + \left(\frac{u - \beta}{\sigma} \right)^2}, \tag{8.7}$$

where β is a coordinate of function maximum, $\mu^T(\beta) = 1$; σ is a parameter of concentration.

The operation of defuzzification is defined in [22] as follows:

$$y_j = \frac{\sum_{p=1}^{q_j} \underline{y}_{jp} \cdot \mu^{e_{jp}}(y_j)}{\sum_{p=1}^{q_j} \mu^{e_{jp}}(y_j)}. \tag{8.8}$$

Relationships (8.6) – (8.8) define the generalized fuzzy model of an object (8.1) as follows:

$$\mathbf{Y} = F_R(\mathbf{X}, \mathbf{R}, \mathbf{B}, \mathbf{\Omega}), \tag{8.9}$$

where $\mathbf{B} = (\beta_1, \beta_2, ..., \beta_K)$ and $\mathbf{\Omega} = (\sigma_1, \sigma_2, ..., \sigma_K)$ are the vectors of β- and σ-parameters for fuzzy terms membership functions in (8.3);

K is the total number of fuzzy terms;

F_R is the operator of inputs-outputs connection, corresponding to formulae (8.6)–(8.8).

8.3 Optimization Problem for Fuzzy Relations Extraction

Let us impose limitations on the knowledge base (8.2) volume in the following form:

$$N \le \overline{N},$$

where $\overline{N}$ is the maximum permissible total number of rules.

So as content and number of linguistic terms a_{il} ($i=\overline{1,n}$, $l=\overline{1,N}$) used in fuzzy knowledge base (8.2) are not known beforehand then we suggest to interpret them on the basis of membership functions (8.7) parameter values ($\beta^{a_{il}}, \sigma^{a_{il}}$). Therefore, knowledge base (8.2) synthesis is reduced to obtaining the matrix of parameters shown in Table 8.2 [22].

Table 8.2. Knowledge base parameters matrix

IF inputs				THEN outputs										
	x_1	...	x_n	y_1			...	y_j			...	y_m		
				e_{11}	...	e_{1q_1}	...	e_{j1}	...	e_{jq_j}	...	e_{m1}	...	e_{mq_m}
				E_1	...			E_k	...					E_M
C_1	$(\beta^{a_{11}}, \sigma^{a_{11}})$	...	$(\beta^{a_{n1}}, \sigma^{a_{n1}})$	r_{11}	...			r_{1k}	...					r_{1M}
...	...	...	...	...	...			...	...					...
C_l	$(\beta^{a_{1l}}, \sigma^{a_{1l}})$	...	$(\beta^{a_{nl}}, \sigma^{a_{nl}})$	r_{l1}	...			r_{lk}	...					r_{lM}
...	...	...	...	...	...			...	...					...
C_N	$(\beta^{a_{1N}}, \sigma^{a_{1N}})$	...	$(\beta^{a_{nN}}, \sigma^{a_{nN}})$	r_{N1}	...			r_{Nk}	...					r_{NM}

This problem can be formulated as follows. It is necessary to find such a matrix (Table 8.2), which satisfies the limitations imposed on knowledge base volume and provides the least distance between model and experimental outputs of the object:

$$\sum_{s=1}^{L}[F_R(\hat{\mathbf{X}}_s, \mathbf{R}, \mathbf{B}, \mathbf{\Omega}) - \hat{\mathbf{Y}}_s]^2 = \min_{\mathbf{R},\mathbf{B},\mathbf{\Omega}} . \tag{8.10}$$

If R_0 is a solution of the optimization problem (8.10), then R_0 is the exact solution of the composite system of fuzzy relational equations:

$$\hat{\boldsymbol{\mu}}^A(\hat{\mathbf{X}}_s) \circ \mathbf{R} = \hat{\boldsymbol{\mu}}^B(\hat{\mathbf{X}}_s), \tag{8.11}$$

where the experimental input and output matrices

$$\hat{\boldsymbol{\mu}}^A = \begin{bmatrix} \hat{\mu}^{C_1}(\hat{\mathbf{X}}_1) & \ldots & \hat{\mu}^{C_N}(\hat{\mathbf{X}}_1) \\ \ldots & \ldots & \ldots \\ \hat{\mu}^{C_1}(\hat{\mathbf{X}}_L) & \ldots & \hat{\mu}^{C_N}(\hat{\mathbf{X}}_L) \end{bmatrix}, \quad \hat{\boldsymbol{\mu}}^B = \begin{bmatrix} \hat{\mu}^{E_1}(\hat{\mathbf{X}}_1) & \ldots & \hat{\mu}^{E_M}(\hat{\mathbf{X}}_1) \\ \ldots & \ldots & \ldots \\ \hat{\mu}^{E_1}(\hat{\mathbf{X}}_L) & \ldots & \hat{\mu}^{E_M}(\hat{\mathbf{X}}_L) \end{bmatrix}$$

are obtained for the given training data.

Following [2], the system (8.11) has a solution set $S(\hat{\boldsymbol{\mu}}^A, \hat{\boldsymbol{\mu}}^B)$, which is determined by the unique maximal solution $\overline{\mathbf{R}}$ and the set of minimal solutions $S^*(\hat{\boldsymbol{\mu}}^A, \hat{\boldsymbol{\mu}}^B) = \{\underline{\mathbf{R}}_I,\ I = \overline{1,T}\}$:

$$S^*(\hat{\boldsymbol{\mu}}^A, \hat{\boldsymbol{\mu}}^B) = \bigcup_{\underline{\mathbf{R}}_I \in S^*} [\underline{\mathbf{R}}_I, \overline{\mathbf{R}}] \ . \tag{8.12}$$

Here $\overline{\mathbf{R}} = [\overline{r}_{lk}]$ and $\underline{\mathbf{R}}_I = [\underline{r}_{lk}^I]$ are the matrices of the upper and lower bounds of the fuzzy relations r_{lk}, where the union is taken over all $\underline{\mathbf{R}}_I \in S^*(\hat{\boldsymbol{\mu}}^A, \hat{\boldsymbol{\mu}}^B)$.

The problem of solving fuzzy relational equations (8.11) is formulated as follows [19 – 21]. Fuzzy relation matrix $\mathbf{R} = [r_{lk}]$, $l = \overline{1,N}$, $k = \overline{1,M}$, should be found which satisfies the constraints $r_{lk} \in [0,1]$ and also provides the least distance between model and experimental outputs of the object; that is, the minimum value of the criterion (8.10).

Following [19 – 21], formation of the intervals (8.12) is accomplished by way of solving a multiple optimization problem (8.10) and it begins with the search for its null solution $\mathbf{R}_0 = [r_{lk}^0]$, where $r_{lk}^0 \le \overline{r}_{lk}$, $l = \overline{1,N}$, $k = \overline{1,M}$. The upper bound ($\overline{r}_{lk}$) is found in the range $[r_{lk}^0, 1]$. The lower bound ($\underline{r}_{lk}^I$) for $I = 1$ is found in the range $[0, r_{lk}^0]$, and for $I > 1$ in the range $[0, \overline{r}_{lk}]$, where the minimal solutions $\underline{\mathbf{R}}_J$, $J < I$, are excluded from the search space.

Let $\mathbf{R}(t) = [r_{lk}(t)]$ be some t-th solution of optimization problem (8.10), that is $F(\mathbf{R}(t)) = F(\mathbf{R}_0)$, since for all $\mathbf{R} \in S(\hat{\boldsymbol{\mu}}^A, \hat{\boldsymbol{\mu}}^B)$ we have the same value of criterion (8.10). While searching for upper bounds $\overline{r}_{lk}$ it is suggested that $r_{lk}(t) \ge r_{lk}(t-1)$, and while searching for lower bounds $\underline{r}_{lk}^I$ it is suggested that $r_{lk}(t) \le r_{lk}(t-1)$ (Fig. 8.1).

The definition of the upper (lower) bounds follows the rule: if $\mathbf{R}(t) \ne \mathbf{R}(t-1)$, then $\overline{r}_{lk}$ ($\underline{r}_{lk}^I$)$= r_{lk}(t)$. If $\mathbf{R}(t) = \mathbf{R}(t-1)$, then the search for the interval solution $[\underline{\mathbf{R}}_I, \overline{\mathbf{R}}]$ is stopped. Formation of intervals (8.12) will go on until the condition $\underline{\mathbf{R}}_I \ne \underline{\mathbf{R}}_J$, $J < I$, has been satisfied.

The hybrid genetic and neuro approach is proposed for solving optimization problem (8.10).

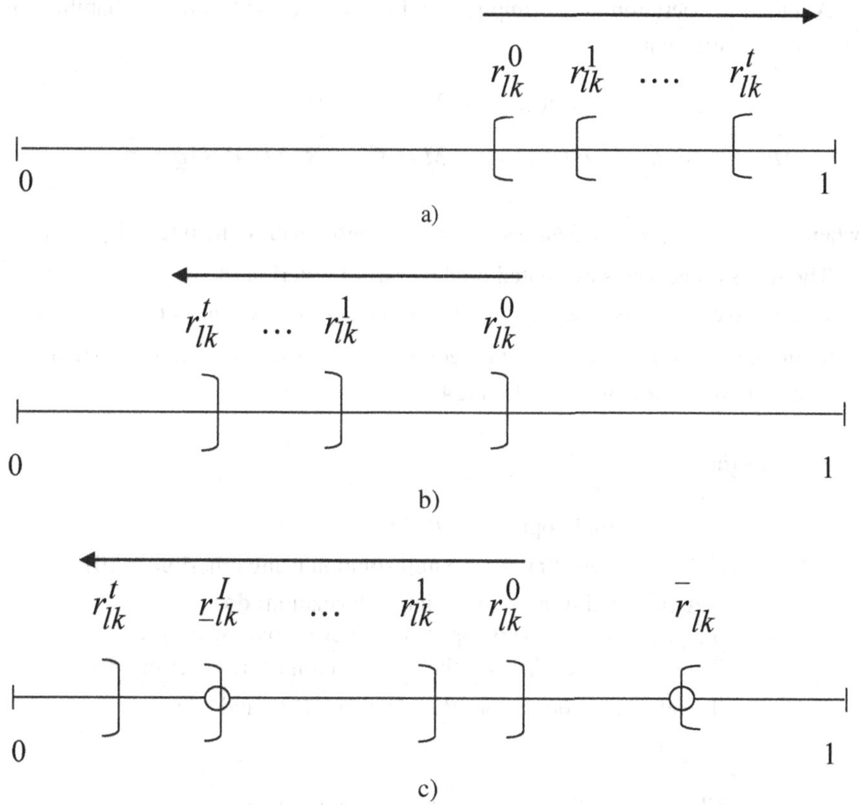

Fig. 8.1. Search for the upper (a) and lower bounds of the intervals for $I = 1$ (b) and $I > 1$ (c)

8.4 Genetic Algorithm for Fuzzy Relations Extraction

To describe the chromosome for the parameters matrix (Table 8.2), we use the string shown in Fig. 8.2, where C_l is the code of IF-THEN rule with number l, $l = \overline{1,N}$. The chromosome needed in the genetic algorithm for solving fuzzy relational equations (8.11) includes only the codes of parameters r_{lk}, $l = \overline{1,N}$, $k = \overline{1,M}$. Parameters of membership functions are defined simultaneously with the null solution.

The crossover operation is defined in Fig. 8.3, and is carried out by way of exchanging chromosomes parts inside each rule C_l ($l = \overline{1,N}$) and inside matrix of rules weights $\mathbf{R}$. The total number of exchange points is equal to $N + 1$.

A mutation operation (Mu) implies random change (with some probability) of chromosome elements:

$$Mu(r_{lk}) = RANDOM([0,1]) \ ,$$

$$Mu(\beta^{a_{il}}) = RANDOM([\underline{x}_i, \bar{x}_i]) \ , \ Mu(\sigma^{a_{il}}) = RANDOM([\underline{\sigma}^{a_{il}}, \bar{\sigma}^{a_{il}}]) \ ,$$

where $RANDOM([\underline{x}, \bar{x}])$ denotes a random number within the interval $[\underline{x}, \bar{x}]$.

The fitness function is evaluated on the basis of criterion (8.10).

If $P(t)$ are chromosomes-parents and $C(t)$ are chromosomes-offsprings on a t-th iteration, then the genetic procedure of optimization will be carried out according to the following algorithm [24, 25]:

```
begin
  t:=0;
  To set the initial population P(t);
  To evaluate the P(t) for the null solution using criterion (8.10);
  while (no condition of null solution formation) do
    To generate the C(t) by operation of cross-over with P(t);
    To evaluate the C(t) for the null solution using criterion (8.10);
    To select the population P(t+1) from P(t) and C(t);
    t:=t+1;
  end
  while (no condition of interval set formation) do
    To generate the C(t) by operation of cross-over with P(t);
    To evaluate the C(t) for the bounds of intervals (8.12) using criterion (8.10);
    To select the population P(t+1) from P(t) and C(t);
    t:=t+1;
  end
end
```

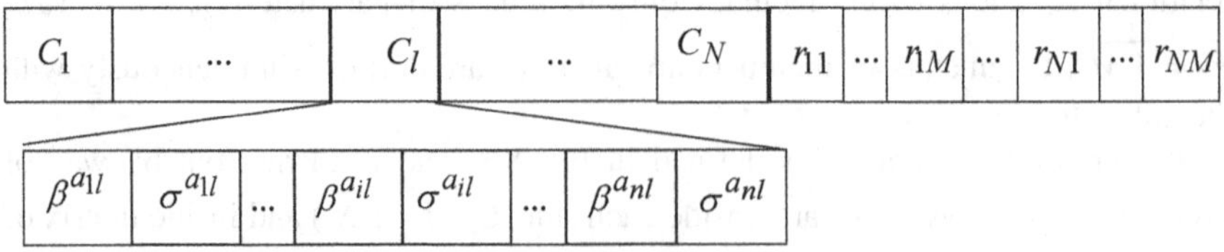

Fig. 8.2. Coding of parameters matrix

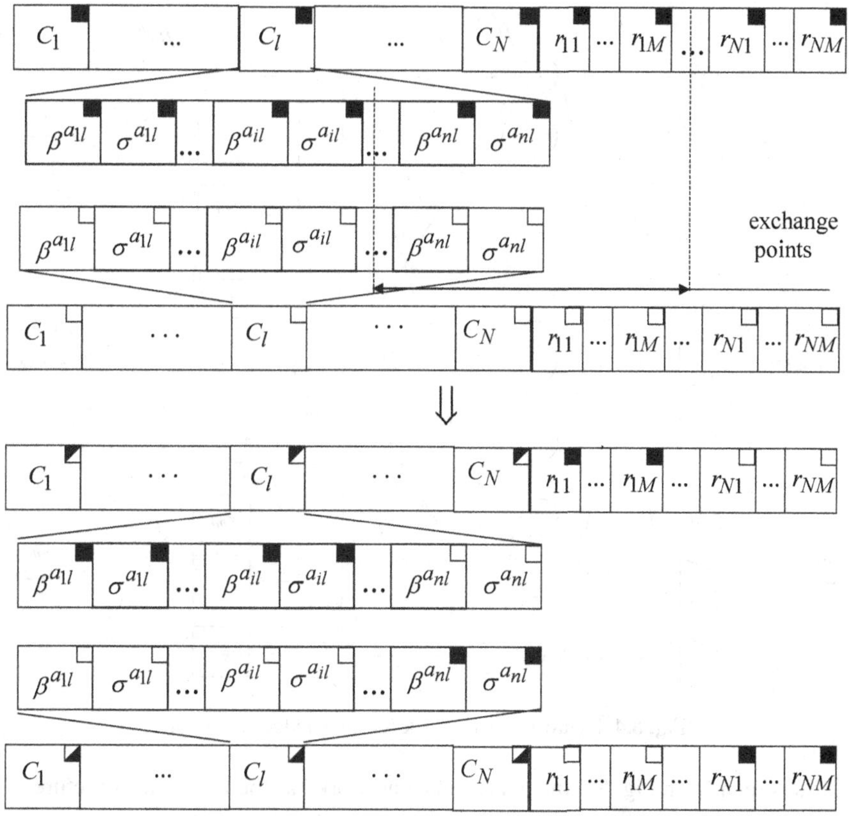

Fig. 8.3. Crossover operation (■, □ - chromosomes-parents, ◪, ◩ - chromosomes-offsprings)

8.5 Neuro-fuzzy Network for Fuzzy Relations Extraction

Let us impose limitations on the knowledge base (8.2) volume in the following form:

$$k_1 \le \overline{k_1}\ ,\ k_2 \le \overline{k_2}\ ,\ ...,\ k_n \le \overline{k_n}\ ,$$

where $\overline{k_i}$ is the maximum permissible total number of fuzzy terms describing a variable x_i , $i = \overline{1,n}$.

This allows embedding system (8.2) into the special neuro-fuzzy network, which is able to extract knowledge [16, 21]. The neuro-fuzzy network for knowledge extraction is shown in Fig. 8.4, and the nodes are presented in Table 3.1.

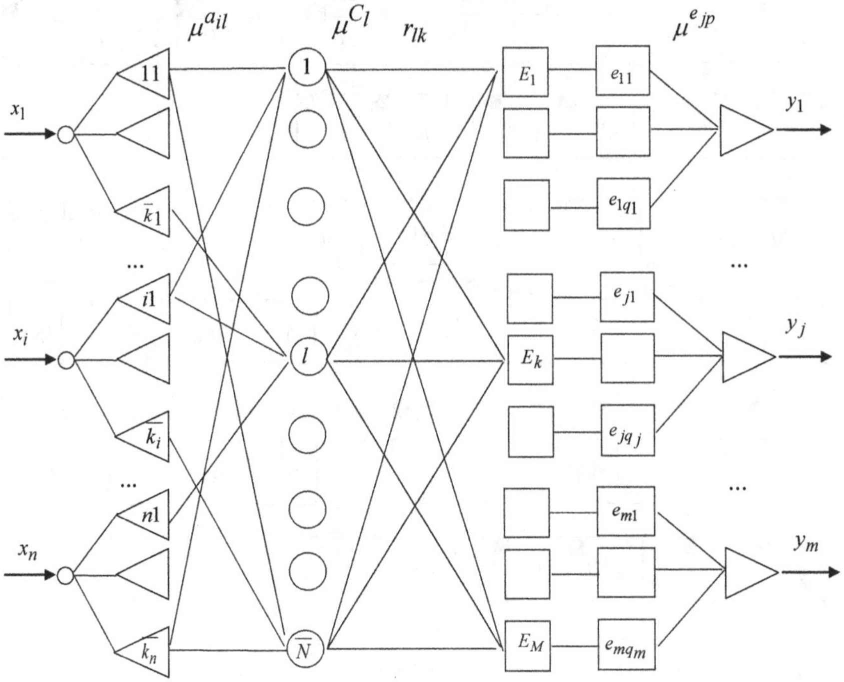

Fig. 8.4. Neuro-fuzzy network for knowledge extraction

As is seen from Fig. 8.4 the neuro-fuzzy network has the following structure:

layer 1 for object identification inputs (the number of nodes is equal to *n*),

layer 2 for fuzzy terms used in the knowledge base (the number of nodes is equal to $\bar{k}_1 + \bar{k}_2 + ... + \bar{k}_n$),

layer 3 for strings-conjunctions (the number of nodes is equal to $\bar{k}_1 \cdot \bar{k}_2 \cdot ... \cdot \bar{k}_n$),

layer 4 for fuzzy rules making classes (the layer is fully connected, the number of nodes is equal to the number of output classes M),

layer 5 for a defuzzification operation for each output.

To train the parameters of the neuro-fuzzy network, the recurrent relations

$$r_{lk}(t+1) = r_{lk}(t) - \eta \frac{\partial \varepsilon_t}{\partial r_{lk}(t)};$$

$$\beta^{a_{il}}(t+1) = \beta^{a_{il}}(t) - \eta \frac{\partial \varepsilon_t}{\partial \beta^{a_{il}}(t)}; \quad \sigma^{a_{il}}(t+1) = \sigma^{a_{il}}(t) - \eta \frac{\partial \varepsilon_t}{\partial \sigma^{a_{il}}(t)}, \quad (8.13)$$

are used which minimize the criterion

$$\varepsilon_t = \frac{1}{2}(\hat{y}_t - y_t)^2,$$

where $\hat{y}_t$ and y_t are the experimental and the model outputs of the object at the t-th step of training;

$r_{lk}(t)$ are fuzzy relations at the t-th step of training;

$\beta^{a_{il}}(t)$, $\sigma^{a_{il}}(t)$ are parameters for the fuzzy terms membership functions at the t-th step of training.

η is a parameter of training [26].

The partial derivatives appearing in recurrent relations (8.13) can be obtained according to the results from Section 7.8.

8.6 Computer Simulations

Experiment 1

The aim of the experiment is to generate the system of IF-THEN rules for the target "input (x) – output (y)" model presented in Fig. 8.5.

$$y = \frac{(1.8x+0.8)(5x-1.1)(4x-2.9)(3x-2.1)(9.5x-9.5)(3x-0.05)+20}{80}. \quad (8.14)$$

The training data in the form of the interval values of input and output variable is presented in Table 8.3.

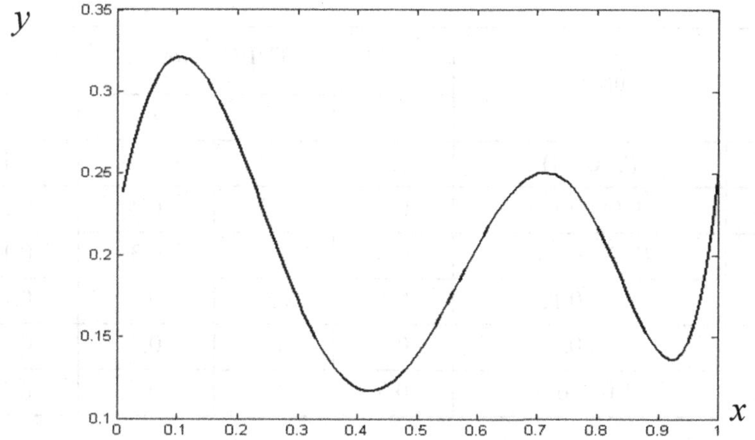

Fig. 8.5. Input-output model-generator

Table 8.3. Training data ($\hat{x}_s$, $\hat{y}_s$)

s	Input	Output
	x_1	y_1
1	[0, 0.1]	[0.22, 0.32]
2	[0.1, 0.2]	[0.32, 0.27]
3	[0.2, 0.3]	[0.27, 0.17]
4	[0.3, 0.4]	[0.17, 0.12]
5	[0.4, 0.5]	[0.12, 0.14]
6	[0.5, 0.6]	[0.14, 0.21]
7	[0.6, 0.7]	[0.21, 0.25]
8	[0.7, 0.8]	[0.25, 0.22]
9	[0.8, 0.9]	[0.22, 0.14]
10	[0.9, 1.0]	[0.14, 0.25]

The total number of fuzzy terms for the input variable is limited to six. The total number of classes for the output variable is limited to four.

The classes for output variable evaluation are formed as follows:

$$[\,\underline{y},\overline{y}\,] = \underbrace{[\,0.10, 0.15\,)}_{e_{11}} \cup \underbrace{[\,0.15, 0.20)\,]}_{e_{12}} \cup \underbrace{[\,0.20, 0.25\,]}_{e_{13}} \cup \underbrace{[\,0.25, 0.35\,]}_{e_{14}},$$

The null solution $\mathbf{R}_0$ presented in Table 8.4 together with the parameters of the knowledge matrix is obtained using the genetic algorithm.

Table 8.4. Fuzzy relational matrix (null solution)

IF input x		THEN output y			
		e_{11}	e_{12}	e_{13}	e_{14}
C_1	(0, 0.14)	0.3	0.9	0.7	0.1
C_2	(0.09, 0.14)	0.2	0.2	0.4	0.9
C_3	(0.40, 0.12)	0.8	0.3	0.3	0.1
C_4	(0.72, 0.12)	0.1	0.3	0.9	0.2
C_5	(0.92, 0.11)	0.9	0.6	0.2	0.3
C_6	(1.0, 0.07)	0.3	0.9	0.6	0.1

The obtained null solution allows us to arrange for the genetic search for the solution set of the system (8.11), where the matrices $\hat{\boldsymbol{\mu}}^A(\hat{x}_s)$ and $\hat{\boldsymbol{\mu}}^B(\hat{x}_s)$ for the training data take the following form:

$\hat{\boldsymbol{\mu}}^A =$

	$\hat{\mu}^{C_1}$	$\hat{\mu}^{C_2}$	$\hat{\mu}^{C_3}$	$\hat{\mu}^{C_4}$	$\hat{\mu}^{C_5}$	$\hat{\mu}^{C_6}$
$\hat{x}_1$	[0.67, 1.0]	[0.75, 1.0]	[0.09, 0.14]	[0.03, 0.04]	[0.01, 0.02]	[0, 0.01]
$\hat{x}_2$	[0.33, 0.67]	[0.62, 0.98]	[0.14, 0.26]	[0.04, 0.05]	0.02	0.01
$\hat{x}_3$	[0.18, 0.33]	[0.31, 0.62]	[0.26, 0.59]	[0.05, 0.08]	[0.02, 0.03]	0.01
$\hat{x}_4$	[0.11, 0.18]	[0.17, 0.31]	[0.59, 1.0]	[0.08, 0.17]	[0.03, 0.04]	0.01
$\hat{x}_5$	[0.07, 0.11]	[0.10, 0.17]	[0.59, 1.0]	[0.17, 0.30]	[0.04, 0.06]	[0.01, 0.02]
$\hat{x}_6$	[0.05, 0.07]	[0.07, 0.10]	[0.25, 0.59]	[0.30, 0.50]	[0.06, 0.11]	[0.02, 0.03]
$\hat{x}_7$	[0.04, 0.05]	[0.05, 0.07]	[0.17, 0.26]	[0.50, 0.97]	[0.11, 0.20]	[0.03, 0.05]
$\hat{x}_8$	[0.03, 0.04]	[0.04, 0.05]	[0.08, 0.17]	[0.69, 1.0]	[0.20, 0.46]	[0.05, 0.11]
$\hat{x}_9$	[0.02, 0.03]	[0.03, 0.04]	[0.05, 0.08]	[0.33, 0.69]	[0.46, 0.97]	[0.11, 0.33]
$\hat{x}_{10}$	0.02	[0.02, 0.03]	[0.04, 0.05]	[0.16, 0.33]	[0.70, 1.0]	[0.33, 1.0]

$\hat{\boldsymbol{\mu}}^B =$

	$\hat{\mu}^{E_1}$	$\hat{\mu}^{E_2}$	$\hat{\mu}^{E_3}$	$\hat{\mu}^{E_4}$
$\hat{x}_1$	0.30	[0.67, 0.90]	[0.67, 0.70]	[0.75, 0.90]
$\hat{x}_2$	0.30	[0.33, 0.67]	[0.40, 0.67]	[0.62, 0.90]
$\hat{x}_3$	[0.30, 0.59]	[0.30, 0.33]	[0.31, 0.40]	[0.31, 0.62]
$\hat{x}_4$	[0.59, 0.80]	0.30	[0.30, 0.31]	[0.17, 0.31]
$\hat{x}_5$	[0.59, 0.80]	0.30	0.30	[0.17, 0.20]
$\hat{x}_6$	[0.26, 0.59]	0.30	[0.30, 0.50]	0.20
$\hat{x}_7$	[0.17, 0.26]	0.30	[0.50, 0.90]	0.20
$\hat{x}_8$	[0.20, 0.46]	[0.30, 0.46]	[0.69, 0.90]	[0.20, 0.30]
$\hat{x}_9$	[0.46, 0.90]	[0.46, 0.60]	[0.33, 0.69]	0.30
$\hat{x}_{10}$	[0.70, 0.90]	[0.60, 0.90]	[0.33, 0.60]	0.30

The complete solution set for the fuzzy relation matrix is presented in Table 8.5, where input x is described by fuzzy terms *Low (L), lower than Average (lA), Average (A), higher than Average (hA), lower than High (lH), High (H)*; output y is described by fuzzy terms *higher than Low (hL), lower than Average (lA), Average (A), High (H).*

Table 8.5. Fuzzy relational matrix (complete solution set)

IF input x		THEN output y			
		hL	lA	A	H
C_1	L	0.30	0.90	0.70	[0, 0.75]
C_2	lA	0.30	0.30	0.40	0.90
C_3	A	0.80	0.30	0.30	[0, 0.20]
C_4	hA	[0, 0.26]	0.30	[0.69, 0.90]	0.20
C_5	lH	0.90	0.60	[0.33, 0.60] ∪ [0, 0.60]	0.30
C_6	H	[0, 0.70]	[0.60, 0.90]	[0, 0.60] ∪ [0.33, 0.60]	[0, 0.30]

The obtained solution provides the approximation of the object shown in Fig. 8.6.

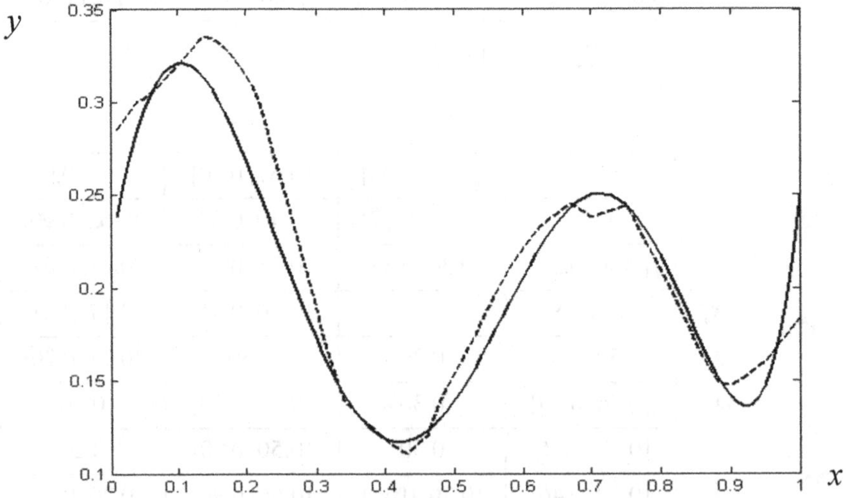

Fig. 8.6. Input-output model extracted from data

The resulting solution can be linguistically interpreted as the set of the two possible rules bases (See Table 8.6), which differ in the fuzzy terms describing output y in rule 6 with overlapping weights.

Table 8.6. System of IF-THEN rules

Rule	IF x	THEN y
1	L	lA
2	lA	H
3	A	hL
4	hA	A
5	lH	hL
6	H	hL or lA

Experiment 2

The aim of the experiment is to generate the system of IF-THEN rules for the target "two inputs (x_1, x_2) – two outputs (y_1, y_2)" model presented in Fig. 8.7:

$$y_1 = f_1(x_1, x_2) = \frac{1}{10}(2z-0.9)\ (7z-1)\ (17z-19)\ (15z-2)\,, \tag{8.15}$$

$$y_2 = f_2(x_1, x_2) = -\frac{1}{2}y_1 + 1\,,$$

where $z = \dfrac{(x_1-3.0)^2 + (x_2-2.5)^2}{40}$.

The training data in the form of the interval values of input and output variables is presented in Table 8.7.

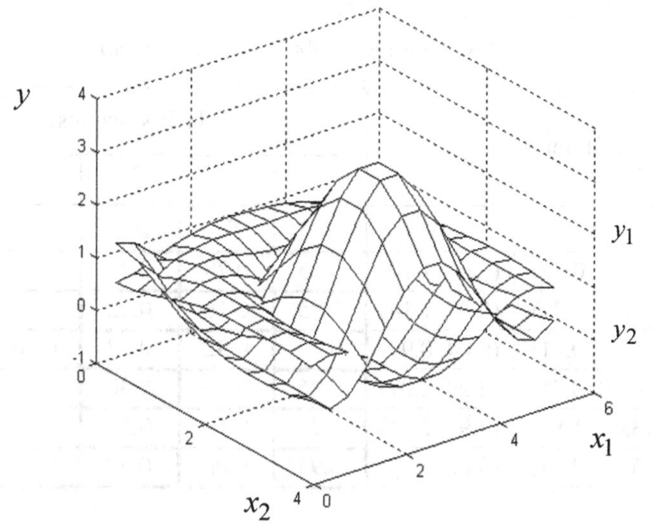

Fig. 8.7. Inputs-outputs model-generator

Table 8.7. Training data ($\hat{\mathbf{X}}_s$, $\hat{\mathbf{Y}}_s$)

s	Inputs		Outputs	
	x_1	x_2	y_1	y_2
1	[0.2, 1.2]	[0.3, 1.6]	[0, 1.0]	[0.5, 1.0]
2	[0.2, 1.2]	[1.3, 4.0]	[0, 0.8]	[0.6, 1.0]
3	[0.7, 3.0]	[0.3, 1.6]	[0, 2.3]	[-0.15, 1.0]
4	[0.7, 3.0]	[1.3, 4.0]	[0, 3.4]	[-0.7, 1.0]
5	[3.0, 5.3]	[0.3, 1.6]	[0, 2.3]	[-0.15, 1.0]
6	[3.0, 5.3]	[1.3, 4.0]	[0, 3.4]	[-0.7, 1.0]
7	[4.8, 5.8]	[0.3, 1.6]	[0, 1.0]	[0.5, 1.0]
8	[4.8, 5.8]	[1.3, 4.0]	[0, 0.8]	[0.6, 1.0]

The total number of fuzzy terms for input variables is limited to three. The total number of combinations of input terms is limited to six.

The classes for output variables evaluation are formed as follows:

$$[\underline{y}_1, \overline{y}_1] = \underbrace{[0, 0.2)}_{e_{11}} \cup \underbrace{[0.2, 1.2)}_{e_{12}} \cup \underbrace{[1.2, 3.4]}_{e_{13}},$$

$$[\underline{y}_2, \overline{y}_2] = \underbrace{[-0.7, 0)}_{e_{21}} \cup \underbrace{[0, 1.2]}_{e_{22}}.$$

The null solution $\mathbf{R}_0$ presented in Table 8.8 together with the parameters of the knowledge matrix is obtained using the genetic algorithm.

Table 8.8. Fuzzy relational matrix (null solution)

IF inputs			THEN outputs				
			y_1			y_2	
	x_1	x_2	e_{11}	e_{12}	e_{13}	e_{21}	e_{22}
C_1	(0.03, 0.72)	(0.01, 1.10)	0.15	0.78	0.24	0.52	0.48
C_2	(3.00, 1.77)	(0.02, 1.14)	0.85	0.16	0.02	0.76	0.15
C_3	(5.96, 0.71)	(0.04, 0.99)	0.10	0.92	0.27	0.50	0.43
C_4	(0.00, 0.75)	(2.99, 2.07)	0.86	0.04	0.30	0.80	0.30
C_5	(3.02, 1.80)	(2.97, 2.11)	0.21	0.11	0.10	0.15	0.97
C_6	(5.99, 0.74)	(3.02, 2.10)	0.94	0.08	0.30	0.75	0.30

The obtained null solution allows us to arrange for the genetic search for the solution set of the system (8.11), where the matrices $\hat{\boldsymbol{\mu}}^A(\hat{\mathbf{X}}_s)$ and $\hat{\boldsymbol{\mu}}^B(\hat{\mathbf{X}}_s)$ for the training data take the following form:

$\hat{\boldsymbol{\mu}}^A =$

	$\hat{\mu}^{C_1}$	$\hat{\mu}^{C_2}$	$\hat{\mu}^{C_3}$	$\hat{\mu}^{C_4}$	$\hat{\mu}^{C_5}$	$\hat{\mu}^{C_6}$
$\hat{\mathbf{X}}_1$	[0.16, 0.74]	[0.16, 0.52]	0	[0.33, 0.61]	[0.28, 0.52]	0
$\hat{\mathbf{X}}_2$	[0.21, 0.46]	[0.21, 0.46]	0	[0.35, 0.90]	[0.28, 0.52]	0
$\hat{\mathbf{X}}_3$	[0, 0.50]	[0.16, 0.74]	0	[0, 0.50]	[0.33, 0.61]	0
$\hat{\mathbf{X}}_4$	[0, 0.46]	[0.21, 0.46]	0	[0, 0.50]	[0.37, 0.95]	0
$\hat{\mathbf{X}}_5$	0	[0.16, 0.74]	[0, 0.50]	0	[0.33, 0.61]	[0, 0.50]
$\hat{\mathbf{X}}_6$	0	[0.21, 0.46]	[0, 0.46]	0	[0.34, 0.95]	[0, 0.50]
$\hat{\mathbf{X}}_7$	0	[0.16, 0.52]	[0.16, 0.74]	0	[0.28, 0.52]	[0.33, 0.61]
$\hat{\mathbf{X}}_8$	0	[0.21, 0.46]	[0.21, 0.46]	0	[0.28, 0.52]	[0.35, 0.90]

$\hat{\boldsymbol{\mu}}^B =$

	$\hat{\mu}^{E_1}$	$\hat{\mu}^{E_2}$	$\hat{\mu}^{E_3}$	$\hat{\mu}^{E_4}$	$\hat{\mu}^{E_5}$
$\hat{\mathbf{X}}_1$	[0.33, 0.61]	[0.16, 0.74]	[0.30, 0.52]	[0.33, 0.61]	[0.30, 0.52]
$\hat{\mathbf{X}}_2$	[0.35, 0.86]	[0.21, 0.46]	[0.30, 0.52]	[0.35, 0.80]	[0.30, 0.52]
$\hat{\mathbf{X}}_3$	[0.21, 0.74]	[0.16, 0.50]	[0.33, 0.61]	[0.16, 0.74]	[0.33, 0.61]
$\hat{\mathbf{X}}_4$	[0.21, 0.46]	[0.16, 0.46]	[0.37, 0.95]	[0.21, 0.50]	[0.37, 0.95]
$\hat{\mathbf{X}}_5$	[0.21, 0.74]	[0.16, 0.50]	[0.33, 0.61]	[0.16, 0.74]	[0.33, 0.61]
$\hat{\mathbf{X}}_6$	[0.21, 0.50]	[0.16, 0.46]	[0.34, 0.95]	[0.21, 0.50]	[0.34, 0.95]
$\hat{\mathbf{X}}_7$	[0.33, 0.61]	[0.16, 0.74]	[0.30, 0.52]	[0.33, 0.61]	[0.30, 0.52]
$\hat{\mathbf{X}}_8$	[0.35, 0.90]	[0.21, 0.46]	[0.30, 0.52]	[0.35, 0.75]	[0.30, 0.52]

The complete solution set for the fuzzy relation matrix is presented in Table 8.9, where input x_1 is described by fuzzy terms *Low (L)*, *Average (A)*, *High (H)*; input

x_2 is described by fuzzy terms *Low (L), High (H)*; output y_1 is described by fuzzy terms *higher than Low* (*hL*), *lower than Average (lA), High (H)*; output y_2 is described by fuzzy terms *Low (L), lower than Average (lA)*.

Table 8.9. Fuzzy relational matrix (complete solution set)

IF inputs			THEN outputs				
			y_1			y_2	
	x_1	x_2	*hL*	*lA*	*H*	*lA*	*L*
C_1	*L*	*L*	[0, 0.21]	[0.74, 1.0]	[0, 0.30]	[0.33, 0.61]	[0, 0.52]
C_2	*A*	*L*	[0.74, 1.0]	[0, 0.16] ∪0.16	[0, 0.30]	[0.74, 1.0]	[0, 0.30]
C_3	*H*	*L*	[0, 0.21]	[0.74, 1.0]	[0, 0.30]	[0.33, 0.61]	[0, 0.52]
C_4	*L*	*H*	0.86	[0, 0.16]	0.30	0.80	0.30
C_5	*A*	*H*	0.21	0.16∪ [0, 0.16]	[0.95, 1.0]	[0, 0.16]	[0.97, 1.0]
C_6	*H*	*H*	[0.90, 1.0]	[0, 0.16]	0.30	0.75	0.30

The obtained solution provides the approximation of the object shown in Fig. 8.8.

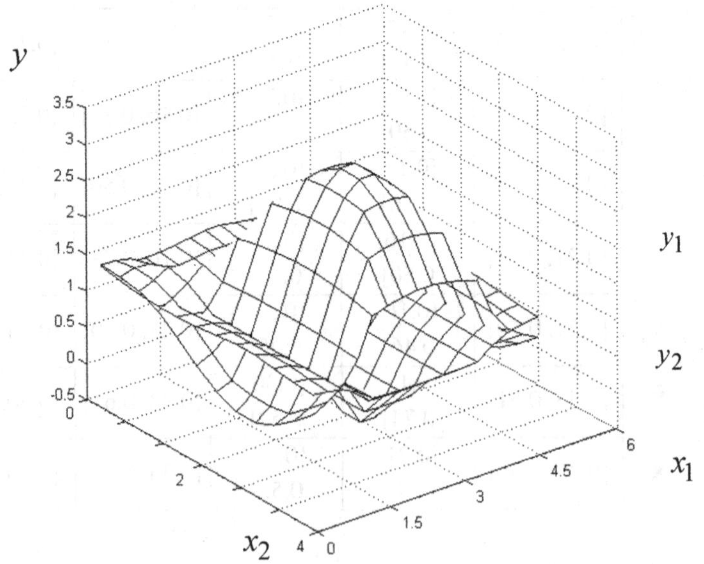

Fig. 8.8. Inputs-outputs model extracted from data

The resulting solution can be linguistically interpreted as the set of the four possible rules bases (See Table 8.10), which differ in the fuzzy terms describing output y_2 in rule 1 and rule 3 with overlapping weights.

Table 8.10. System of IF-THEN rules

Rule	IF inputs		THEN outputs	
	x_1	x_2	y_1	y_2
1	*L*	*L*	*lA*	*lA* or *L*
2	*A*	*L*	*hL*	*lA*
3	*H*	*L*	*lA*	*lA* or *L*
4	*L*	*H*	*hL*	*lA*
5	*A*	*H*	*H*	*L*
6	*H*	*H*	*hL*	*lA*

8.7 Example 7: Fuzzy Relations Extraction for Heart Diseases Diagnosis

The aim is to generate the system of IF-THEN rules for diagnosis of heart diseases. Input parameters are (variation ranges are indicated in parentheses):

x_1 – aortic valve size (0.75 – 2.5 cm^2);

x_2 – mitral valve size (1 – 2 cm^2);

x_3 – tricuspid valve size (0.5 – 2.7 cm^2);

x_4 – lung artery pressure (65 – 100 mm Hg).

Output parameters are:

y_1 – left ventricle size (11–14 mm);

y_2 – left auricle size (40–70 mm);

y_3 – right ventricle size (36–41 mm);

y_4 – right auricle size (38–45 mm).

The training data obtained in the Vinnitsa Clinic of Cardiology is represented in Table 8.11 [27].

In current clinical practice, the number of combined heart diseases (aortic-mitral, mitral-tricuspid, mitral with lung hypertension etc.) is limited to six ($\overline{N}=6$).

The classes for output variables evaluation are formed as follows:

$$[\underline{y}_1, \overline{y}_1] = \underbrace{[11, 12)}_{e_{11}} \cup \underbrace{[13, 14]}_{e_{12}}, \quad [\underline{y}_2, \overline{y}_2] = \underbrace{[41, 50)}_{e_{21}} \cup \underbrace{[50, 70]}_{e_{22}},$$

$$[\underline{y}_3, \overline{y}_3] = \underbrace{[36, 38)}_{e_{31}} \cup \underbrace{[38, 41]}_{e_{32}}, \quad [\underline{y}_4, \overline{y}_4] = \underbrace{[38, 40)}_{e_{41}} \cup \underbrace{[40, 45]}_{e_{42}}.$$

Table 8.11. Training data

s	Input parameters				Output parameters			
	x_1	x_2	x_3	x_4	y_1	y_2	y_3	y_4
1	0.75-2	2	2	65-69	12-14	41-44	36	38
2	2.0-2.5	2	2	65-69	11-13	40-41	36	38
3	2.0-2.5	1-2	2	71-80	11	40	38-40	40-45
4	2.0-2.5	2	2	71-80	11	50-70	37-38	38-40
5	2.0-2.5	2	0.5-2	72-90	11-12	60-70	40-41	40-45
6	2.0-2.5	1-2	2-2.7	80-90	11-12	40	40-41	38
7	2.0-2.5	2	2	80-100	11	50-60	36	38
8	2.0-2.5	1-2	2-2.7	80-100	11	40	40-41	38-40

In clinical practice these classes correspond to the types of diagnoses e_{j1} *low inflation* and e_{j2} *dilation* (*hypertrophy*) of heart sections $y_1 \div y_4$. The aim of the diagnosis is to translate a set of specific parameters $x_1 \div x_4$ into decision e_{jp} for each output $y_1 \div y_4$.

The null solution $\mathbf{R}_0$ presented in Table 8.12 together with the parameters of the knowledge matrix is obtained using the genetic algorithm.

Table 8.12. Fuzzy relational matrix (null solution)

IF inputs				THEN outputs							
x_1	x_2	x_3	x_4	y_1		y_2		y_3		y_4	
				e_{11}	e_{12}	e_{21}	e_{22}	e_{31}	e_{32}	e_{41}	e_{42}
(0.75, 1.30)	(2.00, 0.63)	(2.35, 0.92)	(65.54, 8.81)	0.21	0.95	0.76	0.16	0.95	0.10	0.90	0.10
(2.50, 0.95)	(2.00, 0.65)	(2.44, 1.15)	(64.90, 9.57)	0.40	0.63	0.93	0.15	0.90	0.12	0.85	0.06
(2.52, 1.04)	(1.00, 0.82)	(2.32, 0.88)	(69.32, 10.23)	0.92	0.20	0.86	0.08	0.31	0.75	0.14	0.82
(2.55, 0.98)	(2.00, 0.72)	(2.36, 0.90)	(95.07, 21.94)	0.90	0.15	0.24	0.59	0.55	0.02	0.64	0.26
(2.51, 1.10)	(1.92, 0.75)	(0.50, 0.90)	(100.48, 26.14)	0.85	0.18	0.12	0.95	0.10	0.90	0.21	0.93
(2.55, 0.96)	(1.00, 0.94)	(2.30, 1.20)	(95.24, 22.46)	0.80	0.37	0.76	0.31	0.22	0.88	0.75	0.14

The obtained null solution allows us to arrange for the genetic search for the solution set of the system (8.11), where the matrices $\hat{\boldsymbol{\mu}}^A(\hat{\mathbf{X}}_s)$ and $\hat{\boldsymbol{\mu}}^B(\hat{\mathbf{X}}_s)$ for the training data take the following form:

$\hat{\boldsymbol{\mu}}^A =$

	$\hat{\mu}^{C_1}$	$\hat{\mu}^{C_2}$	$\hat{\mu}^{C_3}$	$\hat{\mu}^{C_4}$	$\hat{\mu}^{C_5}$	$\hat{\mu}^{C_6}$
$\hat{\mathbf{X}}_1$	[0.62, 0.94]	[0.32, 0.74]	[0.30, 0.40]	[0.09, 0.31]	[0.07, 0.35]	[0.08, 0.29]
$\hat{\mathbf{X}}_2$	[0.35, 0.62]	[0.74, 0.90]	0.40	[0.09, 0.31]	[0.07, 0.35]	[0.08, 0.29]
$\hat{\mathbf{X}}_3$	[0.21, 0.54]	[0.2, 0.52]	[0.22, 0.56]	[0.31, 0.72]	0.35	[0.29, 0.77]
$\hat{\mathbf{X}}_4$	[0.21, 0.54]	[0.2, 0.52]	[0.22, 0.40]	[0.31, 0.72]	0.35	[0.29, 0.41]
$\hat{\mathbf{X}}_5$	[0.1, 0.54]	[0.08, 0.52]	[0.07, 0.56]	[0.31, 0.86]	[0.35, 0.89]	[0.29, 0.41]
$\hat{\mathbf{X}}_6$	[0.1, 0.21]	[0.08, 0.21]	[0.07, 0.22]	[0.72, 0.86]	[0, 0.35]	[0.41, 0.85]
$\hat{\mathbf{X}}_7$	[0, 0.21]	[0, 0.21]	[0, 0.22]	[0.72, 0.90]	0.35	0.41
$\hat{\mathbf{X}}_8$	[0, 0.21]	[0, 0.21]	[0, 0.22]	[0.72, 0.90]	[0, 0.35]	[0.41, 1.0]

$\hat{\boldsymbol{\mu}}^B =$

	$\hat{\mu}^{E_1}$	$\hat{\mu}^{E_2}$	$\hat{\mu}^{E_3}$	$\hat{\mu}^{E_4}$	$\hat{\mu}^{E_5}$	$\hat{\mu}^{E_6}$	$\hat{\mu}^{E_7}$	$\hat{\mu}^{E_8}$
$\hat{\mathbf{X}}_1$	[0.32, 0.40]	[0.62, 0.94]	[0.62, 0.76]	[0.16, 0.35]	[0.62, 0.94]	[0.30, 0.40]	[0.62, 0.90]	[0.30, 0.40]
$\hat{\mathbf{X}}_2$	0.40	0.63	[0.74, 0.90]	[0.16, 0.35]	[0.74, 0.90]	0.40	[0.74, 0.85]	0.40
$\hat{\mathbf{X}}_3$	[0.35, 0.77]	[0.21, 0.54]	[0.29, 0.76]	[0.35, 0.59]	[0.31, 0.55]	[0.35, 0.77]	[0.31, 0.75]	[0.35, 0.56]
$\hat{\mathbf{X}}_4$	[0.35, 0.72]	[0.21, 0.54]	[0.29, 0.54]	[0.35, 0.59]	[0.31, 0.55]	[0.35, 0.41]	[0.31, 0.64]	[0.35, 0.4]
$\hat{\mathbf{X}}_5$	[0.35, 0.89]	[0.10, 0.54]	[0.29, 0.56]	[0.35, 0.89]	[0.31, 0.55]	[0.35, 0.89]	[0.31, 0.64]	[0.35, 0.89]
$\hat{\mathbf{X}}_6$	[0.72, 0.86]	0.37	[0.41, 0.76]	0.59	0.55	[0.41, 0.85]	[0.64, 0.75]	[0.26, 0.35]
$\hat{\mathbf{X}}_7$	[0.72, 0.90]	0.37	0.41	0.59	0.55	0.41	0.64	0.35
$\hat{\mathbf{X}}_8$	[0.72, 0.90]	0.37	[0.41, 0.76]	0.59	0.55	[0.41, 0.88]	[0.64, 0.75]	[0.26, 0.35]

The complete solution set for the fuzzy relation matrix is presented in Table 8.13, where, according to current clinical practice, the valve sizes $x_1 \div x_3$ are described by fuzzy terms *stenosis* (*S*) and *insufficiency* (*I*); pressure x_4 is described by fuzzy terms *normal* (*N*) and *lung hypertension* (*H*).

The obtained solution provides the results of diagnosis presented in Table 8.14 for 57 patients. Heart diseases diagnosis obtained an average accuracy rate of 90% after 10000 iterations of the genetic algorithm (100 min on Intel Core 2 Duo P7350 2.0 GHz).

The resulting solution can be linguistically interpreted as the set of the four possible rules bases (See Table 8.15), which differ in the fuzzy terms describing outputs y_1 and y_3 in rule 3 with overlapping weights.

Table 8.13. Fuzzy relational matrix (complete solution set)

IF inputs				THEN outputs							
x_1	x_2	x_3	x_4	y_1		y_2		y_3		y_4	
				L	*D*	*L*	*D*	*L*	*D*	*L*	*D*
S	*I*	*I*	*N*	[0, 0.40]	[0.94, 1.0]	0.76	0.16	[0.94, 1.0]	[0, 0.30]	0.90	[0, 0.30]
I	*I*	*I*	*N*	0.40	0.63	[0.90, 1.0]	[0, 0.35]	[0.90, 1.0]	[0, 0.30]	0.85	[0, 0.30]
I	*S*	*I*	*N*	[0.40, 1.0]	[0, 0.54]	[0.56, 1.0]	[0, 0.35]	[0, 0.55]	[0.40, 1.0]	[0, 0.31]	[0.56, 1.0]
I	*I*	*I*	*H*	[0.90, 1.0]	[0, 0.37] ∪0.37	[0, 0.41]	0.59	0.55	[0, 0.41]	0.64	0.26 ∪ [0, 0.26]
I	*I*	*S*	*H*	[0.89, 1.0]	[0, 0.54]	[0, 0.56]	[0.89, 1.0]	[0, 0.55]	[0.89, 1.0]	[0, 0.31]	[0.89, 1.0]
I	*S*	*I*	*H*	[0.77, 0.90]	0.37 ∪ [0, 0.37]	0.76	[0, 0.59]	[0, 0.55]	[0.85, 1.0]	0.75	[0, 0.26] ∪0.26

Table 8.14. Genetic algorithm efficiency characteristics

Output parameter	Type of diagnosis	Number of cases	Probability of the correct diagnosis
y_1	e_{11}	20	17 / 20 = 0.85
	e_{12}	37	34 / 37 = 0.92
y_2	e_{21}	26	23 / 26 = 0.88
	e_{22}	31	28 / 31 = 0.90
y_3	e_{31}	28	25 / 28 = 0.89
	e_{32}	29	27 / 29 = 0.93
y_4	e_{41}	40	37 / 40 = 0.92
	e_{42}	17	15 / 17 = 0.88

Table 8.15. System of IF-THEN rules

Rule	IF inputs				THEN outputs			
	x_1	x_2	x_3	x_4	y_1	y_2	y_3	y_4
1	*S*	*I*	*I*	*N*	*D*	*L*	*L*	*L*
2	*I*	*I*	*I*	*N*	*D*	*L*	*L*	*L*
3	*I*	*S*	*I*	*N*	*L* or *D*	*L*	*L or D*	*D*
4	*I*	*I*	*I*	*H*	*L*	*D*	*L*	*L*
5	*I*	*I*	*S*	*H*	*L*	*D*	*D*	*D*
6	*I*	*S*	*I*	*H*	*L*	*L*	*D*	*L*

References

1. Di Nola, A., Sessa, S., Pedrycz, W., Sancez, E.: Fuzzy Relational Equations and Their Applications to Knowledge Engineering, p. 430. Kluwer, Dordrecht (1989)
2. Peeva, K., Kyosev, Y.: Fuzzy Relational Calculus Theory, Applications and Software. World Scientific Publishing Company, Singapore (2004) CD-ROM, http://mathworks.net
3. Higashi, M., Klir, G.J.: Identification of fuzzy relation systems. IEEE Trans. Syst., Man, and Cybernetics 14, 349–355 (1984)
4. Pedrycz, W.: An identification algorithm in fuzzy relational systems. Fuzzy Sets Syst. 13, 153–167 (1984)
5. Pedrycz, W.: Fuzzy models and relational equations. Mathematical Modeling 9(6), 427–434 (1987)
6. Zadeh, L.: The Concept of Linguistic Variable and It's Application to Approximate Decision Making. Mir, Moscow (1976)
7. Yager, R.R., Filev, D.P.: Essentials of Fuzzy Modeling and Control. John Willey & Sons, New York (1994)
8. Pedrycz, W., Gomide, F.: An introduction to fuzzy sets / Analysis and Design. – A Bradford Book, p. 465. The MIT Press (1998)
9. Bourke, M.M., Grant Fisher, D.: Calculation and application of a minimum fuzzy relational matrix. Fuzzy Sets and Syst. 74(2), 225–236 (1995)
10. Bourke, M.M., Grant Fisher, D.: Identification algorithms for fuzzy relational matrices, Part 1: Non-optimizing algorithms. Fuzzy Sets and Systems 109(3), 305–320 (2000)
11. Yi, S.Y., Chung, M.J.: Identification of fuzzy relational model and its application to control. Fuzzy Sets and Systems 59(1), 25–33 (1993)
12. Branco, P.J., Dente, J.A.: A fuzzy relational identification algorithm and its application to predict the behaviour of a motor drive system. Fuzzy Sets and Systems 109(3), 343–354 (2000)
13. Wong, C.H., Shah, S.L., Fisher, D.G.: Fuzzy relational predictive identification. Fuzzy Sets and Systems 113(3), 417–426 (2000)
14. Bourke, M.M., Fisher, D.G.: Identification algorithms for fuzzy relational matrices. Part 2: Optimizing algorithms. Fuzzy Sets and Systems 109(3), 321–341 (2000)
15. Blanco, A., Delgado, M., Requena, I.: Identification of fuzzy relational equations by fuzzy neural networks. Fuzzy Sets and Systems 71(2), 215–226 (1995)
16. Chakraborty, D., Pal, N.R.: Integrated Feature Analysis and Fuzzy Rule-Based System Identification in a Neuro-Fuzzy Paradigm. IEEE Transactions on Systems, Man and Cybernetics – Part B: Cybernetics 31(3), 391–400 (2001)
17. Ciaramella, A., Tagliaferri, R., Pedrycz, W., Di Nola, A.: Fuzzy relational neural network. International Journal of Approximate Reasoning 41(2), 146–163 (2006)
18. Rotshtein, A., Posner, M., Rakytyanska, H.: Cause and Effect Analysis by Fuzzy Relational Equations and a Genetic Algorithm. Reliability Engineering & System Safety 91(9), 1095–1101 (2006)
19. Rotshtein, A., Rakytyanska, H.: Fuzzy Relation-based Diagnosis. Automation and Remote Control 68(12), 2198–2213 (2007)
20. Rotshtein, A., Rakytyanska, H.: Diagnosis Problem Solving using Fuzzy Relations. IEEE Transactions on Fuzzy Systems 16(3), 664–675 (2008)
21. Rotshtein, A., Rakytyanska, H.: Adaptive Diagnostic System based on Fuzzy Relations. Cybernetics and Systems Analysis 45(4), 623–637 (2009)

22. Rotshtein, A., Mityushkin, Y.: Extraction of Fuzzy Knowledge Bases from Experimental Data by Genetic Algorithms. Cybernetics and Systems Analysis 37(4), 501–508 (2001)
23. Rotshtein, A.P., Rakytyanska, H.B.: Fuzzy Genetic Object Identification: Multiple Inputs/Multiple Outputs Case. In: Hippe, Z.S., Kulikowski, J.L., Mroczek, T. (eds.) Human - Computer Systems Interaction. AISC, vol. 99, pp. 375–394. Springer, Heidelberg (2012)
24. Gen, M., Cheng, R.: Genetic Algorithms and Engineering Design, p. 352. John Wiley & Sons, New York (1997)
25. Cordon, O., Herrera, F., Hoffmann, F., Magdalena, L.: Genetic Fuzzy Systems. Evolutionary Tuning and Learning of Fuzzy Knowledge Bases. World Scientific, New York (2001)
26. Tsypkin, Y.Z.: Information Theory of Identification, p. 320. Nauka, Moscow (1984) (in Russian)
27. Frid, A., Grines, C.: Cardiology. Practice, Moscow (1996) (in Russian)

Chapter 9
Applied Fuzzy Systems

Data processing not only in physics and engineering, but also in medicine, biology, sociology, economics, sport, art, and military affairs, amounts to the different statements of identification problems. Fuzzy logic is mistakenly perceived by many specialists in mathematical simulation as a mean of only approximate decisions making in medicine, economics, art, sport and other different from physics and engineering humanitarian domains, where the high level of accuracy is not required. Therefore, one of the main goals of the authors is to show that it is possible to reach the accuracy of modeling, which does not yield to strict quantitative correlations, by tuning fuzzy knowledge bases. Only objects with discrete outputs for the direct inference and discrete inputs for the inverse inference were considered in the previous chapters. Such a problem corresponds to the problem of automatic classification arising in particular from medical and technical diagnosis. The main idea which the authors strive to render is that while tuning the fuzzy knowledge base it is possible to identify nonlinear dependencies with the necessary precision.

The use of the fuzzy expert information about the nonlinear object allows us to decrease the volume of experimental researches that gives the significant advantage in comparison with the known methods of identification with the growth of the number of the input variables of the object. Besides that, the fuzzy knowledge base easily interprets the structure of the object, while it is not always possible at the use of known methods.

Numerous examples considered in this chapter testify to wide possibilities of the intellectual technologies of modeling in the different domains of human activity.

9.1 Dynamic System Control

A dynamic system is traditionally considered as one quantitative description of which can be given by the language of differential or other equations [1]. Classical automatic control theory suggests that such equations can be constructed from the laws of physics, mechanics, thermodynamics, and electromagnetism [2]. Construction of dynamic equations requires a deep understanding of the processes and needs good physico-mathematical training [3]. On the other hand, a person can control a complicated object without compiling or solving any equations. We recall for example how easily a driver parks an automobile. Even a novice sitting for the first time in the driver seat can control an automobile by executing the verbal commands from his instructor sitting next to him.

A.P. Rotshtein et al.: Fuzzy Evidence in Identif., Forecast. and Diagn., STUDFUZZ 275, pp. 259–313.
springerlink.com

A unique feature of man is his capacity to learn and to evaluate the observed parameters in natural language: *low* velocity, *large* distance, and so on. Fuzzy set theory makes it possible to formalize natural language statements. Here we show that one can adjust a fuzzy knowledge base and use it to control a dynamic object no less effectively than with classical control theory. This section is written on the basis of work [4].

9.1.1 Control Object

We consider an inverted pendulum (Fig. 9.1), i. e., a rod fixed on a trolley that can oscillate in the longitudinal vertical plane.

The task of the control system is to maintain the inverted pendulum in the vertical position by displacing the trolley. A more ordinary form of this task is to maintain a rod on a finger in the vertical position. In [2] it has been shown that this is the class of problems in simulating the motion of a rocket, a supersonic aircraft, or a set of barges pushed by a tug, all of which are objects in which the centre of mass does not coincide with the point of application of the force.

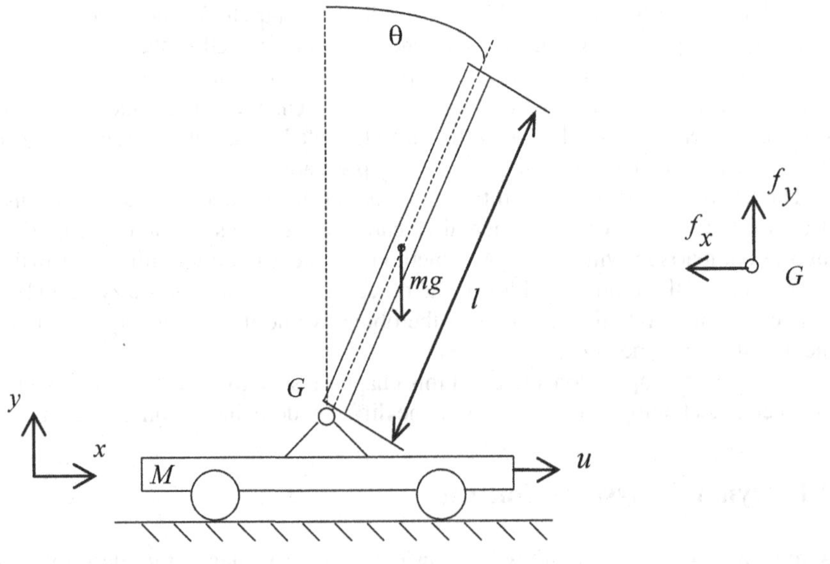

Fig. 9.1. Inverted pendulum

Before we consider the differential equations describing the motion of the pendulum, we note that the rod or the finger is kept vertical by applying simple rules:

If the angle of deviation from the vertical is large, one needs rapid movement in the same direction;

If the angle of deviation is small, one makes a small movement in the same direction;

If the angle of deviation is zero, no movement is made.

9.1.2 Classical Control Model

Following [5], we introduce the following symbols in Fig. 9.1: l – pendulum length, m – pendulum mass, M – trolley mass, g – acceleration due to gravity, u –control for supply to trolley, f_x and f_y – horizontal and vertical components of the forces acting on the pendulum, θ – the angular deviation of the pendulum from vertical, and I – the second moment of the pendulum in the plane of oscillation, which for a rectilinear thin rod is given by $I = \frac{ml^2}{3}$.

The equation of motion for an inverted pendulum as a control object may be written as follows [5]:

turning moment about the point G

$$I\ddot{\theta} = f_x l \cos\theta + f_y l \sin\theta \ ;$$

displacement of the projection of G on the y axis

$$f_y - mg = m\frac{d^2}{dt^2}(l\cos\theta) = -ml(\ddot{\theta}\sin\theta + \dot{\theta}^2\cos\theta) \ ;$$

displacement of the projection of G on the x axis

$$f_x = m\frac{d^2}{dt^2}(x - l\sin\theta) = m\ddot{x} - ml(\ddot{\theta}\cos\theta - \dot{\theta}^2\sin\theta) \ ;$$

and displacement of the trolley parallel to the x axis

$$u - f_x = M\ddot{x} \ ,$$

in which $\dot{\theta}$ is the rate of change in angle θ, $\ddot{\theta}$ is the angular acceleration of the pendulum, and $\ddot{x}$ is the acceleration of the trolley along the x axis.

A linear approximation is used for these equations subject to the condition that θ varies over a fairly narrow range ($\cos\theta \approx 1$, $\sin\theta \approx \theta$, $\theta\dot{\theta} \approx 0$, $\dot{\theta}^2 \approx 0$), which gives us the differential equation of motion as:

$$\ddot{\theta} = \frac{3g(M+m)}{(4M+m)l}\theta + \frac{3u}{(4M+m)l} \ . \tag{9.1}$$

To maintain the pendulum vertical with an ordinary control system with feedback, we represent the control variable as:

$$u = \alpha\theta + \beta\dot{\theta} \ , \tag{9.2}$$

which corresponds to a proportional-differential regulator having proportionality coefficients α and β.

To provide stability, we take the coefficients as:

$$\alpha = -10 \ , \ \beta = -2 \ ,$$

which gives negative values for the roots

$$\lambda_1 = -2.98 \ , \ \lambda_2 = -16.99$$

in the characteristic equation

$$\lambda^2 - \frac{3\beta}{(4M+m)l}\lambda - \frac{3g(M+m)+3\alpha}{(4M+m)l} = 0,$$

corresponding to (9.1).

To keep it vertical, we can thus use the control input

$$u = -10\theta - 2\dot{\theta} \tag{9.3}$$

in which the equation for the stable motion is:

$$\ddot{\theta} = -\frac{6}{(4M+m)l}\dot{\theta} + \frac{3g(M+m)-30}{(4M+m)l}\theta. \tag{9.4}$$

Table 9.1 gives the behaviour of θ (in rad) and $\dot{\theta}$ (rad/sec) from (9.4) with various initial conditions: γ_1, γ_2 and γ_3. In solving equation (9.4) we have used the following parameter values:

$$m = 0.035 \text{ kg},\ M = 0.5 \text{ kg},\ l = 30 \text{ cm},\ g = 9.8 \text{ m/sec}^2.$$

In what follows, Table 9.1 will be used as the training set for adjusting the fuzzy control model.

Table 9.1. Behavior of an inverted pendulum under regulator control

	γ_1		γ_2		γ_3	
t	θ	$\dot{\theta}$	θ	$\dot{\theta}$	θ	$\dot{\theta}$
0.0	0.175	0.0000	0.105	0.0000	0.035	0.0000
0.1	0.150	-0.3523	0.090	-0.2114	0.030	-0.0705
0.2	0.115	-0.3261	0.069	-0.1957	0.023	-0.0652
0.3	0.086	-0.2540	0.052	-0.1524	0.017	-0.0508
0.4	0.064	-0.1908	0.039	-0.1145	0.013	-0.0382
0.5	0.048	-0.1421	0.029	-0.0852	0.010	-0.0284
0.6	0.035	-0.1056	0.021	-0.0633	0.007	-0.0211
0.7	0.026	-0.0784	0.016	-0.0470	0.005	-0.0157
0.8	0.020	-0.0582	0.012	-0.0349	0.004	-0.0116
0.9	0.015	-0.0432	0.009	-0.0259	0.003	-0.0086
1.0	0.011	-0.0321	0.006	-0.0193	0.002	-0.0064
1.1	0.008	-0.0238	0.005	-0.0143	0.002	-0.0048
1.2	0.006	-0.0177	0.004	-0.0106	0.001	-0.0035
1.3	0.004	-0.0131	0.003	-0.0079	0.001	-0.0026
1.4	0.003	-0.0098	0.002	-0.0059	0.001	-0.0020
1.5	0.002	-0.0072	0.001	-0.0043	0.000	-0.0014
1.6	0.002	-0.0054	0.001	-0.0032	0.000	-0.0011
1.7	0.001	-0.0040	0.001	-0.0024	0.000	-0.0008
1.8	0.001	-0.0030	0.001	-0.0018	0.000	-0.0006
1.9	0.001	-0.0022	0.000	-0.0013	0.000	-0.0004
2.0	0.001	-0.0016	0.000	-0.0010	0.000	-0.0003

9.1.3 Fuzzy Control Model

The dependence of the control u on the variables θ and $\dot{\theta}$ is represented as a knowledge base formed from 25 expert rules as follows:

$$\text{IF } \theta = A_i \text{ AND } \dot{\theta} = B_i \text{, THEN } u = C_j \text{, } i = \overline{1,5},\ j = \overline{1,7}.$$

These rules form a 5×5 matrix:

Deviation angle, θ	Rate of change, $\dot{\theta}$: *hN*	*N*	*Z*	*P*	*hP*
hN	*vhN*	*vhN*	*hN*	*N*	*Z*
N	*vhN*	*hN*	*N*	*Z*	*P*
Z	*hN*	*N*	*Z*	*P*	*hP*
P	*N*	*Z*	*P*	*hP*	*vhP*
hP	*Z*	*P*	*hP*	*vhP*	*vhP*

(9.5)

where variables θ and $\dot{\theta}$ are evaluated by means of five terms:

$A_1 = B_1$ = *high negative (hN)*, $A_2 = B_2$ = *negative (N)*, $A_3 = B_3$ = *zero (Z)*, $A_4 = B_4$ = *positive (P)*, $A_5 = B_5$ = *high positive (hP)*.

and variable u is evaluated by means of seven terms:

C_1 - *very high negative (vhN)*, C_2 - *high negative (hN)*, C_3 - *negative (N)*, C_4 - *zero (Z)*, C_5 - *positive (P)*, C_6 - *high positive (hP)*, C_7 - *very high positive (vhP)*.

As the training set for tuning the control model (9.5), we use the Table 9.1 data and equation (9.3). The task of adjustment consists in selecting parameters for the membership functions in the terms A_i and B_i ($i = \overline{1,5}$) and rule weights in (9.5) such as to produce the minimum discrepancy between the theoretical equations (knowledge base (9.5)) on the one hand and the experimental equations (Table 9.1 and formula (9.3)) on the other.

The adjustment is performed by the method described in Section 3. The obtained membership functions are presented in Fig. 9.2. The weights of the fuzzy rules after adjustment correspond to the elements in the following matrix:

Deviation angle, θ	Rate of change, $\dot{\theta}$: *hN*	*N*	*Z*	*P*	*hP*
hN	0.9837	0.3490	0.7902	0.8841	0.9015
N	0.3490	0.9111	0.3901	0.7509	0.2199
Z	0.7902	0.3901	0.7981	0.6381	0.5594
P	0.8841	0.7509	0.6381	0.3690	0.5114
hP	0.9015	0.2199	0.5594	0.5114	0.8708

Fig. 9.3 compares the behaviour of θ for the classical model and the fuzzy model with various initial conditions ($\gamma_1, \gamma_2, \gamma_3$); after the fuzzy control system is adjusted, it provides the same results as a traditional proportional-differential regulator.

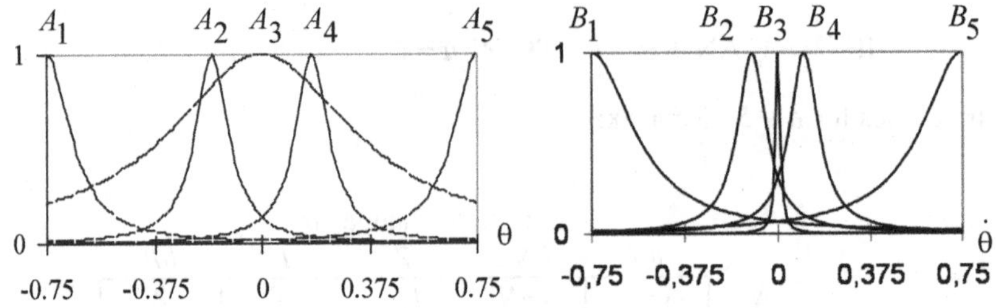

Fig. 9.2. Membership functions for fuzzy levels of variables θ and $\dot{\theta}$ evaluation

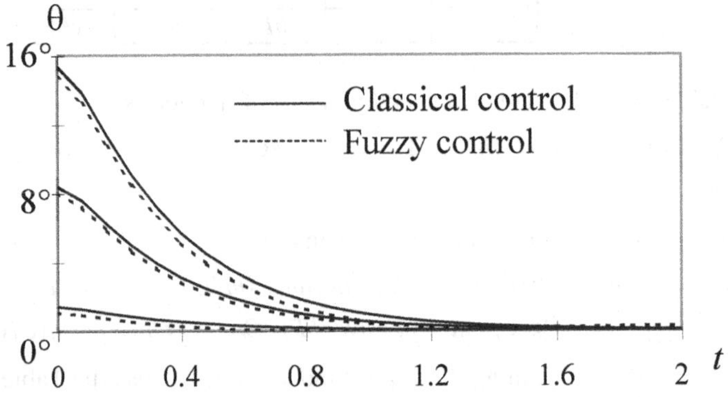

Fig. 9.3. Comparison of fuzzy and classical control systems after tuning

9.1.4 Connection with Lyapunov's Functions

It is shown here, that Lyapunov's functions known in stability theory can be used to synthesize fuzzy rules for control of a dynamic system.

The second or direct Lyapunov's method [3] allows us to study the stability of solutions of the nonlinear differential equations without solving these equations. The stability criterion was developed by Lyapunov on the basis of the following simple physical conception of equilibrium position: equilibrium position of the system is asymptotically stable, if all the trajectories of the process, beginning fairly near from the equilibrium point, stretch in such a way, that a properly defined "energetic" function is converged to the minimum, where position of the local minimum of energy corresponds to this point of equilibrium.

Let us consider the application of this criterion relative to the generalized nonlinear equation:

$$\dot{x} = f(x) \ , \ \ x(0) = x_0 \ , \tag{9.6}$$

where x is the vector of the system condition.

We assume, that $f(0)=0$ and function f is continuous in the neighbourhood of the origin of coordinates.

Definition of Lyapunov's function. Function $V(x)$ is called Lyapunov's function (an energetic function) of system (9.6), if:

1) $V(0)=0$,
2) $V(x)>0$ for all $x \neq 0$ in the neighbourhood of the origin of coordinates,
3) $\dfrac{\partial V(x)}{\partial t}<0$ along the trajectory of system (9.6).

The main result, obtained by Lyapunov, was formulated as the theorem of stability.

Lyapunov's Theorem of Stability. The equilibrium position $x=0$ of system (9.6) is asymptotically stable, if Lyapunov's function $V(x)$ of the system exists.

We stress that Lyapunov's method requires derivation of the system dynamics equations. We are interested in the case with a lack of such equations.

Let us consider the inverted pendulum (Fig. 9.1) in the assumption, that only the following a priori information is known:

a) the system condition is defined by the coordinates $x_1=\theta$ and $x_2=\dot{\theta}$;

b) $\dot{x}_2$ is proportional to control *u*, i.e., if *u* increases (decreases), then $\dot{x}_2$ increases (decreases).

To apply Lyapunov's theorem to the inverted pendulum, the following function is selected as a Lyapunov's function candidate:

$$V(x_1,x_2)=\frac{1}{2}(x_1^2+x_2^2) \ . \tag{9.7}$$

If $V(0,0)=0$ and $V(x_1,x_2)>0$ then to assign $V(x_1,x_2)$ as a Lyapunov's function, it is necessary to provide the condition:

$$\frac{\partial V(x_1,x_2)}{\partial t}=x_1\dot{x}_1+x_2\dot{x}_2=x_1x_2+x_2\dot{x}_2<0 \ . \tag{9.8}$$

A fuzzy knowledge base about control $u=u(x_1,x_2)$ can be formulated as the condition of inequality (9.8) implementation. We consider three cases:

IF x_1 and x_2 have the opposite signs, then $x_1x_2<0$ and inequality (9.8) will be implemented for $x_2\dot{x}_2=0$.

IF x_1 and x_2 are positive, then (9.8) will be implemented for $\dot{x}_2<-x_1$.

IF x_1 and x_2 are negative, then (9.8) will be implemented for $\dot{x}_2>-x_1$.

Using the above mentioned reasoning and priori information relative to the fact that x_2 is *proportional* to u, we obtain four fuzzy rules for stable control the inverted pendulum:

IF x_1 *positive* AND x_2 *negative*, THEN u *zero*,

IF x_1 *negative* AND x_2 *positive*, THEN u *zero*,

IF x_1 *positive* AND x_2 *positive*, THEN u *high negative*,

IF x_1 *negative* AND x_2 *negative*, THEN u *high positive*.

Adjustment of this knowledge base consists of the selection of membership functions for the corresponding terms.

The essential differences between the classical and fuzzy control systems are given in Table 9.2.

Table 9.2. Control System Comparison

System type	Advantages	Disadvantages
Classical	If there is a model that adequately describes the dynamics, one can operate without adjusting it	Difficult to derive differential equations adequately describing the dynamics in the presence of nonlinear perturbations
Fuzzy	Differential equations not necessary, and dynamic model is readily written in terms of linguistic rules	Requires linguistic model adjustment

9.2 Inventory Control

Minimization of the inventory storage cost in enterprises and trade firms stocks including raw materials, stuffs, supplies, spare parts and products, is the most important problem of management. It is accepted that the theory of inventory control relates to operations research [6]. The models of this theory [7, 8] are built according to the classical scheme of mathematical programming: goal function is minimizing storage cost; controllable variables are time moments needed to order (or distribute) corresponding quantity of the needed stocks. Construction of such models requires definite assumptions, for example, of orders flows, time distribution laws and others. Therefore, complex optimization models may produce solutions that are quite inadequate to the real situation.

On the other hand, experienced managers very often make effective administrative decisions on the common sense and practical reasoning level. Therefore, the approach based on fuzzy logic can be considered as a good alternative to the classical inventory control models. This approach elaborated in works [9 – 12] requires neither complex mathematical models construction nor search for optimal solutions on the basis of such models. It is based on a simple comparison of the demand for the stock of the given item at the actual time moment with the quantity of the stock available in the warehouse. Dependent upon this, inventory action is formed consisting of increasing or decreasing corresponding stocks and materials.

"Quality" of a control fuzzy model strongly depends on the "quality" of fuzzy rules and "quality" of membership functions describing fuzzy terms. The more successfully the fuzzy rules and membership functions are selected, the more adequate the control action will be. However, no one can guarantee that the result of fuzzy logical inference will coincide with the correct (i.e. the most rational) control. Therefore, the problem of the adequate fuzzy rules and membership functions construction should be considered as the most actual one while developing control systems on fuzzy logic.

In this chapter it is suggested to build the fuzzy model of stocks and materials control on the grounds of the general method of nonlinear dependencies identification by means of fuzzy knowledge bases [13]. The proposed method is special due to the tuning stage of the fuzzy inventory control model using "demand - supply" training data. Owing to this tuning stage it is possible to select such fuzzy rules weights and such membership functions forms which provide maximal proximity of the results of fuzzy logical inference to the correct managerial decisions.

To substantiate for the expediency to use this fuzzy approach relative to inventory control, we resort to help of analogy with the classical problem of a dynamic system (turned-over pendulum) control which can be successfully solved using fuzzy logic [4].

9.2.1 *Analogy with Turned-Over Pendulum*

The approach to inventory control suggested here is similar to turned-over pendulum control with the aim of retaining it in a vertical position by pushing the cart to the left or to the right (Fig. 9.4). A rather habitual version of such a problem is demonstrated by vertically retaining a stick on the finger. The simplest rules for the problem solution can be represented in the following way:

IF the angle of deflection of the stick from the vertical position is *big*,
THEN the finger should *quickly* move in the same direction to keep the stick up;
IF the angle of deflection of the stick from the vertical position is *small*,
THEN the finger should *slowly* move in the same direction to keep the stick up;
IF the angle of deflection of the stick is equal to zero,
THEN the finger *should stay motionless.*

Keeping the speed of the car constant by the driver takes place in analogy to it; if the speedometer needle drops down, then the driver presses the accelerator down; if the speedometer needle goes up, then the driver reduces the speed. It is known that experienced driver retains some given speed (for example, 90 km/hour) in spite of the quickly changing nonlinear road relief.

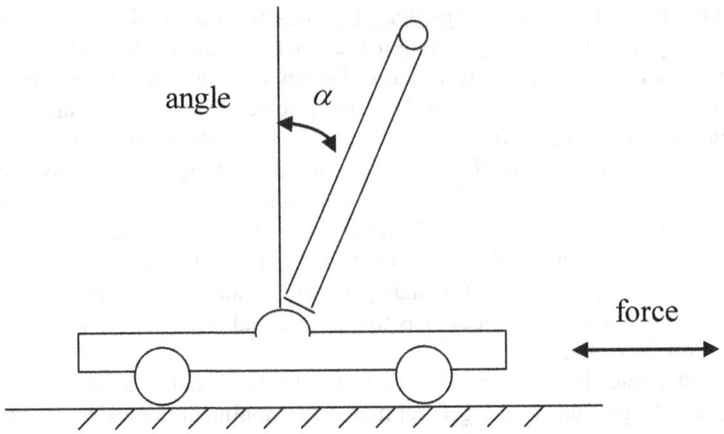

Fig. 9.4. Control system of the turned-over pendulum

Returning to the inventory control system it is not difficult to understand that the actions of the manager must be similar to the actions of the car driver regulating of the vehicle's speed.

9.2.2 Method of Identification

The method of nonlinear objects identification by fuzzy knowledge bases [14] serves as the theoretical basis for the definition of the dependency between control actions and the current state of the control system. The method is based on the principle of fuzzy knowledge bases two-stage tuning. According to this principle the construction of the "inputs – output" object model can be performed in two stages which, in analogy with classical methods [15], can be considered as stages of structural and parametrical identification.

The first stage is traditional for fuzzy expert systems [16]. Formation and rough tuning of the object model by knowledge base construction using available expert information is accomplished at this stage. The higher the professional skill level of an expert, the higher the adequacy of the built fuzzy model at the rough tuning stage will be. However, as was mentioned in the introduction, no one can guarantee the coincidence of the results of fuzzy logic inference (theory) and correct practical decisions (experiment). Therefore, the second stage is needed, at which fine tuning of the model is done by way of training it using experimental data.

The essence of the fine tuning stage consists in finding such fuzzy IF-THEN rules weights and such fuzzy terms membership functions parameters which minimize the difference between desired (experimental) and model (theoretical) behaviour of the object. Fine tuning stage is formulated as nonlinear optimization problem which can be effectively solved by some combination of genetic algorithms and neural networks [14].

9.2.3 Fuzzy Model of Control

Let us present the inventory control system in the form of the object with two inputs ($x_1(t)$, $x_2(t)$) and single output ($y(t)$), where:

$x_1(t)$ is *demand*, i.e. the number of units of the stocks of the given brand, which is needed at time moment *t*;

$x_2(t)$ is *stock quantity-on-hand*, i.e. the number of units of the stocks of the given brand, which is available in the warehouse at moment *t*;

$y(t)$ is an *inventory action* at moment *t*, consisting in increasing – decreasing the stocks of the given brand.

System state parameters $x_1(t)$, $x_2(t)$ and inventory action $y(t)$ are considered as linguistic variables [17], which are estimated with the help of verbal terms on five and seven levels:

$$x_1(t) = \begin{cases} \text{falling } (F) \\ \text{decreased } (D) \\ \text{steady } (S) \\ \text{increased } (I) \\ \text{rising up } (R) \end{cases} \qquad x_2(t) = \begin{cases} \text{minimal } (M) \\ \text{low } (L) \\ \text{adequately sufficient (A)} \\ \text{high } (H) \\ \text{excessive } (E) \end{cases}$$

$$y(t) = \begin{cases} d_1 \text{ – to decrease the stock sharply} \\ d_2 \text{ – to decrease the stock moderately} \\ d_3 \text{ – to decrease the stock minimally} \\ d_4 \text{ – do nothing} \\ d_5 \text{ – to increase the stock minimally} \\ d_6 \text{ – to increase the stock moderately} \\ d_7 \text{ – to increase the stock sharply} \end{cases}$$

Let us note that term "adequately sufficient" in variable $x_2(t)$ estimation depicts the rational quantity of the stock on the common sense level, and does not pretend to be contained within the mathematically strong concept of optimality which envisages the presence of goal function, controllable variables and area of constraints.

Functional dependency

$$y(t) = f(x_1(t), x_2(t)) \tag{9.9}$$

is defined by the table presented in Fig. 9.5.

This table is defined in an expert manner and depicts the complete sorting out of the $(5\times5=25)$ terms combinations in the triplets $\langle x_1(t),\ x_2(t),\ y(t)\rangle$.

Grouping these triplets by inventory actions types, we shall form a fuzzy knowledge base, presented in Table. 9.3.

This fuzzy knowledge base defines a fuzzy model of the object in the form of the following rules, e.g.:

IF demand is falling AND stock is excessive, OR demand is falling AND stock is high, OR demand is decreased AND stock is excessive,

THEN it is necessary to decrease the stock sharply.

$x_2(t)$					
E	d_1	d_1	d_2	d_3	d_4
H	d_1	d_2	d_3	d_4	d_5
A	d_2	d_3	d_4	d_5	d_6
L	d_3	d_4	d_5	d_6	d_7
M	d_4	d_5	d_6	d_7	d_7
	F	D	S	I	R $x_1(t)$

Fig. 9.5. Dependency between state parameters and inventory actions

Fuzzy logical equations correspond to the fuzzy knowledge base (Table 9.3). They establish the connection between membership functions of the variables in correlation (9.9). Let $\mu^j(u)$ be membership function of variable u to term j. Let us go on from the fuzzy knowledge base (Table 9.3) to the system of fuzzy logical equations:

$$\mu^{d_1}(y)=\mu^F(x_1)\cdot\mu^E(x_2)\vee\mu^F(x_1)\cdot\mu^H(x_2)\vee\mu^D(x_1)\cdot\mu^E(x_2);$$
$$\mu^{d_2}(y)=\mu^F(x_1)\cdot\mu^A(x_2)\vee\mu^D(x_1)\cdot\mu^H(x_2)\vee\mu^S(x_1)\cdot\mu^E(x_2);$$
$$\mu^{d_3}(y)=\mu^F(x_1)\cdot\mu^L(x_2)\vee\mu^D(x_1)\cdot\mu^A(x_2)\vee\ \mu^S(x_1)\cdot\mu^H(x_2)\vee\mu^I(x_1)\cdot\mu^E(x_2);$$
$$\mu^{d_4}(y)=\mu^F(x_1)\cdot\mu^M(x_2)\vee\mu^D(x_1)\cdot\mu^L(x_2)$$
$$\vee\mu^S(x_1)\cdot\mu^A(x_2)\vee\mu^I(x_1)\cdot\mu^H(x_2)\ \vee\mu^R(x_1)\cdot\mu^E(x_2);$$
$$\mu^{d_5}(y)=\mu^D(x_1)\cdot\mu^M(x_2)\vee\mu^S(x_1)\cdot\mu^L(x_2)\ \vee\mu^I(x_1)\cdot\mu^A(x_2)\vee\mu^R(x_1)\cdot\mu^H(x_2);$$
$$\mu^{d_6}(y)=\mu^S(x_1)\cdot\mu^M(x_2)\vee\mu^I(x_1)\cdot\mu^L(x_2)\vee\mu^R(x_1)\cdot\mu^A(x_2);$$
$$\mu^{d_7}(y)=\mu^I(x_1)\cdot\mu^M(x_2)\vee\mu^R(x_1)\cdot\mu^M(x_2)\vee\mu^R(x_1)\cdot\mu^L(x_2), \qquad (9.10)$$

where (•) is operation AND (*min*); $\vee$ is operation OR (*max*).

The algorithm of decision making on the basis of fuzzy logical equations consists of the following:

1°. To fix the demand $x_1(t)$ and stock quantity-on-hand $x_2(t)$ values at the time moment $t=t_0$.

Table 9.3. Fuzzy knowledge base

IF		THEN
Demand $x_1(t)$	*Stock quantity-on-hand* $x_2(t)$	*Inventory action* $y(t)$
F *F* *D*	*E* *H* *E*	d_1
F *D* *S*	*A* *H* *E*	d_2
F *D* *S* *I*	*L* *A* *H* *E*	d_3
F *D* *S* *I* *R*	*M* *L* *A* *H* *E*	d_4
D *S* *I* *R*	*M* *L* *A* *H*	d_5
S *I* *R*	*M* *L* *A*	d_6
I *R* *R*	*M* *M* *L*	d_7

2°.To define the membership degrees of $x_1(t)$ and $x_2(t)$ values to the corresponding terms with the help of membership functions.

3°. To calculate the membership degree of the inventory action $y(t)$ at the time $t = t_0$ to each of the $d_1, d_2, ..., d_7$ decisions classes with the help of fuzzy logical equations.

4°. The term with maximal membership function, obtained at step 3° should be considered as inventory action $y(t)$ at the time $t=t_0$. For obtaining the quantitative

$y(t)$ value at the time $t=t_0$ it is necessary to perform the "defuzzification" operation, i.e. to go on from the fuzzy term to a crisp number. According to [14] this operation can be performed as follows. Range $[\underline{y}, \overline{y}]$ of the variable $y(t)$ change is divided into 7 classes:

$$y(t) \in [\underline{y}, \overline{y}] = \underbrace{[\underline{y}, y_1)}_{d_1} \cup \underbrace{[y_1, y_2)}_{d_2} \cup \ldots \cup \underbrace{[y_6, \overline{y}]}_{d_7}.$$

The crisp value of the inventory action $y(t)$ at the time $t=t_0$ is defined by formula:

$$y(t) = \frac{\underline{y}\mu^{d_1}(y) + y_1\mu^{d_2}(y) + \ldots + y_6\mu^{d_7}(y)}{\mu^{d_1}(y) + \mu^{d_2}(y) + \ldots + \mu^{d_7}(y)}. \tag{9.11}$$

9.2.4 Fuzzy Model Tuning

Relations (9.10), (9.11) define the functional dependency (9.9) in the following form

$$y(t) = F(x_1(t), x_2(t), \mathbf{W}, \mathbf{B}_1, \mathbf{C}_1, \mathbf{B}_2, \mathbf{C}_2),$$

where $\mathbf{W} = (w_1, w_2, \ldots, w_{25})$ is the vector of weights in the fuzzy knowledge base (Table 9.3);

$\mathbf{B}_1 = (b_1^{vD}, b_1^{D}, b_1^{St}, b_1^{I}, b_1^{vI})$, $\mathbf{B}_2 = (b_2^{vL}, b_2^{L}, b_2^{S}, b_2^{B}, b_2^{vB})$ are the vectors of centers for variables $x_1(t)$ and $x_2(t)$ membership functions to the corresponding terms;

$\mathbf{C}_1 = (c_1^{vD}, c_1^{D}, c_1^{St}, c_1^{I}, c_1^{vI})$, $\mathbf{C}_2 = (c_2^{vL}, c_2^{L}, c_2^{S}, c_2^{B}, c_2^{vB})$ are the vectors of concentration parameters for variables $x_1(t)$ and $x_2(t)$ membership functions to the corresponding terms;

F is the operator of "inputs – output" connection corresponding to formulae (9.10), (9.11).

It is assumed that some training data sample in the form of M pairs of experimental data can be obtained on the ground of successful decisions about inventory control

$$\langle \hat{x}_1(t), \hat{x}_2(t), \hat{y}(t) \rangle, \ t = \overline{1, M},$$

where $\langle \hat{x}_1(t), \hat{x}_2(t) \rangle$ are the inventory control system state parameters at time moment t, $\hat{y}(t)$ is the inventory action at time moment t.

The essence of the inventory control model tuning consists of such membership functions parameters (b-, c-) and fuzzy rules weights (w-) finding, which provide for the minimum distance between theoretical and experimental data:

$$\sum_{t=1}^{M} [F(\hat{x}_1(t), \hat{x}_2(t), \mathbf{W}, \mathbf{B}_1, \mathbf{C}_1, \mathbf{B}_2, \mathbf{C}_2) - \hat{y}(t)]^2 = \min_{\mathbf{W}, \mathbf{B}_i, \mathbf{C}_i}, \quad i = 1, 2. \tag{9.12}$$

It is expedient to solve the nonlinear optimization problem (9.12) by a combination of the genetic algorithm and gradient methods.

9.2.5 Example of Fuzzy Model Tuning

Fuzzy model of inventory control was constructed for the district food-store house, selling some definite kind of agricultural production (buckwheat). The ranges of the input and output variables change consisted of:

$$x_1(t) \in [0, 200]*10^2 \text{ kg}; \quad x_2(t) \in [70, 170]*10^2 \text{ kg}; \quad y(t) \in [-100, 100]*10^2 \text{ kg}.$$

Inventory control at the enterprise is done once per day. Therefore $t \in [1...365]$ days. The triplets $\langle$ demand $x_1(t)$, stock quantity-on-hand $x_2(t)$, inventory action $y(t)$ $\rangle$ values, corresponding to the experienced manager actions, for which the demand for the produce was satisfied while the permissible produce inventory level in store was minimal where taken as training data sample. Training data sample is presented in Fig. 9.6,a-c in the form of the dynamics of the input and output variables change on time t according to 2001 year data. For example, at moments $t = 120$ and $t = 230$ the control consisted of stock quantity-on-hand increasing by $25*10^2$ kg and reducing by $15*10^2$ kg, respectively. Thus the produce remainder in store after control $\varepsilon(t) = x_2(t) + y(t) - x_1(t)$ consists of $2*10^2$ kg and $53*10^2$ kg, respectively. These values do not exceed the permissible inventory level, which is equal to $70*10^2$ kg. The dynamics of the produce remainder after control $\varepsilon(t)$ change, presented in Fig. 9.6,d is indicative of the control stability, i.e. of the tendency of index $\varepsilon(t)$ approaching a zero value. Membership functions of fuzzy terms for variables $x_1(t)$ and $x_2(t)$, and also their parameters (*b*- , *c*-) before and after training are presented in Fig. 9.7, 9.8 and Tables 9.4, 9.5 respectively. Rules weights included in the fuzzy knowledge base before and after training are presented in Table 9.6.

Table 9.6. Rules weights before (after) training

$x_2(t)$ \ $x_1(t)$	*F*	*D*	*S*	*I*	*R*
E	1 (0.954)	1 (0.755)	1 (0.999)	1 (0.967)	1 (0.578)
H	1 (0.986)	1 (0.711)	1 (0.897)	1 (0.679)	1 (0.953)
A	1 (0.695)	1 (0.538)	1 (0.854)	1 (0.968)	1 (0.680)
L	1 (0.842)	1 (0.943)	1 (0.799)	1 (0.869)	1 (0.947)
M	1 (0.857)	1 (0.851)	1 (0.859)	1 (0.995)	1 (0.867)

Comparison of model and reference control before and after fuzzy model training is presented in Fig. 9.9 and 9.10. Comparison of the produce remainder $\varepsilon(t)$ value in store after control before and after fuzzy model training is shown in Fig. 9.11 and 9.12.

The proposed approach can find application in the automated management systems of enterprises and trade firms. Further development of this approach can be done in the direction of creating adaptive inventory control models, which are tuned with the acquisition of new experimental data about successful decisions. Besides that with the help of supplementary fuzzy knowledge bases factors influencing the demand and quantity-on-hand values (seasonal prevalence, purchase and selling praises, delivery cost, plant-supplier power and others) can be taken into account.

Table 9.4. Membership functions parameters of variable $x_1(t)$ fuzzy terms before (after) training

Linguistic assessments of $x_1(t)$ variable	Parameter	
	b	c
falling (*F*)	0 (1.95)	70 (44.11)
decreased (*D*)	50 (30.54)	70 (42.85)
steady (*S*)	100 (105.77)	70 (35.68)
increased (*I*)	150 (170.04)	70 (40.12)
rising up (*R*)	200 (199.43)	70 (47.55)

Table 9.5. Membership functions parameters of variable $x_2(t)$ fuzzy terms before (after) training

Linguistic assessments of $x_2(t)$ variable	Parameter	
	b	c
minimal (*M*)	70 (75.46)	35 (18.76)
low (*L*)	95 (85.12)	35 (22.12)
adequately sufficient (A)	120 (125.15)	35 (16.75)
high (*H*)	145 (157.99)	35 (14.54)
excessive (*E*)	170 (168.63)	35 (12.69)

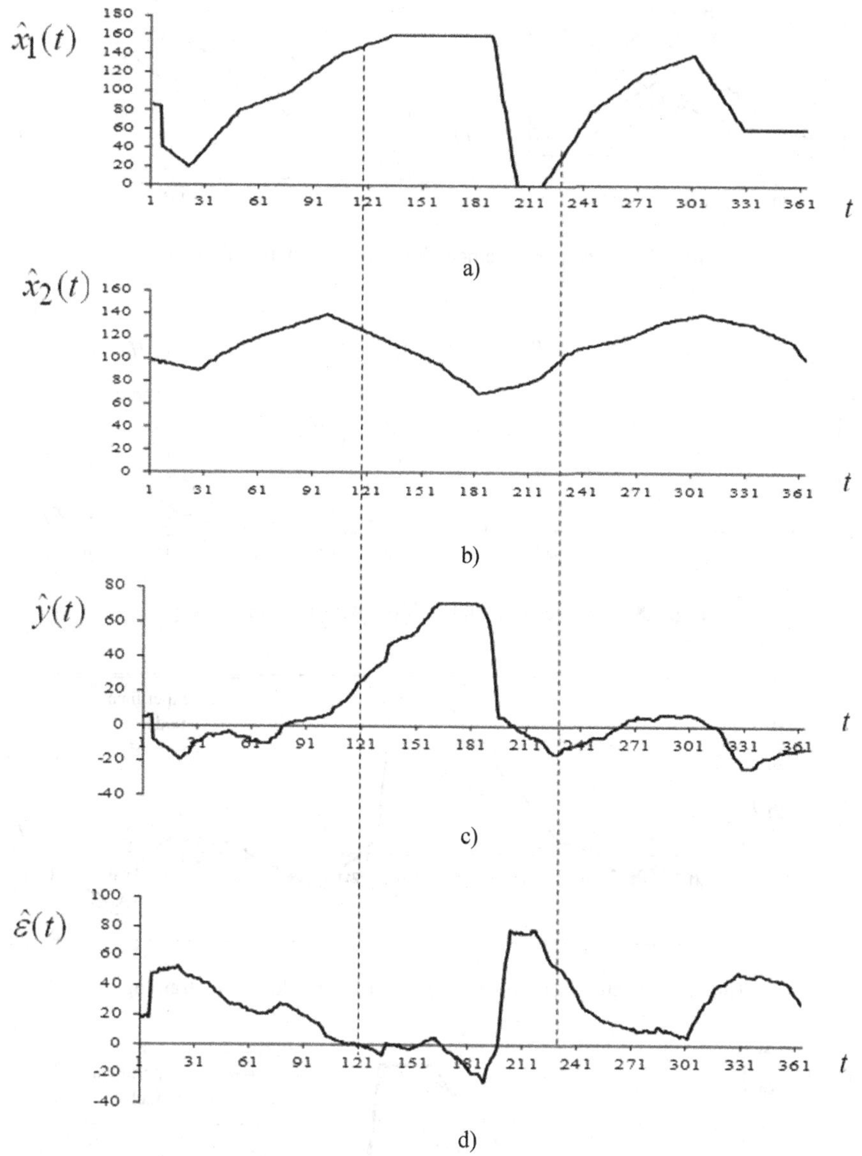

Fig. 9.6. Training data a) change of the demand for the produce in 2001 b) stock quantity-on-hand change in 2001 c) inventory action in 2001 d) change of the produce remainder in store after control in 2001

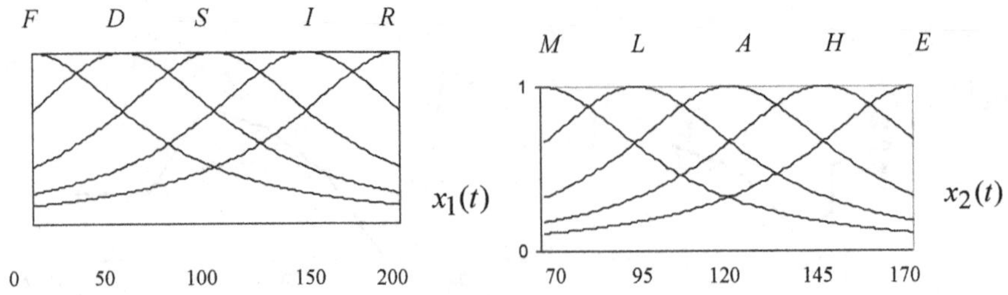

Fig. 9.7. Fuzzy terms membership functions before training

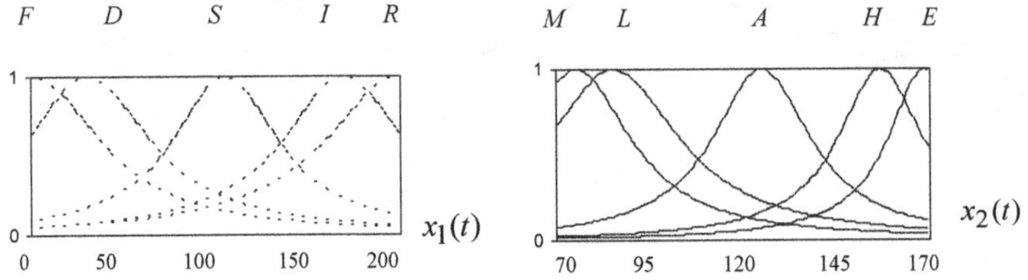

Fig. 9.8. Fuzzy terms membership functions after training

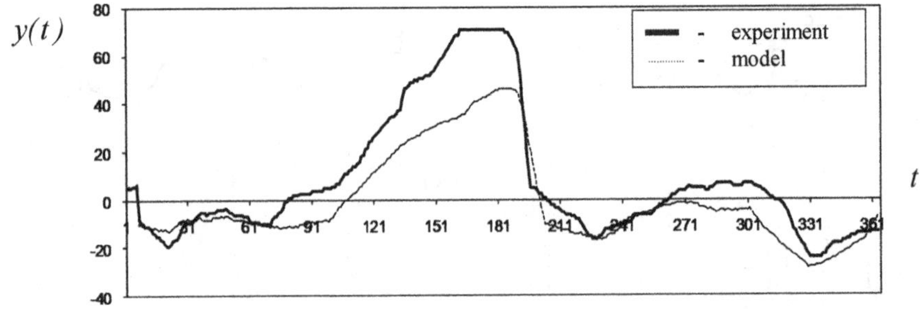

Fig. 9.9. Inventory action generated by fuzzy model before training

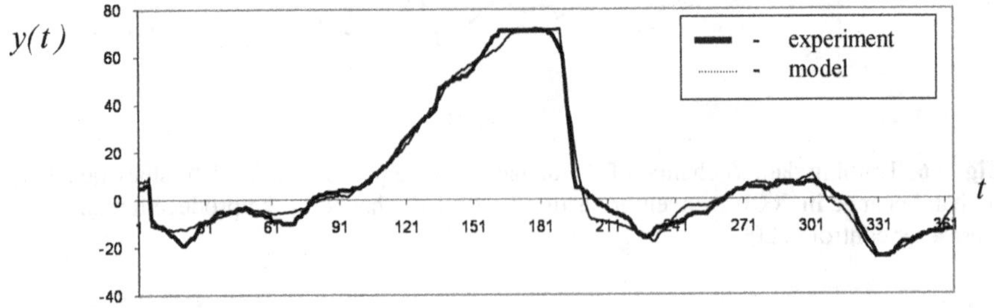

Fig. 9.10. Inventory action generated by fuzzy model after training

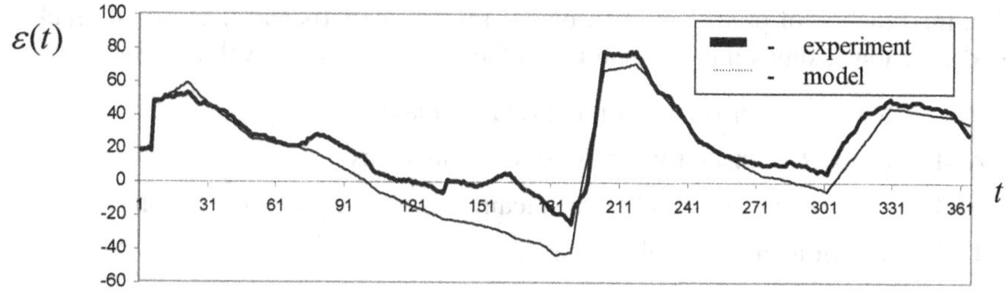

Fig. 9.11. Produce remainder in store after control before fuzzy model training

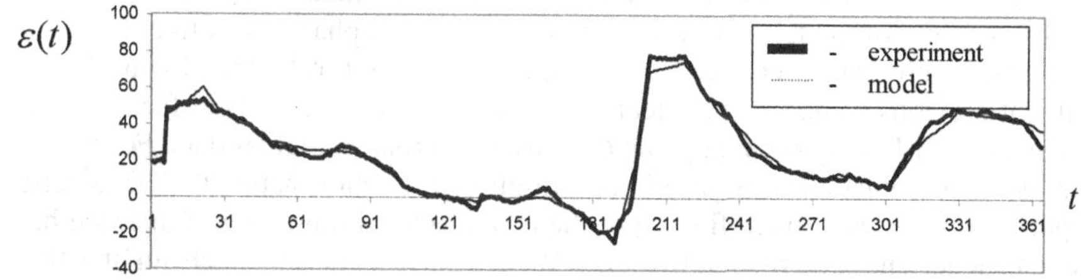

Fig. 9.12. Produce remainder in store after control after fuzzy model training

9.3 Prediction of Football Games Results

The possibilities of the method of non-linear dependencies identification by fuzzy IF-THEN rules [14] are illustrated by an example of the problem of forecasting the results of football games, which is a typical representative of complex forecasting problems that require adaptive model tuning.

Football is a most popular sport attracting hundreds of millions of fans. Prediction of football matches results arouses interest from two points of view: the first one is demonstration of the power of different mathematical methods [18, 19], the second one is the desire of earning money by predicting beforehand any winning result. Models and PC–programs of sport prediction are already being developed for many years (see, for example, http://dmiwww.cs.tut.fi/riku). Most of them use stochastic methods of uncertainty description: regressive and autoregressive analysis [20 – 22], Bayessian approach in combination with Markov chains and the Monte-Carlo method [23 – 26]. The specific features of these models are: sufficiently great complexity, a lot of assumptions, and the need for a great number of statistical data. Besides that, the models cannot always be easily interpreted. Some several years passed before some models using neural networks for the results of football games prediction appeared [27 – 29]. They can be considered as universal approximators of non-linear dependencies trained by experimental data. These models also need a lot of statistical data and do not allow us to define the physical meaning of the weights between neurons after training.

In the practice of prediction making the football experts and fans usually make good decisions using simple reasoning on the common sense level, for example:

IF team T_1 constantly won in previous matches
AND team T_2 constantly lost in previous matches
AND in previous matches between teams T_1 and T_2 team T_1 won,
THEN win of team T_1 should be expected.

Such expressions can be considered as concentration of accumulated experts' experiences and can be formalized using fuzzy logic. That is why it is quite natural to apply such expressions as a support for building a model of prediction.

The process of modeling has two phases. In the first phase we define the fuzzy model structure, which connects the football game result to be found with the results of previous games for both teams. The second phase consists of fuzzy model tuning, i.e., of finding optimal parameters using tournament tables data. For tuning we use a combination of a genetic algorithm and a neural network. The genetic algorithm provides a rough finding of the area of global minimum of distance between model and experimental results. We use the neural approach for the fine model parameters tuning and for their adaptive correction while new experimental data is appearing.

9.3.1 The Structure of the Model

The aim of modeling is to calculate the result of match between teams T_1 and T_2, which is characterized as the difference of scored and lost goals y. We assume that $y \in [\underline{y}, \overline{y}] = [-5, 5]$. For prediction model building we will define the value of y on the following five levels:

d_1 is a big loss (*BL*), $y = -5, -4, -3$;
d_2 is a small loss (*SL*), $y = -2, -1$;
d_3 is a draw (*D*), $y=0$;
d_4 is a small win (*SW*), $y = 1, 2$;
d_5 is a big win (*BW*), $y = 3, 4, 5$.

Let us suppose that the football game result (y) is influenced by the following factors:

$x_1, x_2, ..., x_5$ are the results of five previous games for team T_1;
$x_6, x_7, ..., x_{10}$ are the results of five previous games for team T_2;
x_{11}, x_{12} are the results of two previous games between teams T_1 and T_2.

It is obvious, that values of factors $x_1, x_2, ..., x_{12}$ are changing in the range from –5 to 5.

The hierarchical interconnection between output variable y and input variables $x_1, x_2, ..., x_{12}$ is represented as a tree shown in Fig. 9.13.

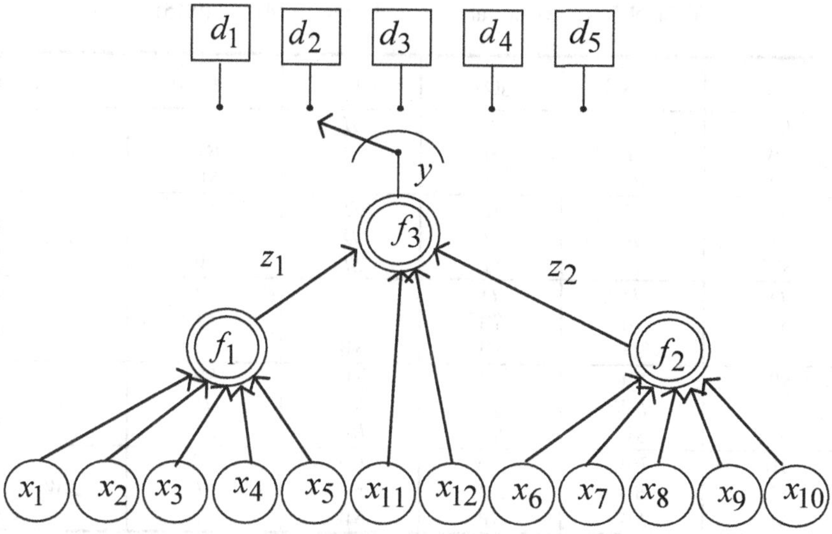

Fig. 9.13. Structure of the Prediction Model

This tree is equal to the system of correlations

$$y = f_3(z_1, z_2, x_{11}, x_{12}), \tag{9.13}$$

$$z_1 = f_1(x_1, x_2, ..., x_5), \tag{9.14}$$

$$z_2 = f_2(x_6, x_7, ..., x_{10}), \tag{9.15}$$

where z_1 (z_2) is the football game prediction for team T_1 (T_2) based on the previous results $x_1, x_2, ..., x_5$ ($x_6, x_7, ..., x_{10}$).

The variables $x_1, x_2, ..., x_{12}$, as well as z_1 (z_2) will be considered as linguistic variables [17], which can be evaluated using above mentioned fuzzy terms: *BL, SL, D, SW* and *BW*.

To describe the correlations (9.13) - (9.15) we shall use the expert matrices of knowledge (Tables 9.7, 9.8). These matrices correspond to fuzzy IF-THEN rules received on the common sense and practical reasoning level. An example of one of these rules for Table 9.7 is given below:

IF (x_{11} =*BW*) AND (x_{12} =*BW*) AND (z_1 =*BW*) AND (z_2 =*BL*)
OR (x_{11} =*SW*) AND (x_{12} =*BW*) AND (z_1 =*SW*) AND (z_2 =*D*)
OR (x_{11} =*BW*) AND (x_{12} =*D*) AND (z_1 =*BW*) AND (z_2 =*SL*)
THEN $y = d_5$.

Table 9.7. Knowledge about correlations (9.14) and (9.15)

$x_1(x_6)$	$x_2(x_7)$	$x_3(x_8)$	$x_4(x_9)$	$x_5(x_{10})$	$z_1(z_2)$
BL	*BL*	*BL*	*BL*	*BL*	
BW	*SL*	*BL*	*SL*	*BW*	*BL*
SW	*BL*	*SL*	*SL*	*SW*	
SL	*SL*	*SL*	*SL*	*SL*	
D	*SL*	*SL*	*D*	*D*	*SL*
SW	*D*	*SL*	*SL*	*SW*	
D	*D*	*D*	*D*	*D*	
SL	*SW*	*SW*	*D*	*SL*	*D*
D	*D*	*SW*	*SW*	*D*	
SW	*SW*	*SW*	*SW*	*SW*	
D	*BW*	*BW*	*SW*	*D*	*SW*
SL	*SW*	*SW*	*BW*	*SL*	
BW	*BW*	*BW*	*BW*	*BW*	
SL	*BW*	*SW*	*BW*	*SL*	*BW*
BL	*SW*	*BW*	*SW*	*BL*	

Table 9.8. Knowledge about correlation (9.13)

x_{11}	x_{12}	z_1	z_2	y
BL	*BL*	*BL*	*BW*	
BW	*D*	*BL*	*D*	d_1
SW	*BL*	*SL*	*SL*	
SW	*SL*	*D*	*SL*	
D	*SL*	*SL*	*D*	d_2
SW	*D*	*SL*	*SL*	
D	*D*	*D*	*D*	
SL	*SW*	*SW*	*D*	d_3
SL	*D*	*SW*	*SW*	
SL	*SW*	*SW*	*BW*	
D	*BW*	*BW*	*SW*	d_4
SL	*SW*	*SW*	*BW*	
BW	*BW*	*BW*	*BL*	
SW	*BW*	*SW*	*D*	d_5
BW	*D*	*BW*	*SL*	

9.3.2 Fuzzy Model of Prediction

Using the generalized fuzzy approximator [14] and the tree of evidence (Fig. 9.13), the prediction model can be described in the following form:

$$y = F_y(x_1, x_2, ..., x_{12}, \mathbf{W}_1, \mathbf{B}_1, \mathbf{C}_1, \mathbf{W}_2, \mathbf{B}_2, \mathbf{C}_2, \mathbf{W}_3, \mathbf{B}_3, \mathbf{C}_3), \quad (9.16)$$

where F_y is the operator of inputs-output connection, corresponding to correlations (9.13) – (9.15),

$$\mathbf{W}_1 = ((w_1^{11},...,w_1^{13}),...,(w_1^{51},...,w_1^{53}))\,,\ \mathbf{W}_2 = ((w_2^{11},...,w_2^{13}),...,(w_2^{51},...,w_2^{53}))\,,$$

$\mathbf{W}_3 = ((w_3^{11},...,w_3^{13}),...,(w_3^{51},...,w_3^{53}))$ are the vectors of rules weights in the correlations (9.13), (9.14), (9.15), respectively;

$$\mathbf{B}_1 = (b_{1-5}^{BL}, b_{1-5}^{SL}, b_{1-5}^{D}, b_{1-5}^{SW}, b_{1-5}^{BW})\ ,\ \mathbf{B}_2 = (b_{6-10}^{BL}, b_{6-10}^{SL}, b_{6-10}^{D}, b_{6-10}^{SW}, b_{6-10}^{BW})\ ,$$

$\mathbf{B}_3 = (b_{11,12}^{BL}, b_{11,12}^{SL}, b_{11,12}^{D}, b_{11,12}^{SW}, b_{11,12}^{BW})$ are the vectors of centres for variables $x_1, x_2, ..., x_5$, $x_6, x_7, ..., x_{10}$ and x_{11}, x_{12} membership functions to terms *BL, SL,..., BW;*

$$\mathbf{C}_1 = (c_{1-5}^{BL}, c_{1-5}^{SL}, c_{1-5}^{D}, c_{1-5}^{SW}, c_{1-5}^{BW})\,,\ \mathbf{C}_2 = (c_{6-10}^{BL}, c_{6-10}^{SL}, c_{6-10}^{D}, c_{6-10}^{SW}, c_{6-10}^{BW})\,,$$

$\mathbf{C}_3 = (c_{11,12}^{BL}, c_{11,12}^{SL}, c_{11,12}^{D}, c_{11,12}^{SW}, c_{11,12}^{BW})$ are the vectors of concentration parameters for variables $x_1, x_2, ..., x_5$, $x_6, x_7, ..., x_{10}$ and x_{11}, x_{12} membership functions to terms *BL, SL,..., BW*.

In model (9.16) we assume that for all of variables $x_1, x_2, ..., x_5$ fuzzy terms *BL, SL,..., BW* have the same membership functions. Same assumption we made for variables $x_6, x_7, ..., x_{10}$ and variables x_{11}, x_{12} (See. Fig. 9.14).

9.3.3 Genetic and Neuro Tuning

The reasonable results of simulation can be reached by fuzzy rules tuning using tournament tables data. Training data in the form of M pairs of experimental data assumed to be obtained with use of tournament tables

$$\left\langle \hat{\mathbf{X}}_l, \hat{y}_l \right\rangle,\ l = \overline{1, M}\ ,$$

where $\hat{\mathbf{X}}_l = \{(x_1^l, x_2^l, ... x_5^l),\ (x_6^l, x_7^l, ... x_{10}^l),\ (x_{11}^l, x_{12}^l)\}$ are the previous matches results for teams T_1 and T_2 in the experiment number l,

$\hat{y}_l$ is the game result between teams T_1 and T_2 in experiment number l.

The essence of the prediction model tuning consists of such membership functions parameters (*b*-, *c*-) and fuzzy rules weights (*w*-) finding, which provide for the minimum distance between theoretical and experimental results:

$$\sum_{l=1}^{M} (F_y(\hat{x}_1^l, \hat{x}_2^l, ..., \hat{x}_{12}^l, \mathbf{W}_i, \mathbf{B}_i, \mathbf{C}_i) - \hat{y}_l)^2 = \min_{\mathbf{W}_i, \mathbf{B}_i, \mathbf{C}_i}\ ,\quad i = 1, 2, 3\,.$$

To solve this non-linear optimization problem we propose a genetic algorithm and neural network combination. The genetic algorithm provides for a rough off-line finding of the area of global minimum, while the neural network is used for on-line improvement of unknown parameters values.

For the fuzzy model tuning we used the results from tournament tables of the Finland Football Championship characterized by a minimal number of sensations. Our training data included results of 1056 matches for the last 8 years from 1994 to 2001. The results of the fuzzy model tuning are given in Tables 9.9 – 9.12 and in Fig. 9.14.

Table 9.9. Fuzzy rules weights in correlation (9.13)

Genetic algorithm	Neuro-fuzzy network
1.0	0.989
1.0	1.000
1.0	1.000
0.8	0.902
0.5	0.561
0.8	0.505
0.6	0.580
1.0	0.613
0.5	0.948
1.0	0.793
0.9	0.868
0.6	0.510
0.6	0.752
0.5	0.500
0.5	0.500

Table 9.10. Fuzzy rules weights in correlation (9.14)

Genetic algorithm	Neuro-fuzzy network
0.7	0.926
0.9	0.900
0.7	0.700
0.9	0.954
0.7	0.700
1.0	1.000
0.9	0.900
1.0	1.000
0.6	0.600
1.0	1.000
0.7	0.700
1.0	1.000
0.8	0.990
0.5	0.500
0.6	0.600

Table 9.11. Fuzzy rules weights in correlation (9.15)

Genetic algorithm	Neuro-fuzzy network
0.7	0.713
0.8	0.782
1.0	0.996
0.5	0.500
0.5	0.541
0.5	0.500
0.5	0.500
0.5	0.522
0.6	0.814
1.0	0.903
0.6	0.503
1.0	0.677
1.0	0.515
0.5	0.514
1.0	0.999

Table 9.12. *b*- and *c*- parameters of membership functions after tuning

Terms	Genetic Algorithm						Neuro-Fuzzy Network					
	$x_1, x_2, ..., x_5$		$x_6, x_7, ..., x_{10}$		x_{11}, x_{12}		$x_1, x_2, ..., x_5$		$x_6, x_7, ..., x_{10}$		x_{11}, x_{12}	
	b-	*c*-	*b*-	*c*-	*b*-	*c*-	*b*-	*c*-	*b*-	*c*-	*b*-	*c*-
BL	-4.160	9	-5.153	9	-5.037	3	-4.244	7.772	-4.524	9.303	-4.306	1.593
SL	-2.503	1	-2.212	5	-3.405	1	-1.468	0.911	-1.450	5.467	-2.563	0.555
D	-0.817	1	0.487	7	0.807	1	-0.331	0.434	0.488	7.000	0.050	0.399
SW	2.471	3	2.781	9	2.749	7	1.790	1.300	2.781	9.000	2.750	7.000
BW	4.069	5	5.749	9	5.238	3	3.000	4.511	5.750	9.000	3.992	1.234

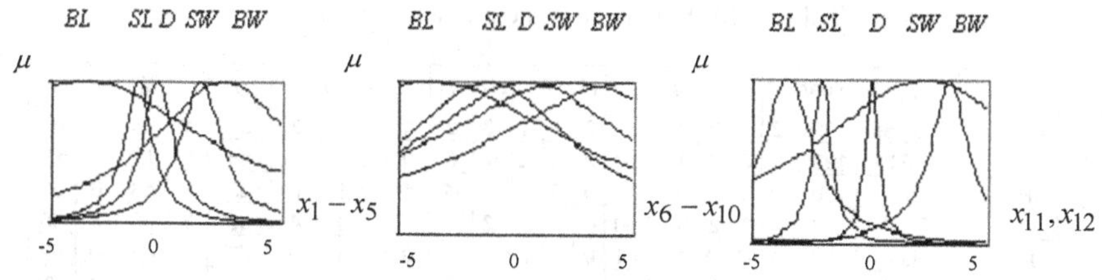

Fig. 9.14. Membership functions after tuning

To test the prediction model we used the results of 350 matches from 1991 to 1993. The fragment of testing data and prediction results are shown in Table 9.13, where:

T_1, T_2 are teams' names,

$\hat{y}$, $\hat{d}$ are real (experimental) results,

y_G, d_G are results of prediction after genetic tuning of the fuzzy model,

y_N, d_N are results of prediction after neural tuning of the fuzzy model.

Symbol * shows no coincidences of theoretical and experimental results.

The efficiency characteristics of fuzzy model tuning algorithms for the testing data are shown in Table. 9.14.

Table 9.14. Tuning algorithms efficiency characteristics

Efficiency characteristics		Genetic Tuning	Neural Tuning
Tuning Time		52 min	7 min
Number of iterations		25000	5000
Probability of correct prediction for different decisions	d_1 – big loss	30 / 35 = 0.857	32 / 35 =0.914
	d_2 – small loss	64 / 84 = 0.762	70 / 84 = 0.833
	d_3 – draw	38 / 49 = 0.775	43 / 49 = 0.877
	d_4 – small win	97 / 126 = 0.770	106 / 126 = 0.841
	d_5 – big win	49 / 56 = 0.875	53 / 56 = 0.946

Table 9.14 shows, that the best prediction results we can receive for the marginal decision classes (the loss and win with big score d_1 and d_5), and the worst results of prediction we can receive for the small loss and small win (d_2 and d_4).

The future improvement of fuzzy prediction model can be done by taking into account some additional factors in fuzzy rules such as: the game on host/guest field, number of injured players, different psychological effects.

Table 9.13. Fragment of the prediction results

№	T_1	T_2	Year	$x1$	$x2$	$x3$	$x4$	$x5$	$x6$	$x7$	$x8$	$x9$	$x10$	$x11$	$x12$	Score	$\hat{y}$	$\hat{d}$	y_G	d_G	y_N	d_N
1	Kuusysi	Reipas	1991	2	1	2	0	1	-1	0	1	-2	-3	2	1	2-0	2	d4	1	d4	1	d4
2	Ilves	PPT	1991	1	3	-1	1	0	0	2	-1	-2	0	0	0	2-1	1	d4	0	d3*	0	d3*
3	Haka	Jaro	1991	-1	2	0	-1	1	1	0	-2	-1	-2	-1	1	1-1	0	d3	0	d3	0	d3
4	MP	OTP	1991	3	1	2	0	2	-1	-2	1	-2	-3	1	3	4-0	4	d5	3	d5	3	d5
5	KuPS	HJK	1991	-1	-3	-4	1	-3	1	0	2	0	0	-2	0	1-3	-2	d2	-1	d2	-1	d2
6	TPS	RoPS	1991	3	1	2	-2	0	2	0	1	-1	1	0	-1	1-0	1	d4	0	d3*	0	d3*
7	PPT	Jaro	1991	0	-5	-1	0	1	1	2	-2	-1	1	1	-3	0-1	-1	d2	-1	d2	-1	d2
8	Haka	Reipas	1991	2	-1	3	1	4	2	-2	0	-1	0	-1	2	3-0	3	d5	2	d4*	2	d4*
9	OTP	Kuusysi	1991	-1	-2	-3	-2	0	1	3	4	-1	2	-2	-1	1-4	-3	d1	-3	d1	-3	d1
10	HJK	TPS	1991	1	1	1	0	2	0	1	-1	2	-3	0	2	2-0	2	d4	2	d4	2	d4
11	MyPa	Jaro	1992	-3	1	2	1	0	2	1	-2	-1	0	-2	0	0-0	0	d3	0	d3	0	d3
12	Jazz	Ilves	1992	2	2	1	-1	0	3	4	-1	0	1	1	-1	2-1	1	d4	0	d3*	1	d4
13	Haka	RoPS	1992	-2	-2	0	1	1	-1	1	1	1	0	1	3	1-1	0	d3	1	d4*	1	d4*
14	HJK	Oulu	1992	2	3	0	0	1	0	-5	1	-2	-1	-1	2	4-0	4	d5	2	d4*	3	d5
15	MP	Kuusysi	1992	0	1	-2	-1	-1	3	1	2	0	1	0	-2	0-3	-3	d1	-3	d1	-3	d1
16	KuPS	HJK	1992	-2	-1	-3	1	-2	4	2	1	2	1	-2	-3	0-5	-5	d1	-4	d1	-4	d1
17	Kuusysi	MP	1992	0	-1	3	2	-1	-3	2	-1	-2	0	1	0	3-1	2	d4	1	d4	1	d4
18	TPS	Haka	1992	-1	2	3	-1	-2	0	-1	0	3	1	-1	1	2-2	0	d3	0	d3	0	d3
19	RoPS	MyPa	1992	-2	-1	2	0	-1	1	-1	1	1	-2	1	-1	1-2	-1	d2	0	d3*	0	d3*
20	Jazz	Ilves	1992	-2	1	-3	5	-1	1	1	-2	0	-1	2	0	1-0	1	d4	1	d4	1	d4
21	TPS	Jaro	1992	-2	-1	2	-1	-3	1	0	2	-1	3	1	-2	0-2	-2	d2	-1	d2	-1	d2
22	Haka	MyPa	1992	1	1	-1	0	1	0	3	2	1	-1	-1	-3	0-1	-1	d2	-2	d2	-2	d2
23	HJK	RoPS	1992	1	2	0	-1	1	-1	2	2	-1	1	0	0	2-1	1	d4	0	d3*	0	d3*
24	MP	Kuusysi	1992	1	-1	-2	-3	1	1	-1	-2	2	3	-2	1	0-2	-2	d2	-1	d2	-1	d2
25	Ilves	Kups	1992	3	0	-2	2	-2	1	1	-1	0	-2	1	0	1-0	1	d4	1	d4	1	d4
26	Haka	HJK	1992	0	-2	-1	-1	0	2	3	-1	0	3	-1	-2	0-3	-3	d1	-3	d1	-3	d1
27	Jaro	MyPa	1992	-1	-1	1	2	1	-3	1	2	1	0	1	1	1-1	0	d3	1	d4*	0	d3
28	RoPS	TPS	1992	-1	1	-1	1	4	-5	-2	3	-1	-2	5	1	2-0	2	d4	2	d4	1	d4
29	MP	Ilves	1992	1	2	-1	1	0	0	1	0	0	-1	1	-2	2-3	-1	d2	-1	d2	-1	d2
30	Kuusysi	KuPS	1992	2	2	0	3	1	-1	-1	1	-3	0	2	3	4-1	3	d5	3	d5	3	d5
31	Jazz	MP	1993	2	2	2	0	3	-2	-1	0	-1	-3	4	3	5-0	5	d5	4	d5	4	d5
32	Kuusysi	TPS	1993	1	-1	0	-1	1	-2	2	0	-1	1	0	1	0-0	0	d3	0	d3	0	d3
33	MyPa	RoPS	1993	-1	-1	2	2	3	2	-1	1	2	-2	3	-1	2-0	2	d4	1	d4	1	d4
34	Haka	HJK	1993	-3	-1	-2	1	0	1	4	1	2	0	-1	-2	1-3	-2	d2	-1	d2	-1	d2
35	Jaro	Ilves	1993	2	0	-1	0	-1	-2	-1	-2	2	1	2	0	2-1	1	d4	1	d4	1	d4
36	Ilves	HJK	1993	1	-2	-1	-1	1	3	1	2	0	1	-1	-1	0-2	-2	d2	-1	d2	-1	d2
37	Jazz	Jaro	1993	2	1	0	1	5	-1	-2	-2	1	-1	2	1	3-0	3	d5	2	d4*	2	d4*
38	MyPa	MP	1993	1	3	1	-1	1	-1	0	2	-1	1	1	0	1-0	1	d4	1	d4	1	d4
39	Kuusysi	Haka	1993	-1	-2	1	1	2	-1	-3	1	-5	2	3	-1	3-1	2	d4	1	d4	1	d4
40	TPS	RoPS	1993	-1	1	-2	1	2	1	2	-1	1	-2	1	1	1-0	1	d4	1	d4	1	d4
41	MP	HJK	1993	-1	-1	0	2	-1	2	3	1	-1	1	-2	1	1-2	-1	d2	0	d3*	0	d3*
42	Kuusysi	Jaro	1993	2	2	-2	1	2	0	-1	2	-2	0	1	2	2-1	1	d4	1	d4	1	d4
43	Jazz	Haka	1993	2	3	2	-1	1	-1	-3	-4	-2	0	2	2	4-0	4	d5	3	d5	3	d5
44	FinnPa	MyPa	1993	-1	1	-2	-1	2	1	-2	-1	1	0	-1	-1	1-2	-1	d2	-1	d2	-1	d2
45	TPS	Ilves	1993	2	1	2	1	-1	2	2	-2	1	-3	0	2	2-0	2	d4	1	d4	1	d4
46	RoPS	Jazz	1993	-1	-1	2	-2	-1	4	1	5	0	2	1	-3	2-5	-3	d1	-3	d1	-3	d1
47	MyPa	Ilves	1993	5	0	2	1	1	-3	-1	-2	1	-2	3	0	5-1	4	d5	3	d5	3	d5
48	TPV	Kuusysi	1993	-2	-1	0	1	0	-1	0	2	-1	0	0	1	0-0	0	d3	0	d3	0	d3
49	RoPS	HJK	1993	-1	-1	1	-2	0	3	1	-2	1	1	-2	1	0-2	-2	d2	0	d3*	-1	d2
50	TPS	Jaro	1993	-1	-1	1	2	2	-2	-1	1	-2	1	3	1	1-0	1	d4	1	d4	1	d4

9.4 Identification of Car Wheels Adhesion Factor with a Road Surface

The task of car wheels adhesion factor (AF) evaluation with a road surface arises with an execution of a technical expert's examination during an investigation of traffic accidents (TA). The objectivity of decision making relative to guilt or innocence of the driver who caused the TA depends on the precision of the AF definition (for example, run over a pedestrian). The existing technique [30, 31] allows determining of only some range of possible AF values depending upon a series of the influencing factors. Therefore, its final evaluation is determined by the auto engineering expert, subjectively taking into account the additional factors and conditions which are not involved in this technique.

Decision making relative to the cause of the accident is very sensitive to the value of AF: the subjective choice of the lower or upper value of AF can decide the fate of the accident participants.

The purpose of this research, the results of which are presented in this chapter, is to develop a mathematical model of AF evaluation taking into account all accessible information about the influencing factors, and at the expense of the AF magnitude improvement to raise a solution's objectivity.

This section is based on materials of [32].

9.4.1 Technique of Identification

The model of AF evaluation was developed on the basis of fuzzy rule-based methodology of identification described in [14]. The model was created in two stages: first – structural identification; second – parametrical identification. At the first stage the structure of the AF dependence upon the influencing factors was built by expert IF-THEN rules. At the second stage we selected such parameters of membership functions and such weights of fuzzy rules which allow us to minimize the difference between model and experimental results.

9.4.2 Structural Identification

The structure of the suggested model is shown in Fig. 9.15 in the form of a tree, whose trailing tops are the factors influencing AF.

The characteristic of the model consists of the fact that it takes into account both of the traditional factors, which are generalized by the integrated index Q, and additionally entered factors: S, H, P, N, V. All the influencing factors shown in Table 9.15 are considered as linguistic variables given using the appropriate universal sets and are estimated by fuzzy terms.

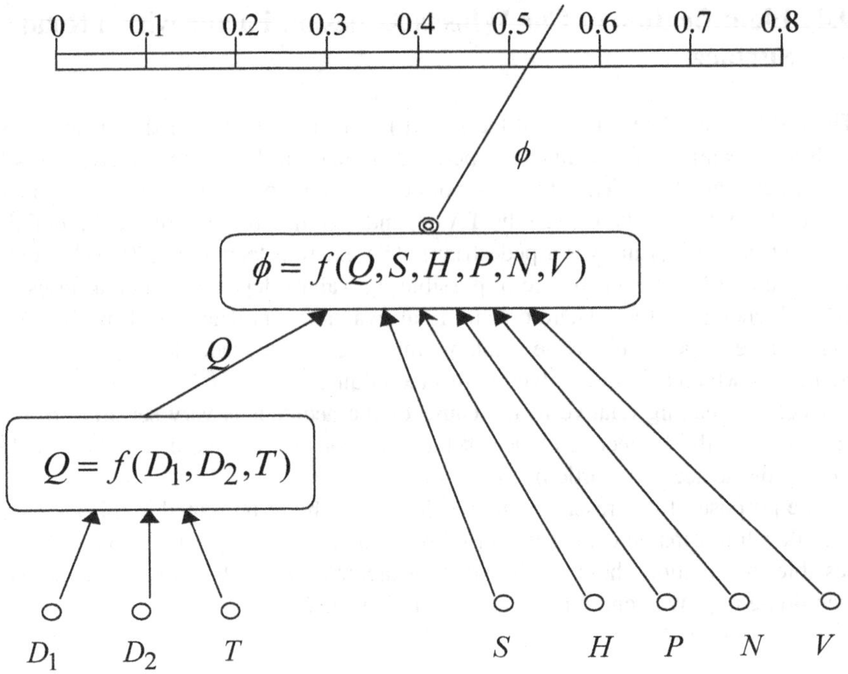

Fig. 9.15. The model structure for AF definition

The integrated index Q included in Table 9.15 depends on the factors: D_1 – road surface type; D_2 – road surface condition; T – tires type. The recommendations for the evaluation of the integrated index Q are given in Table 9.16 according to the known technique [30].

The expert knowledge base necessary for AF evaluation is shown in Table 9.17 (experts V. Rebedailo, A. Kashkanov). The application of the model of fuzzy logic inference to the knowledge base (Table 9.17) allows us to predict AF in some practical range of its modification. However, the exact evaluation of this factor depends on the choice of parameters for the model tuning.

9.4.3 Parametrical Identification

The tuning of the model was realized using training data, which represents the population of pairs "influencing factors – adhesion factor". To provide this training data a specially organized experiment with the automobile "Moskvich – 412" was carried out. In this experiment we used the car braking with different motion speeds on the horizontal road. Values of the factors which influence on AF were registered together with the values of car brake distances and values of brake initial velocity [31].

Table 9.15. The factors influencing AF

Factor	Universal set	Terms for estimations
Q – Integrated index "type of tires – road"	(0 – 9) conditional unit	Low (Q_1), Below average (Q_2), Average (Q_3), Above average (Q_4), High (Q_5)
S – Degree of tires slip	(0 – 100)%	Rolling with slip (S_1), Skid (S_2)
H – Wear of tires	(0 – 100)%	New (H_1), Within admissible range (H_2), Worn tire (H_3)
P – Pressure in tires	(0.1 – 0.325) MPa	Reduced (P_1), Normal (P_2), Higher than normal (P_3)
N – Load on a wheel	(0 – 100)%	Without load (N_1), Average (N_2), Full load (N_3)
V – Velocity of the car	(0 – 130) kms/h	Low (V_1), Below average (V_2), Average (V_3), Above average (V_4), High (V_5)

Table 9.16. Recommendations for evaluation of the integrated index Q

Road surface		Index Q for a type of tires (T)		
Type (D_1)	Condition (D_2)	High pressure	Low pressure	High permeability
Asphalt, Bitumen	Dry	5.63 – 7.88	7.88 – 9	7.88 – 9
	Rain moisture	3.1 – 4.33	4.33 – 4.95	4.33 – 4.95
	Wet	3.94 – 5.06	5.06 – 6.19	5.63 – 6.75
	Covered with a dirt	2.81 – 5.06	2.81 – 4.5	2.81 – 5.06
	Wet snow (t>0°C)	2.1 – 3.4	2.1 – 4.2	2.1 – 4.2
	Ice (t<0°C)	0.9 – 1.69	1.13 – 2.25	0.56 – 1.13
Cobble	Dry	4.5 – 5.63	5.63 – 6.19	6.75 – 7.88
	Wet	2.7 – 3.75	3.75 – 4.43	4.5 – 6.19
Metal	Dry	5.63 – 6.75	6.75 – 7.88	6.75 – 7.88
	Wet	3.38 – 4.5	4.5 – 5.63	4.5 – 6.19
Ground road	Dry	4.5 – 5.63	5.63 – 6.75	5.63 – 6.75
	Rain moisture	2.25 – 4.5	3.38 – 5.06	3.94 – 5.63
	Time of bad roads	1.68 – 2.81	1.68 – 2.81	2.25 – 3.38
Virgin soil in summer: Sand	Dry	2.25 – 3.38	2.48 – 4.5	2.25 – 3.38
	Damp	3.94 – 4.5	4.5 – 5.63	4.5 – 5.63
Clayed soil	Dry	4.5 – 5.63	5.06 – 6.19	4.5 – 5.63
	Humidified up to a plastic state	2.25 – 4.5	2.81 – 4.5	3.38 – 5.06
	Humidified up to a fluid state	1.69 – 2.25	1.69 – 2.81	1.69 – 2.81
Virgin soil in winter: Snow	Mellow	2.25 – 3.38	2.25 – 4.5	2.25 – 4.5
	Smooth	1.69 – 2.25	2.25 – 2.81	3.38 – 5.63

The total volume of the training sample included 60 pairs of "influencing factors – AF" data.

After tuning we received the membership functions shown in Fig. 9.16. Parameters of centres (b) and concentration (c) of the tuned membership functions presented in Table 9.18. Weights of the fuzzy rules obtained after tuning are given in the right side of Table 9.17.

Table 9.17. Fuzzy knowledge base

Q	S	H	P	N	V	ϕ	Weight
Q_1	S_2	H_2	P_2	N_1	V_1		1.000
Q_1	S_1	H_1	P_1	N_3	V_1	ϕ_1	0.700
Q_1	S_1	H_3	P_3	N_2	V_2		0.999
Q_2	S_2	H_2	P_2	N_2	V_3		0.700
Q_1	S_1	H_2	P_1	N_2	V_2	ϕ_2	0.700
Q_2	S_1	H_1	P_3	N_3	V_3		0.998
Q_2	S_1	H_2	P_2	N_3	V_5		0.700
Q_2	S_1	H_1	P_3	N_2	V_3	ϕ_3	0.400
Q_2	S_2	H_2	P_3	N_1	V_2		0.300
Q_2	S_1	H_2	P_2	N_1	V_2		0.400
Q_3	S_2	H_2	P_2	N_2	V_3	ϕ_4	0.997
Q_3	S_1	H_1	P_1	N_1	V_5		0.400
Q_4	S_2	H_1	P_2	N_3	V_2		0.999
Q_3	S_1	H_1	P_3	N_1	V_1	ϕ_5	1.000
Q_4	S_2	H_3	P_2	N_1	V_3		0.400
Q_4	S_2	H_2	P_2	N_1	V_1		0.999
Q_4	S_1	H_2	P_1	N_3	V_2	ϕ_6	0.400
Q_4	S_2	H_1	P_2	N_1	V_3		0.400
Q_4	S_1	H_1	P_2	N_1	V_2		0.699
Q_5	S_1	H_1	P_2	N_3	V_5	ϕ_7	1.000
Q_5	S_2	H_2	P_1	N_2	V_4		1.000
Q_5	S_2	H_2	P_2	N_3	V_2		1.000
Q_5	S_2	H_2	P_2	N_1	V_3	ϕ_8	1.000
Q_5	S_1	H_1	P_2	N_1	V_4		0.600

Table 9.18. Parameters of membership functions after tuning

Term	b	c	Term	b	c	Term	b	c
Q_1	0.90	0.97	H_1	21.36	24.33	N_2	64.48	28.92
Q_2	2.50	0.40	H_2	57.15	38.68	N_3	85.92	20.31
Q_3	4.63	0.59	H_3	90.21	26.55	V_1	10.40	14.74
Q_4	6.23	0.42	P_1	0.14	0.04	V_2	10.40	30.06
Q_5	8.58	0.75	P_2	0.20	0.04	V_3	14.07	42.26
S_1	24.88	41.76	P_3	0.32	0.07	V_4	64.65	5.82
S_2	98.93	41.95	N_1	0.10	38.98	V_5	119.99	13.48

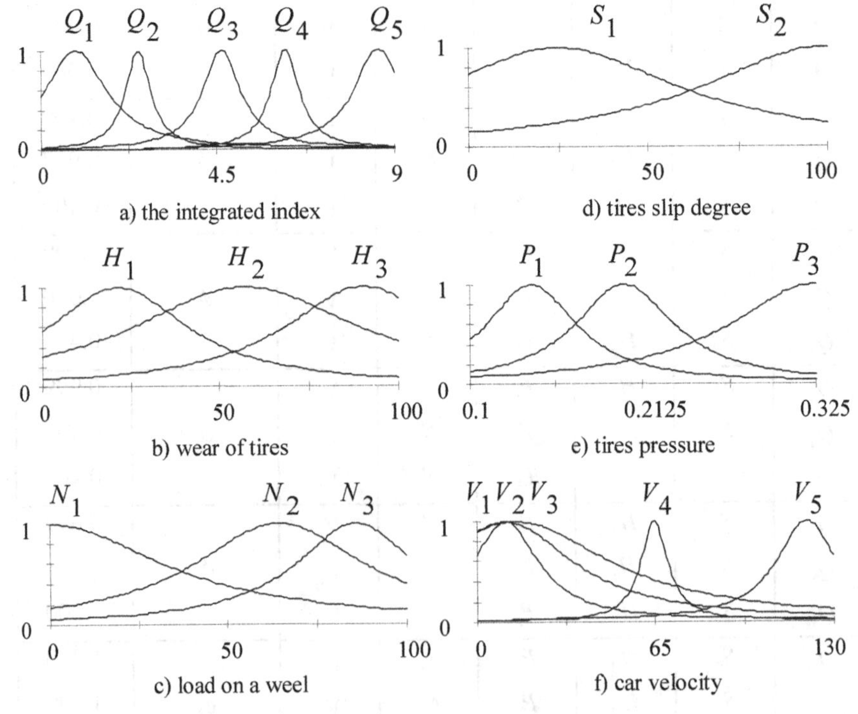

Fig. 9.16. Fuzzy terms membership functions after tuning

Table 9.19. Comparison of decisions

Factors						Adhesion factor		
Q	*S*	*H*	*P*	*N*	*V*	Tabular	1)	2)
6,15	100	62	0,2	15	20	0.45 – 0.55	0,55	0,54
4,45	100	65	0,2	15	60	0.25 – 0.4	0,33	0,35
4,7	100	65	0,18	20	40	0.30 – 0.45	0,39	0,39
3,4	90	45	0,17	95	120	0.22 – 0.40	0,26	0,26
3,7	64	95	0,25	45	72	0.20 – 0.40	0,28	0,29
3,9	84	81	0,27	67	65	0.25 – 0.45	0,32	0,31
8,1	67	72	0,25	20	58	0.60 – 0.70	0,68	0,68
3,4	65	80	0,14	15	15	0.25 – 0.40	0,27	0,28
3,6	40	75	0,18	20	45	0.30 – 0.45	0,34	0,31
3,9	100	35	0,29	45	110	0.20 – 0.40	0,29	0,29
7,4	35	70	0,19	60	90	0.60 – 0.70	0,62	0,62
5,3	30	5	0,26	90	35	0.40 – 0.50	0,45	0,45
8,6	100	60	0,2	15	20	0.70 – 0.80	0,76	0,75
6,15	100	62	0,2	15	40	0.45 – 0.55	0,52	0,52
6,3	100	65	0,18	20	20	0.50 – 0.60	0,56	0,54
4,7	100	65	0,18	20	60	0.30 – 0.45	0,36	0,38
4,8	15	55	0,21	62	32	0.40 – 0.50	0,42	0,41
5	37	15	0,18	17	25	0.40 – 0.50	0,44	0,42
6,8	70	28	0,16	90	52	0.50 – 0.70	0,55	0,54
7,3	41	37	0,2	50	65	0.60 – 0.70	0,62	0,62
6,7	80	55	0,12	56	62	0.50 – 0.60	0,52	0,54
4,8	100	20	0,23	10	80	0.35 – 0.50	0,39	0,38
3,3	50	90	0,3	50	85	0.25 – 0.40	0,24	0,24
2,1	20	55	0,23	70	40	0.15 – 0.20	0,16	0,15
8,6	100	60	0,2	15	40	0.70 – 0.80	0,74	0,74
6,15	100	62	0,2	15	60	0.45 – 0.55	0,48	0,51
6,3	100	65	0,18	20	40	0.50 – 0.60	0,53	0,52
7,2	70	70	0,19	15	60	0.60 – 0.70	0,63	0,62
1,7	35	30	0,16	74	34	0.10 – 0.20	0,16	0,15
1,3	72	35	0,15	70	33	0.08 – 0.15	0,12	0,13
2,25	62	21	0,31	85	64	0.20 – 0.25	0,17	0,18
4,5	32	75	0,19	90	80	0.35 – 0.50	0,35	0,36
7,5	75	25	0,18	71	67	0.60 – 0.70	0,64	0,63
2,6	65	50	0,16	60	55	0.20 – 0.30	0,22	0,20
5	70	20	0,17	100	25	0.40 – 0.50	0,39	0,40
0,7	100	75	0,18	20	10	0.05 – 0.10	0,06	0,06
8,6	100	60	0,2	15	60	0.70 – 0.80	0,70	0,70
4,45	100	65	0,2	15	20	0.25 – 0.40	0,40	0,38
6,3	100	65	0,18	20	60	0.50 – 0.60	0,51	0,52
5,6	100	75	0,2	25	100	0.45 – 0.55	0,46	0,44
2,9	48	25	0,24	51	68	0.20 – 0.40	0,22	0,21
2,85	56	75	0,29	40	40	0.20 – 0.30	0,20	0,22
5,5	53	98	0,18	100	35	0.40 – 0.50	0,42	0,43
5,2	78	20	0,17	38	129	0.40 – 0.55	0,41	0,41
8,2	15	10	0,2	100	115	0.70 – 0.80	0,67	0,66
8,3	100	30	0,17	80	40	0.70 – 0.80	0,71	0,71
4,3	90	10	0,13	10	120	0.35 – 0.40	0,33	0,33
8,6	100	62	0,2	15	80	0.70 – 0.80	0,67	0,68
4,45	100	65	0,2	15	40	0.25 – 0.40	0,36	0,38

1) Experimental
2) On suggested models

The comparison of the model with the experimental results of the AF evaluation shown in Table 9.19 testifies the adequacy of the obtained model for practical use.

9.4.4 Example and Comparison with the Technique in Use Now

The case of the run over a pedestrian by the automobile "GAZ-24" is discussed. The traffic accident protocol information:

- type of road surface (D_1) – asphalt;
- condition of road surface (D_2) – covered by dirt;
- type of tires (T) – low pressure;
- tires slip degree (S) – rolling with slip;
- wear of tires (H) – in admissible limits (about 50%);
- pressure in tires (P) – normal (0.2MPa);
- load on a wheel (N) – low (about 10%);
- car velocity (V) – 55 km/h.

We consider the horizontal road strip. After the run-over and up to the full stoppage automobile GAZ – 24 in the state of employed brakes run the distance of 9.2 m. From the moment when the motion barrier occurred and up to the moment of the pedestrian run-over he walked 5 m with the velocity of 4.5 km/h. The pedestrian was knocked-down by the front part of the car.

The results of the AF calculations as follows:

a) using the conventional technique [30]: $\varphi = 0.25 - 0.4$;

b) using the suggested technique: $\varphi = 0.35$.

The results using all the known information are presented in Table 9.20. The last column of this table shows the significance of the exact AF knowledge for the relevant decision making.

Table 9.20. Calculation results for decision making

Technique	Adhesion factor	Car braking distance	Distance up to the obstacle at the moment of dangerous situation	Decision making about the possibility to avoid collision
In use	0.25	68.8 m	46.2 m	Impossible
	0.4	51.0 m	55.3 m	Possible
Suggested	0.35	55.3 m	53.3 m	Impossible

9.5 Innovative Projects Creditworthiness Evaluation

Estimation of innovation project quality level is an important task of any investment firm. An instant and correct solution of this problem that can generally be accomplished only by specialist economists allows one to manage financial resources optimally. In this connection it is necessary to design computer based information system providing intelligent support for investment firm's personnel in decision making.

The expert system suggested here was developed to the order of Ukraine Innovation Fund. Expert IF-THEN rules were obtained from a group of analysts under the leadership of Vinnitsa Chapter of Ukraine Innovation Fund Director Prof. N. Petrenko.

This chapter is written on the basis of the work [33].

9.5.1 Types of Decisions and Partial Figures of Quality

Innovation project quality estimation is used for making one of the following decisions: d_1 - to finance, d_2 - to finance after retrofit, d_3 - to finance when means are available, d_4 - to reject.

Let us use letter D to designate the integral figure of innovation project quality. To estimate this figure we will use the following information:

X - level of the enterprise-applicant, which is estimated using the following partial figures: x_1 - level of enterprise leader, x_2 - enterprise assets, x_3 - enterprise liabilities, x_4 - enterprise balance profit, x_5 - enterprise debt receivables, x_6 - enterprise indebtedness under credits. To estimate enterprise leader level we take into account the following figures: a_1 - sociability, a_2 - fidelity, a_3 - education, a_4 - leader work experience, a_5 - comfort;

Y - technical economic level of the project, in point for which estimation the following partial figures are used: y_1 - project scale, y_2 - project novelty, y_3 - development trend priority, y_4 - degree of perfection, y_5 - juridical protection, y_6 - ecology level;

V - expected sales level;

Z - financial level of the enterprise-applicant, which is estimated using the following partial figures: z_1 - ratio of internal funds to innovation funds, z_2 - innovation fund means return.

The task of estimation is in bringing one of the decisions $d_1 \div d_4$ into correspondence with some innovation project with known partial figures.

9.5.2 *Fuzzy Knowledge Bases*

A hierarchy diagram of accepted innovation project quality figures is shown in Fig. 9.17 in the form of a fuzzy logic inference tree, to which this system of relations corresponds:

$$D = f_D(X,Y,V,Z) \ , \tag{9.17}$$

$$X = f_X(x_1,x_2,x_3,x_4,x_5,x_6) \ , \tag{9.18}$$

$$x_1 = f_{x_1}(a_1,a_2,a_3,a_4,a_5) \ , \tag{9.19}$$

$$Y = f_Y(y_1,y_2,y_3,y_4,y_5,y_6) \ , \tag{9.20}$$

$$Z = f_Z(z_1,z_2) \ . \tag{9.21}$$

Partial figures in point $x_1 \div x_6$, $a_1 \div a_5$, $y_1 \div y_6$, V, z_1 and z_2, and also enlarged figures X, Y, Z are considered as linguistic variables. To estimate the introduced linguistic variables we will use the unitary scale of qualitative terms: *vL* – very Low, *L* - Low, *lA* – lower than average, *A* - average, *hA* – higher than average, *H* - High, *vH* – very high.

Each of these terms represents some fuzzy set preset using the following membership function model. Using introduced quality terms let us represent relations (9.17) - (9.21), in the knowledge base form by Tables 9.21-9.25.

Table 9.21. Knowledge about relation (9.17)

X	Y	V	Z	D
H	*H*	*H*	*H*	
hA	*H*	*H*	*H*	d_1
H	*H*	*H*	*hA*	
hA	*hA*	*hA*	*hA*	
hA	*H*	*H*	*hA*	d_2
hA	*hA*	*H*	*A*	
H	*H*	*A*	*A*	
H	*A*	*A*	*A*	d_3
H	*A*	*hA*	*A*	
L	*L*	*L*	*L*	d_4
A	*L*	*L*	*L*	

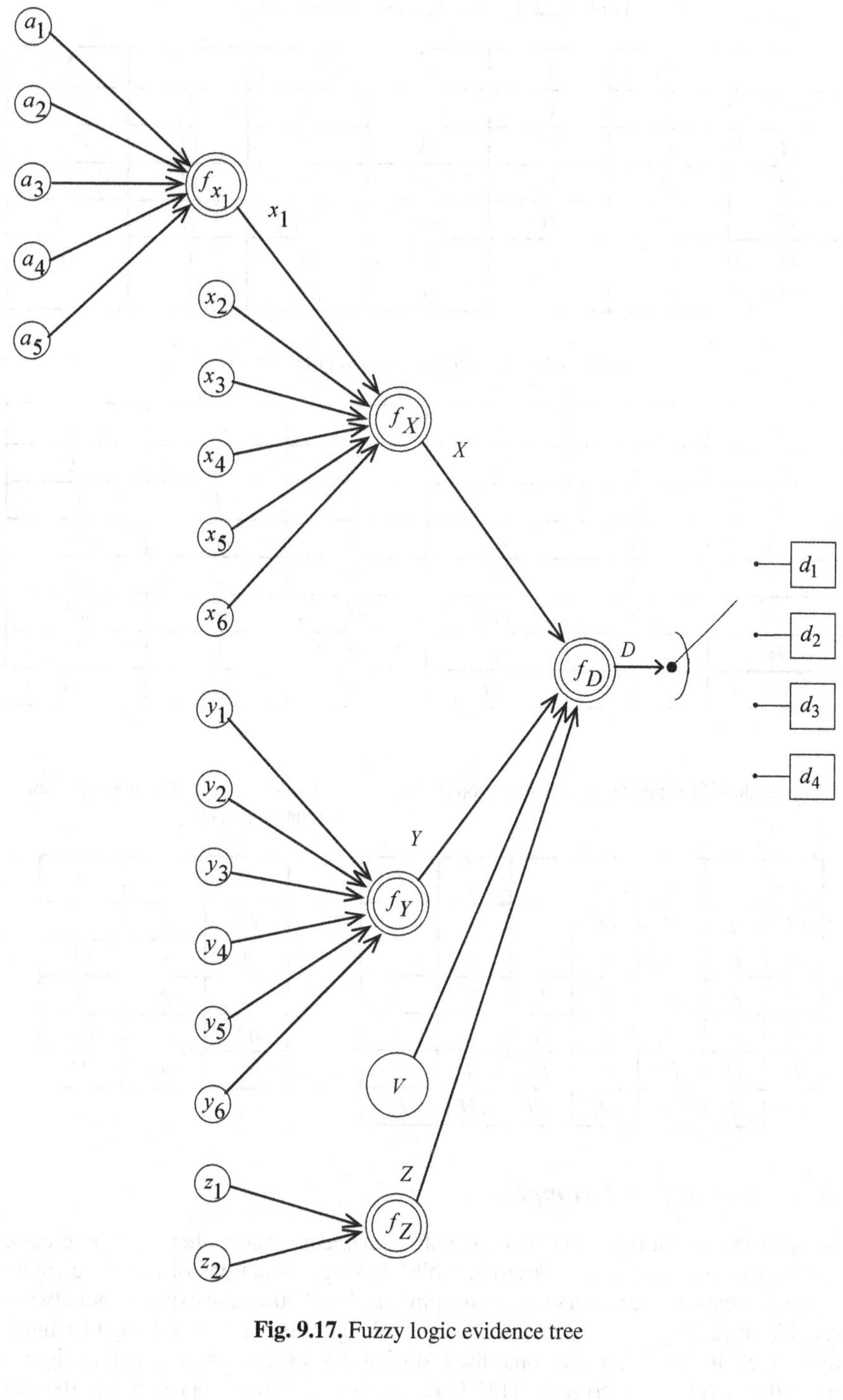

Fig. 9.17. Fuzzy logic evidence tree

Table 9.22. Knowledge about relation (9.18)

x_1	x_2	x_3	x_4	x_5	x_6	X
L	*L*	*L*	*L*	*L*	*H*	*L*
H	*H*	*H*	*lA*	*lA*	*lA*	*lA*
H	*H*	*H*	*A*	*A*	*lA*	*A*
H	*H*	*H*	*hA*	*hA*	*A*	*hA*
H	*H*	*H*	*H*	*H*	*L*	*H*
H	*H*	*H*	*A*	*hA*	*L*	

Table 9.23. Knowledge about relation (9.19)

a_1	a_2	a_3	a_4	a_5	x_1
vL	*vL*	*vL*	*vL*	*vL*	*vL*
L	*L*	*L*	*L*	*L*	*L*
lA	*A*	*lA*	*A*	*lA*	*lA*
A	*A*	*A*	*A*	*A*	*A*
hA	*H*	*hA*	*H*	*A*	*hA*
H	*H*	*H*	*H*	*H*	*H*
vH	*vH*	*vH*	*vH*	*vH*	*vH*

Table 9.24. Knowledge about relation (9.20)

y_1	y_2	y_3	y_4	y_5	y_6	Y
vL	*vL*	*vL*	*vL*	*vL*	*vL*	*L*
L	*L*	*L*	*L*	*L*	*L*	
A	*A*	*L*	*L*	*L*	*A*	*lA*
A	*A*	*A*	*A*	*A*	*A*	*A*
H	*H*	*H*	*H*	*H*	*H*	*hA*
vH	*vH*	*vH*	*vH*	*vH*	*vH*	*H*

Table 9.25. Knowledge about relation (9.21)

z_1	z_2	Z
vL	*vL*	*L*
A	*L*	*lA*
A	*A*	*A*
hA	*H*	*hA*
vH	*vH*	*H*

9.5.3 Evaluation Examples

Some of the partial figures have a qualitative character; that is, they have no precise quantitative measurement. Therefore, while making estimations of the same figure by several experts there can be various opinions. In addition, the expert is not always capable of making an estimation of the partial figure using words though he intuitively feels its level. To overcome these difficulties we can estimate partial figures using the thermometer principle [14]. Convenience of such an approach is in the fact that various sense partial figures are defined as linguistic variables given on the unitary universal set $U = [0, 100]$, which is the scale of a thermometer. Parameters (b) and (c) of membership functions are introduced in Table 9.26.

Table 9.26. Membership functions parameters

Term	vL	L	lA	A	hA	H	vH
b	0.0	16.7	33.3	50.0	66.7	83.3	100
c	15	15	15	15	15	15	15

Examples of three innovation projects' estimations by the suggested fuzzy model are represented in Table 9.27. Results of decision making are well in accordance with expert assessments of quality.

Table 9.27. Examples of innovation projects quality estimation

Partial figure	Project 1	Project 2	Project 3
a_1			
a_2			
a_3			
a_4			
a_5			
x_2			
x_3			
x_4			
x_5			
x_6			
y_1			
y_2			
y_3			
y_4			
y_5			
y_6			
z_1			
z_2			
V			
Decision	To finance with means available	To finance	To finance after retrofit

9.6 System Reliability Analysis

Probabilistic models of reliability of technological processes and systems were considered in [34 – 39]. The application of these models presumes the availability of statistical data on probabilities of correct execution of elements of algorithmic process, i. e., technological operations. To take into account influencing factors, it is expedient to use experiment planning theory and regression models. It is very difficult to provide the equal conditions of experiment reiteration necessary for the statistical methods' correct application while evaluating the probabilities of correct (noncorrect) performance of the system and its elements' functioning process. From the other side an experiment and statistical data processing is too complicated because of the many factors influencing the reliability such as environment task conditions; psychological stress and the degree of fatigue of an operator etc. It is relatively easy and natural to take into account such a factor's influence linguistically, e.g., "if the degree of fatigue of an operator is low, environment task conditions is good, psychological stress is low then human reliability is high".

The active research on fuzzy logic using in reliability theory began in the 10th decade of the last century. The first approaches to the fuzzy reliability theory creation have been proposed in monographs [40, 41]. The overwhelming majority of known works uses for the system reliability analysis the descriptive possibilities of fuzzy logic in combination with probability theory and descriptive possibilities of Boolean algebra [42 – 45]. In this chapter, we consider basic principles, mathematical models and the example of application of the new method of complex systems reliability analysis on the basis of algebra of algorithms [46, 47] and fuzzy logic [48]. Here we present the results of simulation of the bioconversion technological process reliability. In reliability modeling of a technological system it is necessary to take into account not only the structure of the technological process but also the influencing factors, connected with the quality of raw material, the technological equipment and the operator, controlling the process.

This chapter is written on the basis of the works [49, 50].

9.6.1 Basic Principles

The approach proposed in [49, 50] is based on the following principles:

1. *Principle of algorithmization*

This principle, adopted from theory of reliability of man-machine systems [35], envisages construction of the reliability model on the basis of the algorithmic description of the events, connected with the occurring, detecting and removal of the failures (faults, defects, errors) in the system. To depict the algorithm, we use graph-schemes or the language of V.M. Glushkov's algorithmic algebra [46, 47], in which any regular algorithm can be built with the help of the three structures:

a) *linear* (B *-structure*): $A_1A_2 = B$, producing the operator B, which is equivalent to the consecutive performance of the operators A_1 and A_2;

b) *alternative* (C*-structure*): $(A_1 \underset{\alpha}{\vee} A_2) = C$, producing the operator C, such that

$$C = \begin{cases} A_1, \ if \ condition \ \alpha \ is \ true \ (\alpha = 1) \\ A_2, \ if \ condition \ \alpha \ is \ fault \ (\alpha = 0) \end{cases};$$

c) *iterative* (D*-structure*): $\underset{\alpha}{\{A\}} = D$, producing the operator D, which is equivalent to repeated implementation of operator A till the condition α has become true $(\alpha = 1)$.

2. *Principle of fuzzy correctness*

Conception of the crisp boarder between "correct" (1) and "noncorrect" (0) results of the system and its elements functioning lacking underlies this principle. For the formal evaluation of the level of operator A correct performance, we use the multidimensional membership function $\mu_A^1(x_1, x_2, ..., x_n)$, which depends on the measured parameters (input variables). Correctness of each of the parameters is defined by the membership function $\mu^1(x_i)$, which can be interpreted as a parameter x_i values' correctness distribution.

3. *Principle of linguistic evaluation of control quality*

The system functioning process control is accomplished with the help of the checking and correction operations. If the checking operation is performed by a human, then the 1st type error (false alarm or rejection of "good" result) can be connected with the level of "objectivity – preconception" of the inspector, and the 2nd type error (acceptance of defective goods), – with the level of "vigilance – negligence" of the inspector.

This principle envisages the possibility of evaluation of the checking and correction operations using verbal terms: low (average, high) tendency of man-operator to commit the 1st and 2nd type errors; low (average, high) repair quality, etc. Membership functions, necessary for these terms formalization, are formed with the help of extension-compression operations [48], which underlie the idea of Soft Computing – computing with words.

4. *Principle of fuzzy identification*

This principle emphasizes that the problem of system reliability evaluation amounts to the problem of "multiple inputs – single output" object identification with the help of fuzzy knowledge bases [14].

Inputs of the object are the measured parameters of the quality of raw material, equipment and man-operator. Output of the object is the discrete double-throw switch: 1 - correct; 0 - noncorrect. Because of the lack of the crisp border between 1 and 0 results, the degrees of membership of the vector of input parameters to the levels 1 and 0 are calculated during system reliability modeling.

Fuzzy knowledge bases, i.e., IF-THEN rules, necessary for solving the identification problem, are determined by B-, C- and D- structures, from which the algorithmic model of reliability is built.

9.6.2 Fuzzy-Algorithmic Elements

The fuzzy-algorithmic model of system reliability is built using the following elements (Fig. 9.18):

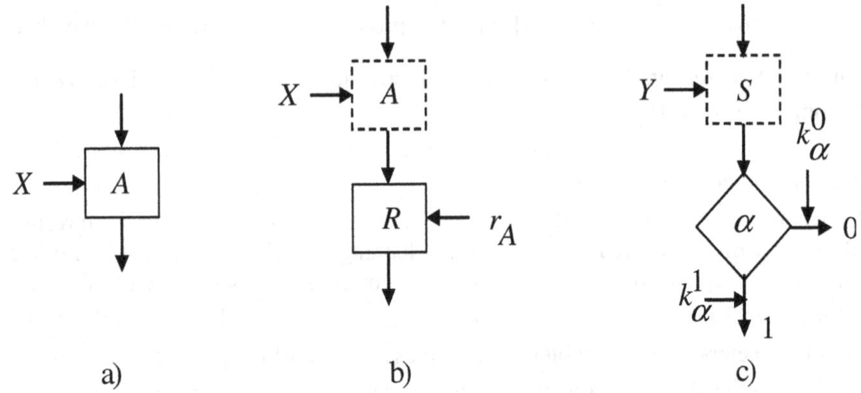

Fig. 9.18. Elements of reliability model

Working operator A (Fig. 9.18a) is the element of the model, describing occurring abnormalities in the system functioning process. Quality of the working operator A performance depends on the vector of measured parameters $\mathbf{X} = (x_1, x_2, ..., x_n)$, where $x_i = x_i(t)$, i.e., parameters values depend on time.

Correctness of the working operator A performance is defined by the formula:

$$\mu_A^1(\mathbf{X}) = \prod_{i=1}^{n} \mu^1(x_i), \tag{9.22}$$

where $\mu_A^1(\mathbf{X})$ is the multidimensional membership function of the vector of parameters $\mathbf{X}$ to the term "correct performance of the operator A",

$\mu^1(x_i)$ is the membership function, which describes the distribution of parameter x_i, $i = 1, 2, ..., n$, values' correctness.

Correction operator R (Fig. 9.18b) is the element of the model, which describes removal of abnormalities, occurred while performing the working operator A.

Different kinds of repair and updating included in the system functioning algorithm can be described by the correcting operator R.

Correctness of the correction operator R performance is defined by the formula:

$$\mu_R^1(\mathbf{X}) = 1 - [1 - \mu_A^1(\mathbf{X})]^{r_A}, \tag{9.23}$$

where $\mu_A^1(\mathbf{X})$ is defined by formula (9.22),

r_A is the parameter, which characterizes the quality of correction:

$r_A = 1, 3, 5, 7, 9$, if the quality of correction is low (1), lower than average (3), average (5), higher than average (7), high (9).

If, for example, a working operator A has correctness $\mu_A^1(\mathbf{X}) = 0.5$, then the correctness of the correcting operator R is increased with the growth of parameter r_A:

r_A	1	3	5	7	9
$\mu_R^1(X)$	0.5	0.875	0.967	0.992	0.9998

Correctness of algorithm AR, i.e., "work (A) – correction (R)", performance is defined by formula:

$$\mu_{AR}^1(\mathbf{X}) = \mu_A^1(\mathbf{X}) + [1 - \mu_A^1(\mathbf{X})] \cdot \mu_R^1(\mathbf{X}), \tag{9.24}$$

from which it is shown, that if $\mu_R^1(\mathbf{X}) = 1$, then $\mu_{AR}^1(\mathbf{X}) = 1$.

Logical condition α (Fig. 9.18c) is the element of the model, which describes correctness checking for the vector of parameters $\mathbf{Y} = (y_1, y_2, ..., y_l)$. This vector of parameters can correspond to the condition of the system components: the raw material, the equipment, the man-operator or the results of functioning process implementation. In particular, the diagnostic and functional checking which are used in reliability theory of man-machine systems [35] can be described by logical condition α.

While performing condition α the two results are possible:

$\alpha = 1$, if all the parameters of vector $\mathbf{Y}$ are correct,
$\alpha = 0$, if at least one of the parameters of vector $\mathbf{Y}$ is noncorrect.

Correctness of condition α performance is defined as follows:

$\mu_\alpha^{11}(\mathbf{Y})$ is the possibility distribution of the condition α performance for result 1, i.e., without 1st type errors, when real correctness (1) is subjectively recognized as true (1),

$\mu_\alpha^{00}(\mathbf{Y})$ is the possibility distribution of the condition α performance for result 0, i.e., without 2nd type errors, when false (0) is subjectively recognized as false (0).

These distributions are defined by the formulae:

$$\mu_\alpha^{11}(\mathbf{Y}) = [\mu_\alpha^1(\mathbf{Y})]^{k_\alpha^1}, \tag{9.25}$$

$$\mu_\alpha^{00}(\mathbf{Y}) = [1 - \mu_\alpha^1(\mathbf{Y})]^{k_\alpha^0}, \tag{9.26}$$

$$\mu_\alpha^1(\mathbf{Y}) = \prod_{i=1}^{l} \mu^1(y_i), \tag{9.27}$$

where $\mu^1(y_i)$ is the correctness distribution of the parameter y_i, $i = 1, 2, ..., l$.

k_α^1 and k_α^0 are the coefficients, describing the tendency of the checking operation α to the 1st and 2nd type errors, respectively ($k_\alpha^1 \ge 1, k_\alpha^0 \ge 1$).

If $k_\alpha^1 = 1$ and $k_\alpha^0 = 1$, then the 1st and 2nd type errors are absent. The increase of these coefficients results in compression of the membership functions in (9.25) and (9.26), and, respectively, lowering down of the level of correctness of checking condition α performance for results 1 and 0. This is equivalent to the growth of the 1st and 2nd type errors levels.

For calculations on the basis of linguistic assessments one can use:

$k_\alpha^1 = 1$, if the 1st type errors are absent (if the inspector is objective),

$k_\alpha^1 = 2$, for small tendency to the 1st type errors (if the inspector is somewhat preconceived),

$k_\alpha^1 = 3$, for sufficient tendency to the 1st type errors (if the inspector is preconceived),

i.e. with the growth of the inspector preconception (or with the lowering down of his/her objectivity) the possibility of the 1st type error is increased.

For the 2nd type errors:

$k_\alpha^0 = 1$, if the 2nd type errors are absent (if the inspector is vigilant),

$k_\alpha^0 = 2$, for small tendency to the 2nd type errors (if the inspector is somewhat negligent),

$k_\alpha^0 = 3$, for sufficient tendency to the 2nd type errors (if the inspector is negligent),

i.e., with the lowering down of the inspector vigilance (or with the growth of his/her negligence) the possibility of the 2nd type error is increased.

9.6.3 *Fuzzy-Algorithmic Structures*

Each of the algorithmic structures produces the mathematical model, which allows us to calculate the correctness of this structure implementation depending on the correctness of the included operators and conditions implementation. Such models are obtained in [50] on the basis of the graphs of events, taking place while performing each of the structures (Fig. 9.19). Necessary formulae are given below.

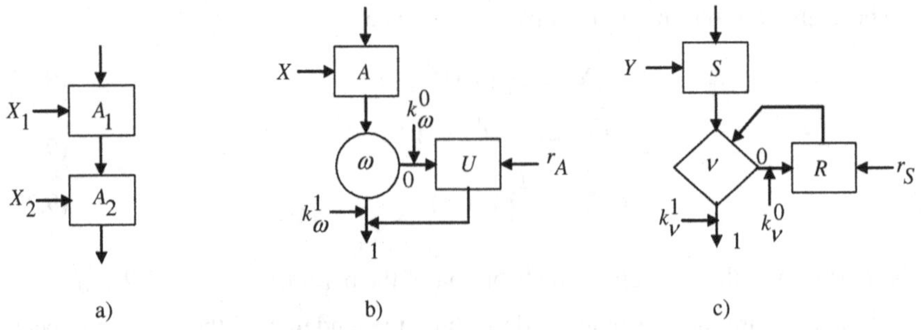

Fig. 9.19. Algorithmic structures

Linear structure (Fig. 9.19a) is given by algorithm

$$B = A_1 A_2 , \tag{9.28}$$

in which the working operators A_1 and A_2 depend on the vectors of parameters $\mathbf{X}_1 = (x_1^1, x_2^1, ..., x_n^1)$ and $\mathbf{X}_2 = (x_1^2, x_2^2, ..., x_n^2)$, respectively.

Fuzzy correctness of the equivalent operator B performance in (9.28) is defined by the formula:

$$\mu_B^1(\mathbf{X}_1, \mathbf{X}_2) = \mu^1(\mathbf{X}_1) \cdot \mu^1(\mathbf{X}_2), \tag{9.29}$$

where $\mu^1(\mathbf{X}_1) = \prod_{i=1}^{n_1} \mu^1(x_i^1)$, $\mu^1(\mathbf{X}_2) = \prod_{j=1}^{n_2} \mu^1(x_j^2)$.

Alternative structure (Fig. 9.19b) is given by algorithm

$$C = A \underset{\omega}{(} E \vee U), \tag{9.30}$$

in which ω is the logical condition, verifying during the checking of the correctness of the working operator A implementation, where

$$\omega = \begin{cases} 1, \textit{ if vector of parameters } \mathbf{X} \textit{ is normal}, \\ 0, \textit{ otherwise}. \end{cases}$$

E is the identical operator, corresponding to the checking operation ω results fixation,

U is the operator correcting the parameters of the working operator A,

k_ω^1 , k_ω^0 and r_A are the parameters of condition ω and operator U implementation quality, respectively.

Structure (9.30) corresponds to the process "work – checking – correction without feedback" [35].

Fuzzy correctness of the equivalent operator C performance in (9.30) is defined by the formula:

$$\mu_C^1(\mathbf{X}, k_\omega^1, k_\omega^0, r_A) = \mu_\omega^1 \cdot \mu_\omega^{11} + [\mu_\omega^1(1 - \mu_\omega^{11}) + (1 - \mu_\omega^1)\mu_\omega^{00}]\mu_U^1 , \tag{9.31}$$

where $\mu_\omega^1 = \prod_{i=1}^{n} \mu^1(x_i)$, $\mu_\omega^{11} = (\mu_A^1)^{k_\omega^1}$, $\mu_\omega^{00} = (1 - \mu_A^0)^{k_\omega^0}$,

$\mu_U^1 = 1 - (1 - \mu_A^1)^{r_A}$,

$\mu^1(x_i)$ is the correctness distribution of the parameter x_i $i = 1, 2, ..., n$,

k_ω^1 , k_ω^0 and r_A are the numbers (1,2,3,…), which define the quality of checking ω and correcting U operators, respectively.

Iterative structure (Fig. 9.19c) is given by algorithm

$$D = \underset{\nu}{S\{R\}}, \tag{9.32}$$

in which ν is the logical condition, verifying during the checking of the parameters of the working operator S, where

$$\nu = \begin{cases} 1, \textit{ if vector of parameters } \mathbf{Y} \textit{ is normal,} \\ 0, \textit{ otherwise.} \end{cases}$$

R is the operator correcting parameters of the working operator S,

k_ν^1, k_ν^0 and r_S are the parameters of condition ν and operator R implementation quality, respectively.

Structure (9.32) describes the process "diagnostics – repair with feedback" [35] when the equipment is diagnosed.

In the general case, operator S corresponds to the equipment functioning, the raw material preparation or the man-operator work.

Fuzzy correctness of the equivalent operator D performance in (9.32) is defined by the formula:

$$\mu_D^1(\mathbf{Y}) = a + ba_1 \cdot \frac{1}{1-b_1}, \tag{9.33}$$

where $a = \mu_\nu^1 \cdot \mu_\nu^{11}$, $a_1 = \mu_R^1 \cdot \mu_\nu^{11}$,

$b = \mu_\nu^1\left(1-\mu_\nu^{11}\right)+\left(1-\mu_\nu^1\right)\mu_\nu^{00}$, $b_1 = \mu_R^1\left(1-\mu_\nu^{11}\right)+\left(1-\mu_R^1\right)\cdot\mu_\nu^{00}$,

$\mu_\nu^1 = \prod_{l=1}^{m} \mu^1(y_l)$, $\mu_\nu^{11} = \left(\mu_\nu^1\right)^{k_\nu^1}$, $\mu_\nu^{00} = \left(1-\mu_\nu^1\right)^{k_\nu^0}$,

$\mu_R^1 = 1-\left(1-\mu_\nu^1\right)^{r_\nu}$,

$\mu^1(y_l)$ is the correctness distribution of the parameter y_l, $l = 1, 2, ..., m$,

k_ν^1, k_ν^0 and r_S are the numbers (1,2,3,...), which define the quality of checking ν and correcting R operators, respectively.

9.6.4 Example of Technological System Reliability Analysis

Let us consider the bioconversion technological process (BCTP), the algorithmic model of which (Fig. 9.20) is defined by the formula:

$$F = \underset{\nu}{S\{R\}} A(E \underset{\omega}{\vee} U), \tag{9.34}$$

where S is the working operator, corresponding to the raw material preparation;

ν is the raw material parameters checking (ho - homogeneity, Ph - hydrogen factor, hu - humidity);

R is the operator of the raw material parameters correction;

A is the working operator, corresponding to the process performance;

ω is the process parameters checking (V – rate of mixing, t^0 -temperature);

E is the identical operator, corresponding to the checking operation ω results fixation;

k_V^1, k_V^0, r_S, k_ω^1, k_ω^0 and r_A are the parameters of the process control quality, shown in Fig. 9.20.

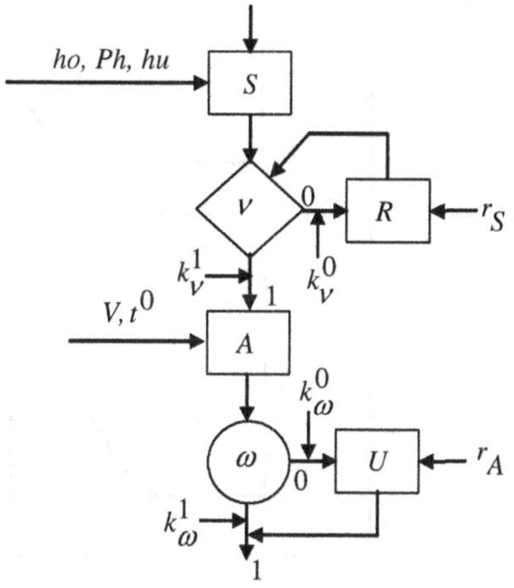

Fig. 9.20. Algorithmic model of the bioconversion process reliability

The parameters correctness distributions are presented in Table 9.28.

Algorithm (9.34) is presented as follows

$$F = D \cdot C\,, \quad D = \underset{V}{S}\{R\}, \quad C = \underset{\omega}{A}\,(E \vee U)\,.$$

Therefore, the problem of reliability analysis is reduced to the consecutive application of the models of B-, C- and D- structures:

$$\mu_F^1(hu, Ph, ho, V, t) = \mu_D^1(hu, Ph, ho) \cdot \mu_C^1(V, t)\,,$$

where $\mu_F^1(\ldots)$ is the process (9.34) performance correctness distribution;

$\mu_D^1(\ldots)$ is the operator D performance correctness distribution calculated by formula (9.33) for $\mathbf{Y} = (hu, Ph, ho)$,

$\mu_C^1(\ldots)$ is the operator C performance correctness distribution calculated by formula (9.31) for $\mathbf{X} = (V, t)$.

The tree of inference, which defines the interconnection of the fuzzy-algorithmic structures in identifying the process (9.34) reliability level, is shown in Fig. 9.21, where *double circles* are the models of B-, C- and D- structures;

single circles are the operators and conditions, appearing in algorithm (9.34);

output arrow is the process performance correctness level, which is defined by the membership function μ_F^1;

input arrows are the variables, influencing the correctness level μ_F^1.

Table 9.28. Parameters correctness distributions

Parameter	Membership function
Homogeneity (*ho, %*)	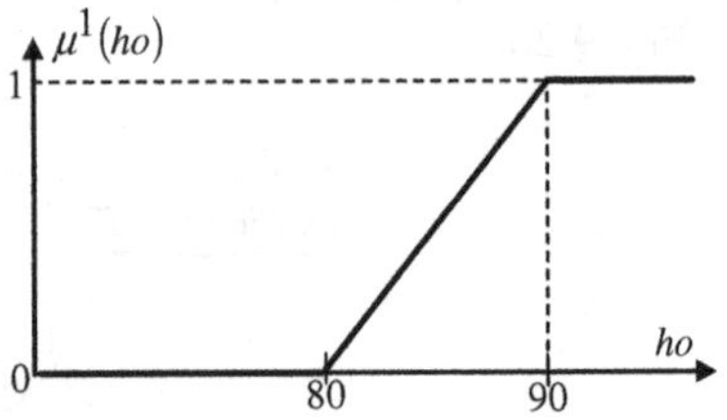
Humidity (*hu, %*)	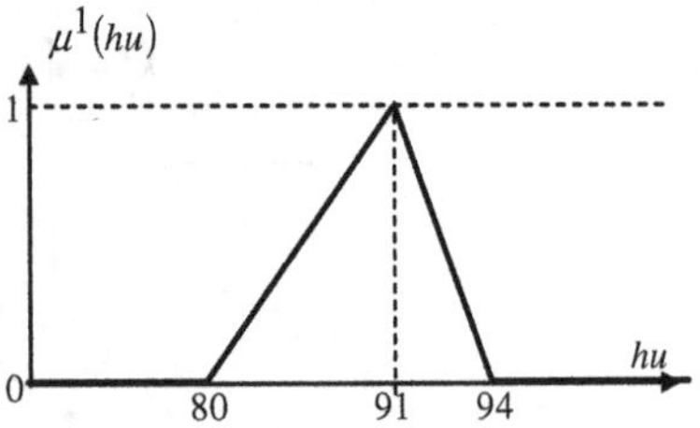
Hydrogen factor (*Ph,* c.u.)	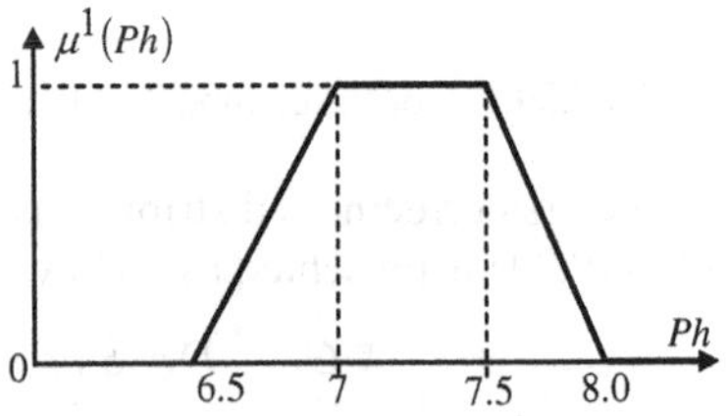
Rate of mixing (*V*, rpm)	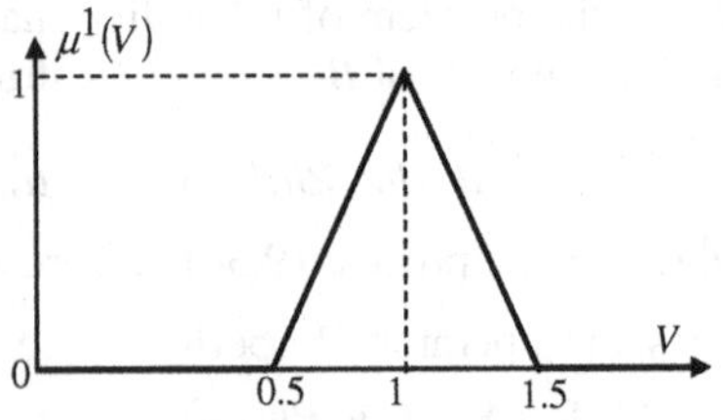
Temperature (t^0 C)	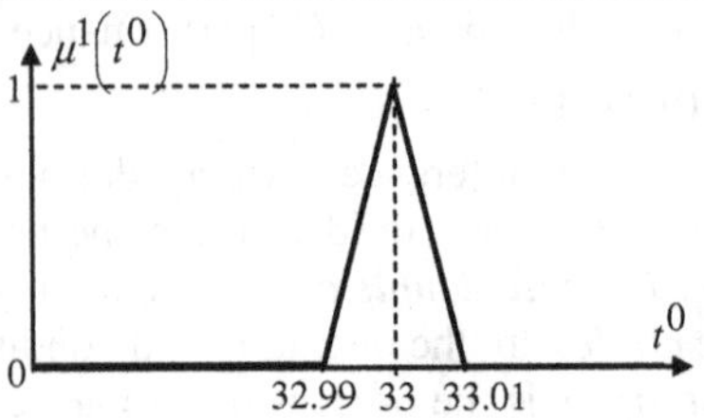

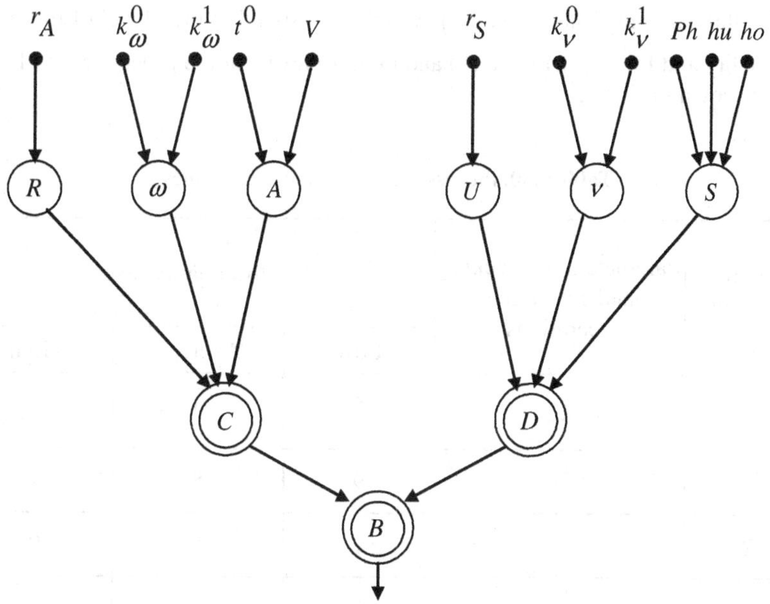

Fig. 9.21. Tree of inference

The aim of simulation consisted of the construction of three-dimensional correctness distributions $\mu_F^1(V, t^0)$ for different combinations of the raw material quality levels (Table 9.29) and the process control quality levels (Table 9.30). The nine three-dimensional distributions were obtained.

Table 9.29. Values of the raw material parameters

Raw material parameters	Quality levels		
	Low	Average	High
ho (%)	83	87	91
hu (%)	83	87	91
Ph (c.u.)	6.6	6.8	7.1

The three distributions $\mu_F^1\left(V,t^0\right)$, which correspond to high level of raw material quality and low (a), average (b) and high (c) levels of the process control quality are shown in Fig. 9.22.

Table 9.30. Processes control parameters values

Control element	Parameters of checking and correcting operations	Control quality levels		
		Low	Average	High
V	k_V^1	5	3	1
	k_V^0	9	5	1
R	r_S	1	5	9
ω	k_ω^1	5	3	1
	k_ω^0	9	5	1
U	r_A	1	5	9

The correctness distributions $\mu_F^1\left(V,t^0\right)$ allow us to obtain the regions of parameters change (V and t^0), which provide the required level of the process performance correctness. Let call them zones (cross-sections) of μ - working capacity, $\mu \in [0, 1]$. Such zones for levels $\mu_F^1(V,t^0) = 0.9, 0.8, 0.7, 0.6$ are presented in Table 9.31. The obtained zones of μ - working capacity provide the possibility of optimization of the system reliability with taking into account the restrictions of the region of permissible parameters change [51].

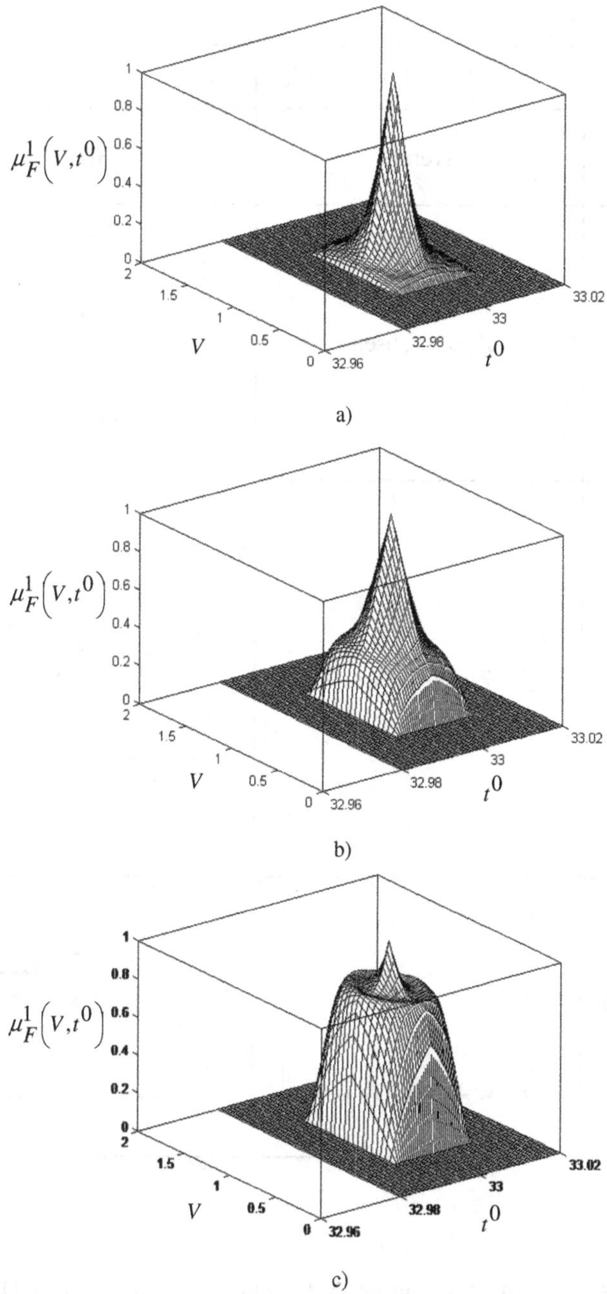

Fig. 9.22. Process performance correctness distributions for low (a), average (b) and high (c) control quality levels

Table 9.31. Working capacity zones for process performance (0.6, 0.7, 0.8, 0.9)-correctness levels

Control quality	Raw material quality		
	Low	Average	High
Low	Zone is absent	Zone is absent	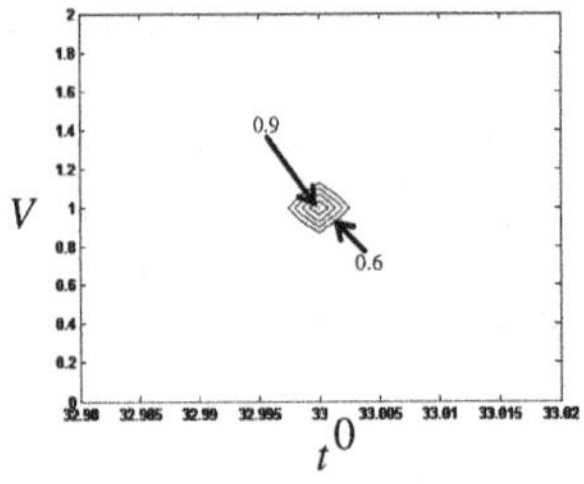
Average	Zone is absent	Zone is absent	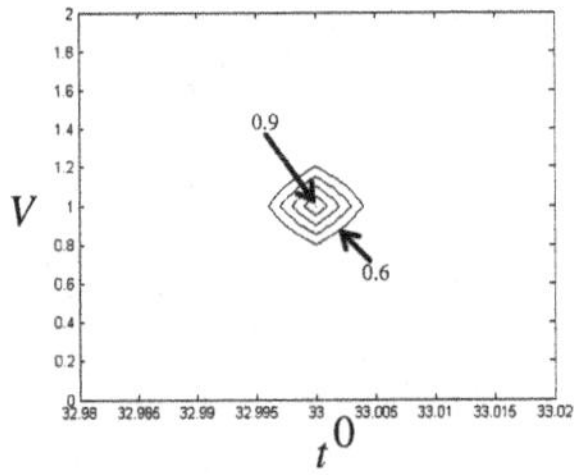
High	Zone is absent	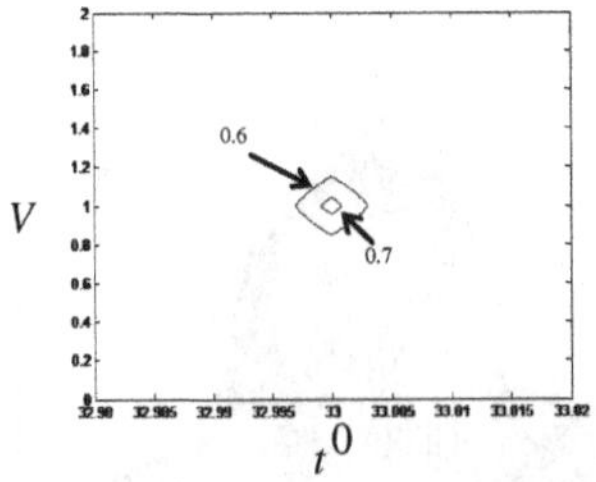	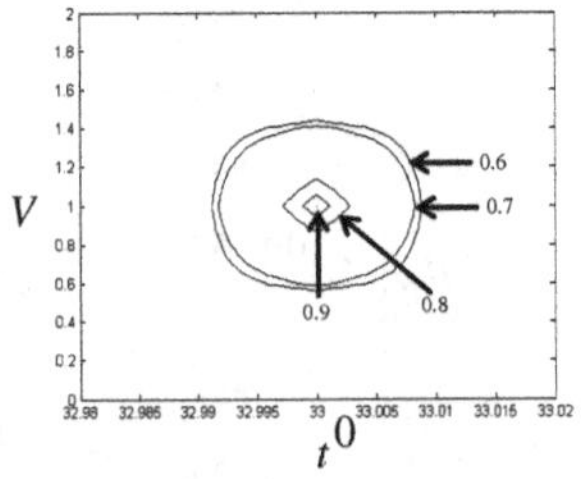

References

1. Butenin, N.V., Neimark, Y.I., Fufaev, N.A.: Introduction to Nonlinear Oscillation Theory, p. 384. Nauka, Moscow (1987) (in Russian)
2. Aleksakov, G.N., Gavrilin, V.V., Fedorov, V.A.: Structural Models of Dynamic processes, p. 62. MIFI, Moscow (1989) (in Russian)

3. Chemodanov, B.K. (ed.): Mathematical principles of Automatic Control Theory, vol. I, p. 366. Vysshaja Shkola, Moscow (1977) (in Russian)
4. Rotshtein, A.P., Shtovba, S.D.: Managing a Dynamic System by means of a Fuzzy Knowledge Base. Automatic Control and Computer Sciences 35(2), 16–22 (2001)
5. Nakano, E.: Introduction to Robotics, p. 334. Mir, Moscow (1988) (in Russian)
6. Gorelik, V.A., Ushakov, I.A.: Operations Research, p. 288. Machine building, Moscow (1986) (in Russian)
7. Ryzgikov, Y.I.: Inventory Control, p. 364. Nauka, Moscow (1969) (in Russian)
8. Rubalskiy, G.B.: Inventory Control under Random Demand, p. 160. Sov. Radio, Moscow (1977) (in Russian)
9. Petrovich, D., Sweeney, E.: Knowledge-based Approach to Treating Uncertainty in Inventory Control. Computer Integrated Manufacturing Systems 7(3), 147–152 (1994)
10. Cox, D.E.: Fuzzy Logic for Business and Industry, p. 280. Charles River Media, Inc., Rockland (1995)
11. Bojadziev, G., Bojadziev, M.: Fuzzy Logic for Business, Finance and Management, p. 252. World Scientific Publishing (1997)
12. Venkatraman, R., Venkatraman, S.: Rule-based System Application for a Technical Problem in Inventory Issue. Advanced Engineering Informatics 14(2), 143–152 (2000)
13. Rotshtein, A.P., Rakytyanska, H.B.: Inventory Control as an Identification Problem based on Fuzzy Logic. Cybernetics and Systems Analysis 42(3), 411–419 (2006)
14. Rotshtein, A.: Intellectual Technologies of Identification: Fuzzy Sets, Genetic Algorithms, Neural Nets, p. 320. UNIVERSUM, Vinnitsa (1999) (in Russian), `http://matlab.exponenta.ru/fuzzylogic/book5/index.php`
15. Tsypkin, Y.Z.: Information Theory of Identification, p. 320. Nauka, Moscow (1984) (in Russian)
16. Terano, T., Asai, K., Sugeno, M. (eds.): Applied Fuzzy Systems. Omsya, Tokyo (1989); Mir, Moscow (1993) (in Russian)
17. Zadeh, L.A.: The Concept of a Linguistic Variable and its Application to Approximate Reasoning. Part 1-3. Information Sciences 8, 199–251 (1975); 9, 301 - 357, 43 - 80 (1976)
18. Ivachnenko, A.G., Lapa, V.G.: Forecasting of Random Processes, p. 416. Kiev, Naukova dumka (1971) (in Russian)
19. Markidakis, S., Wheelwright, S.C., Hindman, R.J.: Forecasting: Methods and Applications, 3rd edn., p. 386. John Wiley & Sons, USA (1998)
20. Mingers, J.: Rule Induction with Statistical Data – A Comparison with Multiple Regression. J. Operation Research Society 38, 347–351 (1987)
21. Willoughby, K.A.: Determinants of Success in the CFL: a Logistic Regression Analysis. In: Proc. of National Annual Meeting to the Decision Sciences, vol. 2, pp. 1026–1028. Decision Sci. Inst., Atlanta (1997)
22. Glickman, M.E., Stern, H.S.: A State-space Model for National Football League Scores. Journal of the American Statistical Association 93, 25–35 (1998)
23. Stern, H.: On Probability of Winning a Football Game. The American Statistician 45, 179–183 (1991)
24. Dixan, M.I., Coles, S.C.: Modelling Association Football Scores and Inefficiencies in the Football betting Market. Applied Statistics 46(2), 265–280 (1997)
25. Koning, R.H.: Balance in Competition in Dutch Soccer. The Statistician 49, 419–431 (2000)

26. Rue, H., Salvensen, O.: Prediction and Retrospective Analysis of Soccer Matches in a League. The Statistician 49, 399–418 (2000)
27. Walczar, S., Krause, J.: Chaos, Neural networks and Gaming. In: Proc. of Third Golden West International Conference, Intelligent Systems, pp. 457–466. Kluwer, Las Vegas (1995)
28. Purucker, M.C.: Neural Network Quarterbacking. IEEE Potentials 15(3), 9–15 (1996)
29. Condon, E.M., Colden, B.L., Wasil, E.A.: Predicting the Success of Nations at the Summer Olympic using Neural Networks. Computers & Operations Research 26(13), 1243–1265 (1999)
30. Galasa, P.V.: Expert Analysis of Traffic Accidents. Expert-service, Kiev (1995) (in Ukrainian)
31. Litvinov, A.S., Farobin, A.E.: Automobile. Theory of Operation Properties. Machine building, Moscow (1989) (in Russian)
32. Rotshtein, A., Rebedailo, V., Kashkanov, A.: Fuzzy Logic-based Identification of Car Wheels Adhesion Factor with a Road Surface. Fuzzy Systems & A.I. Reports and Letters 6(1-3), 53–64 (1997)
33. Rotshtein, A., Shtovba, S., Mostav, I.: Fuzzy Rule Based Innovation Projects Estimation. In: Annual Conference of the North American Fuzzy Information Processing Society - NAFIPS -, vol. 2, pp. 122–126 (2001)
34. Barlow, R.E., Proshan, F.: Statistical Theory of Reliability and Life Testing. Holt, Rinehart and Winston, New York (1975)
35. Gubinsky, A.I.: Reliability and Quality of Ergonomic Systems Functioning. Nauka, Leningrad (1982) (in Russian)
36. Druzginin, G.V.: Analysis of Ergonomic Systems. Energoatomizdat, Moscow (1984) (in Russian)
37. Rotshtein, A.P.: Probabilistic-Algorithmic Models of Man-Machine Systems. Automation (5), 81–87 (1987)
38. Rotshtein, A.P., Kuznetcov, P.D.: Design of Faultless Man-Machine Technologies. Technika, Kiev (1992) (in Russian)
39. Ryabinin, I.A.: Reliability and Safety of Structural-Complex Systems. Polytechnika, Saint Petersburg (2000) (in Russian)
40. Onisawa, T., Kacprzyk, J. (eds.): Reliability and Safety under Fuzziness, p. 390. Physica-Verlag, Berlin (1995)
41. Cai, K.-Y.: Introduction to Fuzzy Reliability, p. 290. Kluwer Academic Publishers, New York (1996)
42. Rotshtein, A.: Fuzzy Reliability Analysis of Man-Machine Systems. In: Onisawa, T., Kasprzyk, J. (eds.) Reliability and Safety Analysis under Fuzziness. Studies in Fuzzyness, vol. 4, pp. 245–270. Phisika-Verlag, Springer (1995)
43. Rotshtein, A.P., Shtovba, S.D.: Fuzzy Reliability of Algorithmic Processes, p. 142. Kontinent – PRIM, Vinnitsa (1997) (in Russian), http://www.vinnitsa.com/shtovba/doc/fuzzy_reliability.djvu
44. Rotshtein, A.P., Shtovba, S.D.: Predicting the Reliability of Algorithmic Processes with Fuzzy Input Data. Cybernetics and System Analysis 34(4), 545–552 (1998)
45. Utkin, L.V., Shubinsky, I.B.: Unconventional Methods of Information Systems Reliability Analysis. Lubavich, Saint Petersburg (1998) (in Russian)
46. Glushkov, V.M.: Automatic Control Theory and Formal Transformations of Micro Programs. Cybernetics (5), 1–10 (1965) (in Russian)

47. Glushkov, V.M., Tceitlin, G.E., Ucshenko, E.L.: Algebra. Languages. Programming. Naukova Dumka, Kiev (1989) (in Russian)
48. Zadeh, L.A.: Fuzzy Sets as a Basic for a Theory of Possibility. Fuzzy Sets and Systems 1, 3–28 (1978)
49. Rotshtein, A.P.: Fuzzy Analysis of Activity Algorithms Reliability. Reliability 3(22), 3–16 (2007)
50. Rotshtein, A.P.: Algebra of Algorithms and Fuzzy Logic in System Reliability Analysis. Journal of Computer and Systems Sciences International 49(2), 253–264 (2010)
51. Rotshtein, A.P., Katelnikov, D.I.: Fuzzy Algorithmic Simulation of Reliability: Control and Correction Resource Optimization. Journal of Computer and Systems Sciences International 49(6), 967–971 (2010)